W9-BRU-817

Elements of Probability and Statistics

INTERNATIONAL SERIES IN DECISION PROCESSES

INGRAM OLKIN, Consulting Editor

A Basic Course in Statistics, 2d ed., T. R. Anderson and M. Zelditch, Jr.
Introduction to Statistics, R. A. Hultquist
Applied Probability, W. A. Thompson, Jr.
Elementary Statistical Methods, 3d ed., H. M. Walker and J. Lev
Reliability Handbook, B. A. Kozlov and I. A. Ushakov (edited by J. T. Rosenblatt and L. H. Koopmans)
Fundamental Research Statistics for the Behavioral Sciences, J. T. Roscoe
Statistics: Probability, Inference, and Decision, Volumes I and II, W. L. Hays and R. L. Winkler
An Introduction to Probability, Decision, and Inference, I. H. LaValle
Elements of Probability and Statistics, S. A. Lippman

FORTHCOMING TITLES

Elementary Probability, C. Derman, L. Gleser, and I. Olkin
Probability Theory, Y. S. Chou and H. Teicher
Statistical Inference, 2d ed., H. M. Walker and J. Lev
Statistics for Psychologists, 2d ed., W. L. Hays
Decision Theory for Business, D. Feldman and E. Seiden
Times Series Analysis, D. R. Brillinger
Statistics Handbook, C. Derman, L. Gleser, G. H. Golub, G. J. Lieberman, I. Olkin, A. Madansky, and M. Sobel
An Introduction to Statistics and Probability for Engineers, C. Derman and P. Kolesar
Statistics for Scientists, M. Zelen
Mathematics Models in Experimental Design, M. Zelen
Time Series, D. R. Brillinger
Applied Multivariate Analysis Including Bayesian Techniques, S. J. Press
Bayesian Statistics, R. L. Winkler
Multivariate Linear Model, J. Finn
Statistics and Computer Programming, I. Olkin and others
A First Course in the Methods of Multivariate Analysis, C. Y. Kramer
Modern Mathematical Methods for Economics and Business, R. E. Miller

SERIES IN QUANTITATIVE METHODS FOR DECISION-MAKING

ROBERT L. WINKLER, Advisory Editor

Statistics: Probability, Inference, and Decision, W. L. Hays and R. L. Winkler
An Introduction to Probability, Decision, and Inference, I. H. LaValle
Elements of Probability and Statistics, S. A. Lippman

FORTHCOMING TITLES
Bayesian Statistics, R. L. Winkler
Modern Mathematical Methods for Economics and Business, R. E. Miller

Elements of Probability and Statistics

Steven A. Lippman
University of California, Los Angeles

Only yesterday the practical things of today were decried as impractical, and the theories which will be practical tomorrow will always be branded as valueless games by the practical men of today.

William Feller

HOLT, RINEHART AND WINSTON, INC.
New York · Chicago · San Francisco · Atlanta
Dallas · Montreal · Toronto · London · Sydney

Library of Congress Catalog Card Number: 74–129485
SBN: 03–083000–1
Printed in the United States of America

1 2 3 4 038 9 8 7 6 5 4 3 2 1

To Nancy

PREFACE

The serious but nonmathematical reader who needs to use probability and statistics in his research has until now had the choice of excellent books beyond his mathematical means or else so-called "cookbooks" that all too often lead him to colossal blunders. This text is intended to alleviate this problem by offering the reader who is not familiar with calculus a firm grounding in the basic concepts of probability and statistics.

This text has been prepared for use in an introductory course for graduate students of business, education, and all of the social sciences, as well as for undergraduates in mathematics and engineering. Moreover, because of the emphasis on the conceptual framework of probability and statistics, the reader should be able to pursue more advanced work after completing this book. Since calculus is almost never used (and the reader can skip its occurrences without loss of continuity), the only mathematical prerequisite to the successful reading of this book is high school algebra. As a matter of fact, however, the author's experience has shown that most undergraduates with no exposure to calculus are likely to find this text too difficult. Thus, the mathematical level seems to be ideally suited to serious students having had one or two semesters of calculus.

This book strongly emphasizes probability theory and offers an abundance of real examples, drawn from such diverse fields as business, physics, education, sociology, geography, and psychology. The examples serve three purposes: they motivate and exemplify the theory, they help instill the flavor and excitement of probability and statistics, and they illustrate the relevance of probabilistic and statistical analysis in our modern world.

Most readers will make best use of this book by beginning with a very quick reading of the appendix on sets and functions to acquaint themselves with the set notation employed throughout the text. Some readers will need to give close attention to this material. First read the three chapters of Part I on probability theory; then begin Part II on statistical inference with the introduction to hypothesis testing given in Chapter 4. At this point, the reader can skip to Chapter 6 on chi-square tests or to Chapter 9 on nonparametric tests, or he can proceed to Chapter 5 on normal tests. Chapter 8 on regression should be studied only after Chapter 7 on estimation, which, in turn, requires familiarity with the material of Chapter 5. These various possibilities are represented schematically below.

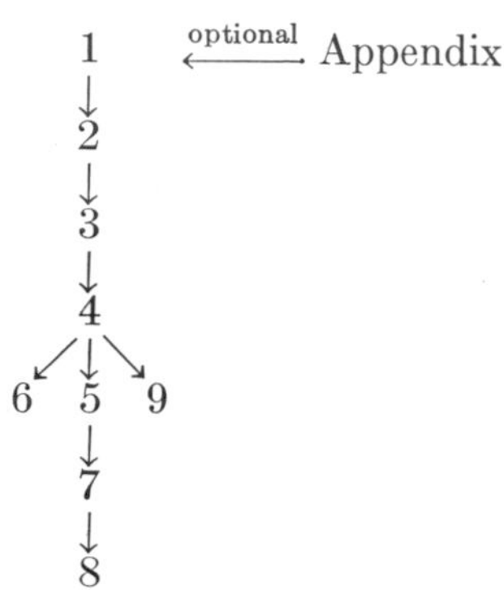

Dependence Relationships among the Chapters

Certain portions of the text that contain more advanced or supplementary material have been written so that they can be omitted without loss of continuity. When these portions comprise an entire section or subsection of a chapter, they are preceded by an asterisk; if they do not comprise an entire section, then they are preceded by "▶▶" and terminated with "▶." Furthermore, we have demarcated the proof of each theorem by beginning it with the word "Proof" and concluding it with the symbol "♦." This method of setting the proofs apart from the rest of the text will enable those who so desire to omit some of the more difficult proofs. Finally the symbol "•" is used to designate the end of an example. Because of all this flexibility, the book can readily be used for a course of study as short as a quarter or as long as a year.

It has been the author's experience that the subject matter of Chapters 1 through 6 can be handled in forty hours of lectures, excluding quiz sections, if all starred material is omitted, if some proofs are omitted, and if no time is spent reviewing the Appendix.

At the end of each subsection are numerous exercises. These exercises—about 400 in all—form an essential part of the book. They serve not only to reinforce and test the reader's grasp of the concepts presented but also to present additional material of a more theoretical nature; for example, most of the material on conditional expectation is to be found in the exercises. Within each subsection, the exercises are listed in order of increasing difficulty. Beginning with simple numerical examples that are straightforward applications of the textual material, they conclude with more thought-provoking problems.

One final word about how to attack this book: read it, think about it, reread it, use it, but most importantly, ENJOY IT.

S. A. L.

Los Angeles, California
December 1970

ACKNOWLEDGEMENTS

During the development of this text, I was fortunate in receiving valuable suggestions from many students including Robert Kris Brown, Herbert Libow, and especially Carlo Brumat who was my most valuable critic. I am grateful to Professors James R. Jackson (UCLA) and J. Morgan Jones (UCLA) for class testing preliminary versions of this text, a task which surely is as trying as debugging a computer program. Thanks are due also to Professors Gerald Lieberman (Stanford) and Harvey Wagner (Yale) for their sage counsel, to Professor Ingram Olkin (Stanford) for editorial advice, and to Mrs. Nelly Williams and Kay Fujimoto for cheerfully typing the many drafts of this text. Not to be forgotten is my wife, Nancy, whose poetry graces Chapter 7 and whose criticism has graced the entire manuscript.

I am indebted to the Literary Executor of the late Sir Ronald A. Fisher, F.R.S., to Dr. Frank Yates, F.R.S., and to Oliver & Boyd, Edinburgh, for permission to reprint Appendix Table C from their book *Statistical Tables for Biological, Agricultural and Medical Research.*

Finally, it is a great pleasure to acknowledge the many fine teachers who gave unsparingly of their time and have had a profound influence on my life and on my thinking. In particular, I would like to thank Professor Saul Hymans (University of Michigan) who first interested me in statistics, Professor Elizabeth Scott (UC Berkeley) who arranged for my special undergraduate major, and most of all Professor Arthur F. Veinott, Jr. (Stanford) who taught me more than I first realized and more than he himself can ever imagine.

CONTENTS

Elements of Probability and Statistics

"A picture of reality drawn in a few sharp lines cannot be expected to be adequate to the variety of all its shades."
(Herman Weyl)

part I

Probability

1

PROBABILITY

1. PROBABILITY MODELS

We are about to study a subject that has fascinated both research workers and laymen alike for over 300 years, namely, probability. Everyone has some notion of probability, and most of us would agree that the word connotes "the likelihood of some event happening." This last phrase, however, is somewhat vague. It is now our intention to make precise just what we mean by the term "probability."

When we toss a "fair" coin, we all would agree that the probability of its landing heads is $\frac{1}{2}$; moreover, on any one toss, the coin will land either heads or tails. Yet, if we tossed this fair coin a great many times, say n, and observed the total number $T(n)$ of tails that appeared,[1] then we would be quite unlikely to observe $T(n) = \frac{1}{2}n$ or even to observe $T(n)$ being within one, two, or even a few of $\frac{1}{2}n$. What we would find, however, is that the difference between $T(n)/n$ and $\frac{1}{2}$ is small. In fact, as n gets larger and larger, $[T(n)/n] - \frac{1}{2}$ gets smaller and smaller.[2] That is, $T(n)/n$, the *relative frequency* of tails, approaches $\frac{1}{2}$ as the number of tosses is increased. This is the notion of probability which we shall entertain and generalize.

Physicists tell us that when an object such as a ball is shot straight up into the air, it will reach a height of $\frac{1}{2}v^2/g$ feet, where g is the acceleration of gravity and v is the initial velocity of the ball. But we know that the ball will not travel *exactly* $\frac{1}{2}v^2/g$ feet, because factors such as friction have

[1] Here the domain of the function T is $\mathfrak{N} = \{1, 2, \ldots\}$.

[2] We shall show this in Section 4 of Chapter 2.

not been taken into account. Nevertheless, the formula

$$\text{height} = \frac{\frac{1}{2}v^2}{g}$$

yields an excellent estimate of the height that the object in question actually will reach. Thus, this formula is an idealization of reality, or what we call a *mathematical model.* Idealizations of this type are standard practice in the sciences and in this text.

Emanating from the fact that a mathematical model takes into account only the most essential features of the real situation being studied are two advantages. First, the precise mathematical statements we are able to make about our model enable us to gain new insights into the true nature of our problem. For example, we see that we need not double the initial velocity to send our ball twice as high, we need only (approximately) increase it by a factor of $\sqrt{2}$. Second the results obtained are applicable to a wide variety of situations that occur under similar conditions. For example, the formula above provides a good approximation almost independently of the initial velocity, type, weight, size, or shape of the projectile being shot into the air.

Like the physicists, we probabilists and statisticians also build mathematical models of real-world situations. Our mathematical models will be called *random experiments.* Coin tossing is typical of the type of problem for which we will build a mathematical model.

DEFINITION

Consider an experiment (hypothetical or real) such that

(i) the outcome is not necessarily known with certainty prior to performing the experiment,
(ii) the set Ω[3] of all possible outcomes is known prior to performing the experiment, and
(iii) the experiment can be repeated under the same conditions.

Then the experiment is said to be a *random experiment,* and Ω is called the *sample space.*

EXAMPLE 1

Suppose we roll an ordinary die (singular of dice). Then the outcome of this experiment is one of the numbers 1, 2, . . . , 6—that is, the outcome is a member of the set $\Omega = \{1, 2, 3, 4, 5, 6\}$. Just as the physicist ignored the

[3] The symbol "Ω" is the upper-case Greek letter omega. We will always denote the sample space by Ω.

force of friction in the example above, we ignore the possibility of the die's landing on its edge. This is part of the process of modeling. That is, we only take into account the most salient features of the problem and we omit the inessential features. (We also ignored such unimportant parts of the description of the outcome as where the die landed, how many times it rolled, and so on.) Consequently, by fiat, condition (ii) is satisfied.

Next, we consider whether or not this experiment can be repeated under the same conditions. In reality, rolling the die a great many times might eventually result in a change, albeit small, in the physical shape of the die. Nevertheless, we shall say that the experiment can be repeated under the same conditions—that is, we claim that the change in conditions that might result is inconsequential.

Finally, it is clear that the outome of the experiment is not known with certainty before the experiment is performed. Hence, we see that rolling a die is a random experiment with sample space $\Omega = \{1, 2, 3, 4, 5, 6\}$. •

EXAMPLE 2

Suppose we toss a coin. Then the outcome of this experiment is a member of the set $\Omega = \{\text{heads, tails}\}$, and it is not known with certainty prior to performing the experiment. Once again, our model ignores the possibility of the coin's landing on its edge. We also ignore the possibility that the coin may get bent after many tosses, and we claim that the experiment can be repeated under the same conditions. Thus, tossing a coin is also seen to be a random experiment. •

EXAMPLE 3

Jack asked Jill to marry him, and he gave her one day to reach a decision. There are two possible outcomes of this experiment: an answer of "yes" or "no." Moreover, Jack didn't know what Jill's answer would be before he asked her; that is, the outcome is not known with certainty before the experiment is performed. However, in no reasonable way would we consider it possible to repeat the experiment: having asked the big question changes their relationship in an essential way. Hence, this is not a random experiment. •

Thus, we shall not attempt to deal with (philosophical) questions about, say, the probability that Jill will say yes to Jack or that the sun will rise tomorrow. (If it did not rise one morning, then we would not be able to repeat the experiment.) Neither will we try to interpret or make precise statements of judgment such as, "Had he known the truth he probably would have forgiven her."

Any subset E of Ω is called an *event*, and E is called an *elementary event* if E contains only one element of Ω. For instance, if $E = \{1, 2, 3\}$,

$F = \{2\}$, and $\Omega = \{1, 2, 3, 4, 5, 6\}$, then E is an event, whereas F is an elementary event.

We now prescribe the way in which probabilities (numbers) can be assigned to events. In particular, we define a function P, called a probability function, whose domain is the set of all subsets of Ω[4] and whose range is contained in $\{x: 0 \leq x \leq 1\}$.

DEFINITION

The function P is said to be a *probability function* on Ω if P satisfies
Axiom 1: $P(E) \geq 0$ for each $E \subset \Omega$,
Axiom 2: $P(\Omega) = 1$,
Axiom 3: If $E_i \cap E_j = \varnothing$ for $i \neq j$, $i, j = 1, 2, \ldots$, then[5]

$$P\left(\bigcup_{i=1}^{\infty} E_i\right) = \sum_{i=1}^{\infty} P(E_i).$$

The number $P(E)$ is called the probability that the outcome of the random experiment is in E or simply the probability of E. Of course, Ω and P differ for different random experiments, but the function P must always have the above properties. These properties can be interpreted roughly as follows. The probability of the occurence of any event is nonnegative, certainty is represented by probability 1, and probabilities add. We refer to the pair (Ω, P) as a *probability model.*

When put in proper perspective, the three axioms satisfied by a probability function seem neither intimidating nor arbitrary but rather understandable and necessary. The proper perspective is attained by considering random experiments where Ω has a finite number of elements, so we can interpret probability as (limiting) relative frequency. Take for example the random experiment wherein a fair die is tossed, and let $A(n)$ and $B(n)$ be the number of 1's and 2's that appear during n rolls of the die so $A(n)/n$ and $B(n)/n$ are the relative frequencies of the events "a 1 appears" and "a 2 appears," respectively. Since we are thinking of probability as meaning (limiting) relative frequency, we have (for n large)

$$P(\{1\}) \approx \frac{A(n)}{n} \quad \text{and} \quad P(\{2\}) \approx \frac{B(n)}{n},$$

[4] In general, it is not always possible to define a probability function on *all* of the subsets of Ω. This, however, is a deep mathematical technicality (related to the topic of measure theory) with which we need not concern ourselves. The more advanced reader will be relieved to know that if Ω is a finite or a countable set (see Exercise 18 of the Appendix), then all difficulties alluded to vanish.

[5] The reader who is uncomfortable with the idea of infinity, written ∞, can simply say to himself "some large number" every time the symbol ∞ appears.

where "$\approx$" means "is very close to." Since $A(n) \geq 0$ and $B(n) \geq 0$ for all n, $A(n)/n$ and $B(n)/n$ are each nonnegative, so Axiom 1 is obviously an essential part of what we mean by probability. Next, observe that

$$P(\{1\} \cup \{2\}) \approx \frac{A(n) + B(n)}{n} = \frac{A(n)}{n} + \frac{B(n)}{n} \approx P(\{1\}) + P(\{2\}),$$

so probabilities really do add, as stated in Axiom 3. Finally, let $C(n)$ be the number of times that a 3, 4, 5, or 6 appears. Then

$$n = A(n) + B(n) + C(n),$$

so

$$P(\Omega) \approx \frac{A(n) + B(n) + C(n)}{n} = 1,$$

expressing the fact that certainty is, in fact, correctly represented by probability 1.

Historically, the concept of probability was restricted to (limiting) relative frequency. The more general axiomatic approach—which we use—was developed only in the last 50 years. Consequently, it is not surprising that the concept of (limiting) relative frequency is a useful pedagogical tool for dealing with probability.

EXAMPLE 4

A color-blind child selects a crayon from a box containing three crayons that differ only in that one is red, one is green, and one is blue. In this experiment, $\Omega = \{\text{red, green, blue}\}$. Since the child is color-blind and the only distinguishing feature of the crayons is their color, it seems reasonable that the probability function P that best describes the likelihood of the three simple events satisfies

$$P(\{\text{red}\}) = \tfrac{1}{3}, \quad P(\{\text{green}\}) = \tfrac{1}{3}, \quad \text{and} \quad P(\{\text{blue}\}) = \tfrac{1}{3}$$

in addition to the three axioms. In view of this and Axiom 3, we have $P(\{\text{red, green}\}) = \frac{2}{3}$, which means that one of the colors red or green will be selected with probability $\frac{2}{3}$. Similarly, $P(\{\text{red, blue}\}) = \frac{2}{3}$, $P(\{\text{green, blue}\}) = \frac{2}{3}$, and $P(\Omega) = 1$. •

EXAMPLE 5

A certain wealthy businessman owns a portfolio of stocks of which 3 percent, 27 percent, and 70 percent are rails, utilities, and industrials, respectively. One of his stocks is chosen at random; that is, each stock is as likely to be chosen as any other. If we only notice which type of stock is chosen, then $\Omega = \{\text{rail, utility, industrial}\}$ and

$$P(\{\text{rail}\}) = \tfrac{3}{100}, \quad P(\{\text{utility}\}) = \tfrac{27}{100}, \quad \text{and} \quad P(\{\text{industrial}\}) = \tfrac{70}{100}. \bullet$$

EXAMPLE 6

Suppose a die is rolled, then $\Omega = \{1, 2, 3, 4, 5, 6\}$. We call the die "fair" if

$$P(\{1\}) = P(\{2\}) = P(\{3\}) = P(\{4\}) = P(\{5\}) = P(\{6\}).^{6}$$

It follows from Axioms 2 and 3 and the above that $P(\{1\}) = \frac{1}{6}$. (Verify this!) More generally, if there are n events that partition Ω, and are such that $P(\text{event } 1) = P(\text{event } 2) = \cdots = P(\text{event } n)$, then $P(\text{event } i) = 1/n$ $(i = 1, 2, \ldots, n)$, and we say these events are *equally likely*. (Does it follow that if there are n possible simple events, each has probability $1/n$?) •

We now state and prove some simple properties of probability functions. In following these proofs, it behooves the reader to draw Venn diagrams of the sets of interest, such as E, $\tilde{E}$, and Ω for Theorem 1. The drawings, of course, have only pedagogical value and are not part of the proofs. Note that the proofs depend only upon the three axioms.

THEOREM 1

If $E \subset \Omega$, then $P(\tilde{E}) = 1 - P(E)$.

PROOF. By Exercise 9 of the Appendix, $E \cup \tilde{E} = \Omega$ and $E \cap \tilde{E} = \varnothing$. Hence, by Axioms 2 and 3 we have

$$1 = P(\Omega) = P(E \cup \tilde{E}) = P(E) + P(\tilde{E}). \blacklozenge$$

One of the more obvious consequences of Theorem 1 is that $P(\varnothing) = 0$, which means that events that cannot occur have probability zero. (Surprisingly, the converse of this is not true, as we shall see in Chapter 3.)

EXAMPLE 7

First, suppose that we rolled a fair die and wanted to know the probability of the event $E \equiv$ a six appears.[7] Then, $P(E) = 1 - P(\tilde{E}) = 1 - \frac{5}{6}$. In this case, the use of Theorem 1 affords us no advantage whatsoever.

Next, suppose that we rolled a fair die 3 times and wanted to find the probability of the event $E \equiv$ at least one six appears. Although in this example we could also find $P(E)$ directly (see Example 8), it is far easier to find $P(E)$ by first finding $P(\tilde{E})$ and then using Theorem 1. In Example 17 we show that $P(\tilde{E}) = (\frac{5}{6})^3$. Thus, $P(E) = 1 - (\frac{5}{6})^3$. •

The usefulness of this approach is better illustrated in more complex problems; refer to Exercises 28, 29, and 32 at this time.

[6] Although we have not yet explicitly defined P on all subsets of Ω, the definition of P on these subsets is implicit (see Exercise 17).

[7] The symbol "$\equiv$" signifies that the objects immediately to the left and to the right of the symbol "$\equiv$" are being defined to be equal.

Just for the moment, think of the relative-frequency interpretation of probability and let $E(n)$ denote the number of times in n trials or experiments that some particular event E has occurred. Then since $E(n) \leq n$, we have that $E(n)/n$, the relative frequency of E, is less than or equal to 1. Consequently, we should expect that $P(E) \leq 1$. This is the assertion of

THEOREM 2

If $E \subset \Omega$, then $P(E) \leq 1$.

PROOF. $P(E) + P(E) = 1$ by Theorem 1, so $P(E) \leq 1$, since $P(\tilde{E}) \geq 0$ by Axiom 1. ♦

The next theorem can be interpreted as saying that the more ways (that is, sample points) there are for an event to happen, the more likely it is to happen. For example, if we roll a fair die 3 times, then the probability of at least one six appearing is larger than the probability of at least two sixes appearing.

THEOREM 3

If $A \subset B \subset \Omega$, then $P(A) \leq P(B)$.

PROOF. Using Exercises 8 and 9 of the Appendix, Axiom 3, and Axiom 1, we obtain

$$P(B) = P(A \cup (B \sim A)) = P(A) + P(B \sim A) \geq P(A). \ ♦$$

THEOREM 4

(General Addition Rule) If $A \subset \Omega$ and if $B \subset \Omega$, then

$$P(A \cup B) = P(A) + P(B) - P(A \cap B).$$

PROOF. Since $A \cup B = A \cup (B \cap \tilde{A})$, we can conclude from Axiom 3 that

$$P(A \cup B) = P(A) + P(B \cap \tilde{A}).$$

Similarly, we have

$$P(B) = P(B \cap A) + P(B \cap \tilde{A}).$$

Combining these two equalities, we obtain the desired result. ♦

If you have drawn a Venn diagram, the truth of Theorem 4 is apparent. We subtracted $P(A \cap B)$ so as not to double-count the probability of the event $A \cap B$.

Of fundamental importance is the following theorem, which is indispensable in solving a great many problems. (See Examples 14, 15, 18, and Exercise 22.)

THEOREM 5

If $E \subset \Omega$ and $A_1, A_2, \ldots$ partition Ω, then

$$P(E) = \sum_{i=1}^{\infty} P(E \cap A_i).$$

PROOF. Using the hypothesis, Law h of the Appendix, and Axiom 3, we have

$$P(E) = P(E \cap \Omega) = P\left[E \cap \left(\bigcup_{i=1}^{\infty} A_i\right)\right]$$

$$= P\left(\bigcup_{i=1}^{\infty} (E \cap A_i)\right) = \sum_{i=1}^{\infty} P(E \cap A_i). \blacklozenge$$

We now present a very simple example to illustrate the use of Theorem 5. This theorem, in conjunction with the concept of conditional probability (to be introduced in Section 2), is essential in solving innumerable more complex and more interesting problems. Refer to Examples 15 and 18 at this time.

EXAMPLE 8

Suppose that we roll a fair die 3 times, and we want to find the probability of the event $E \equiv$ at least one six appears. Consider the events

A_0 = no six appears, A_1 = one six appears,
A_2 = two sixes appear, A_3 = three sixes appear.

They partition Ω. Consequently, it follows from Theorem 5 that

$$P(E) = \sum_{i=0}^{3} P(E \cap A_i) = P(A_1) + P(A_2) + P(A_3).$$

In Example 27 we show that $P(A_1) = 75/6^3$, $P(A_2) = 15/6^3$, and $P(A_3) = 1/6^3$, so $P(E) = 91/6^3 = 1 - (\frac{5}{6})^3$ as stated in Example 7. Theorem 5 is often called the "breakdown rule," since it involves decomposing or breaking down the event E into simpler components, namely, A_1, A_2, and A_3. We shall make frequent use of this theorem later. •

EXERCISES:

1. Describe the sample space of each of the following experiments and decide whether or not they are random experiments.
   (a) A coin is tossed twice.

(b) Two coins are tossed.
(c) A ball is withdrawn from an urn that contains 3 identical black balls and 5 identical white balls.
(d) A basketball player shoots a free throw.
(e) You choose an architect who lives in the city to build your house.
(f) A pair of dice are rolled.
(g) We count the number of misprints on the front page of the local newspaper.
(h) A card is dealt from a deck of 52 cards.
(i) Five cards are dealt from a deck of 52 cards.
(j) The university chooses a new chancellor.

2. Show that the probability of any elementary event is $1/n$ if each of the n elementary events in the sample space is equally likely.

3. Shooting baskets is a hit-or-miss affair; either you hit or you miss. Consequently, it follows from the axioms of probability that $P(\text{hit}) = P(\text{miss}) = \frac{1}{2}$. On the other hand, Lew Alcindor makes 65 percent of his shots. Resolve the apparent contradiction.

4. Consider the random experiment of tossing a fair die once. Find the probability that the outcome is (a) less than 3, (b) more than 3, (c) less than 3 and more than 5, (d) less than 5 and more than 2, (e) either less than 3 and more than 1 or less than 6 and more than 2.

5. The numbers on a roulette wheel are 0, 1, 2, . . ., 36 and 00. The numbers 0 and 00 are green, while the other even numbers are red and the odd numbers are black. Assuming that the wheel is fair, find the probability that the outcome is (a) 13, (b) 00, (c) 13 or 7, (d) 13 or green, (e) 13 and green, (f) 13 or green or black, (g) greater than 17 and black, (h) black, (i) red or black.

6. A card is drawn at random from an ordinary deck of playing cards. Find the probability that the card is (a) the ace of hearts, (b) a heart, (c) a picture card (ace, king, queen, or jack), (d) a heart or a picture card, (e) a 10 or higher.

7. Four fair coins are tossed simultaneously. What is the probability that the outcome will be 2 heads and 2 tails?

8. A pair of fair dice are tossed. For $i = 1, 2, \ldots, 12$, find the probability that the sum of the points showing is i. (First carefully define the sample space and note that it has 36 points, each of which is an ordered pair.)

9. You are a reporter at a national meeting of the Electoral College. There are 302 electors favoring Nixon, 191 electors favoring Humphrey, and 45 electors favoring Wallace. You spot an elector who is standing in the hall. What is the probability that he favors Humphrey?

10. The numbers from 1 to 15 are painted on fifteen identical balls, one number per ball. If one of the balls is drawn at random, what is the probability that the number on it is (a) divisible by 5, (b) even, (c) odd, (d) a perfect square, (e) a 2-digit number?

11. A bag contains 5 times as many dimes as pennies. One coin is drawn at random. What is the probability that it is a dime?

12. Let Ω be the set of all integers between 1 and 2400 inclusive, let A and B be the subsets of Ω which consist of all numbers divisible by 2 and 3, respectively. If an element is chosen from Ω at random, what is the probability that it will not be a member of either A or B?

13. Suppose 50 percent of the graduate students at PU own cars while only 30 percent of the upperclassmen and 15 percent of the lowerclassmen own cars. If PU has 30 graduates, 210 upperclassmen, and 260 lowerclassmen, what is the probability that a student picked at random will own a car?

14. Show that $P(A \cap B) \leq P(A)$ and $P(A \cap B) \leq P(B)$, where $A \subset \Omega$, $B \subset \Omega$, and P is a probability function on Ω.

15. Let P be a probability function on Ω. Find $P(\varnothing)$.

16. Show that

$$P(A \cup B \cup C) = P(A) + P(B) + P(C) + P(A \cap B \cap C) - P(A \cap C) - P(B \cap C) - P(A \cap B).$$

17. Suppose (Ω is finite or countable and) we are given a function f such that $f(\omega) \geq 0$ for each $\omega \in \Omega$ and $\sum_{\omega \in \Omega} f(\omega) = 1$. Show that there is a unique probability function P on Ω such that $P(\{\omega\}) = f(\omega)$ for each $\omega \in \Omega$. [Consequently, in defining a probability function P on Ω we merely need to specify $P(\{\omega\})$ for each $\omega \in \Omega$ and not $P(E)$ for each $E \subset \Omega$.]

18. Suppose that we defined a probability function P' on Ω as in the definition except we replaced Axiom 3 with

Axiom 3′: if $A \cap B = \varnothing$, then $P'(A \cup B) = P'(A) + P'(B)$.

Show that a probability function always satisfies Axiom 3′, and show that

$$P'(A \cup B \cup C) = P'(A) + P'(B) + P'(C) \qquad \text{if } A \cap B = A \cap C = B \cap C = \varnothing.$$

2. CONDITIONAL PROBABILITY

In this section we show how to construct new probability functions from old ones. We now focus attention on the change in probability that takes place when we are given information about the outcome of our random experiment in addition to that information already contained in our probability model (Ω, P). This additional information consists in knowing (with certainty) that the outcome of our random experiment lies in some subset B of Ω.

2.1 Conditional Probability

We begin this subsection with two examples, and then motivate the basic definition of conditional probability.

EXAMPLE 9

Suppose that in rolling a pair of fair dice[8] [colored green (g) and black (b)] we find that the green die came up 6; then we know that the outcome of this experiment is a member of the set $B = \{(6, b): b = 1, 2, \ldots, 6\}$. Prior to obtaining this information, we found that the probability of the dice totaling 11 was $\frac{2}{36}$. In light of the new information, however, it is now evident that this probability is $\frac{1}{6}$, since each point of Ω (and hence each point of B) has the same probability of occurrence. •

EXAMPLE 10

My friend Morgan is a compulsive gambler. Yesterday he played the following game. He rolled a fair die; if it came up 1, 2, 3, 4, 5, or 6, he would win $-\$100$, $-\$50$, $-\$10$, \$20, \$20, or \$120, respectively. I couldn't stand the excitement so I left the room before he rolled the die. After the game was over, he came out with a big smile on his face. Obviously, he had won; that is, the outcome of the experiment of rolling the die was a member of the set $B = \{4, 5, 6\}$. Before he rolled the die, the probability that he would roll a 6 was $\frac{1}{6}$; however, after I saw his smile, I suspected that the probability that he rolled a 6 was $\frac{1}{3}$. •

More generally, suppose we are given the information that some event $B \subset \Omega$ has occurred. We now must assign new probabilities to the points in B in a manner consistent with those probabilities assigned by P. We shall denote our new probability function by P_B. Of course, we must have $P_B(\{\omega\}) = 0$ for each $\omega \notin B$.[9] A consistent assignment of probabilities also requires that (1) the new probabilities add to 1 and that (2) the essential nature of the relationship of the probabilities of the points within B to one another is unchanged, that is, each point in B has proportionately the same probability that it previously had. More

[8] Here, $\Omega = \{(g, b): g, b = 1, 2, 3, 4, 5, 6\}$.

[9] We use ω, the lower-case Greek letter omega, to denote a generic element of Ω.

formally, we require that

$$\sum_{\omega \in B} P_B(\{\omega\}) = 1 \tag{1.1}$$

and

$$P_B(\{\omega\}) = kP(\{\omega\}) \qquad \text{for each } \omega \in B, \tag{1.2}$$

where $k > 0$ is some constant. Combining (1.1) and (1.2), we have

$$1 = \sum_{\omega \in B} P_B(\{\omega\}) = k \sum_{\omega \in B} P(\{\omega\}) = kP(B).$$

Hence, $k = 1/P(B)$, so

$$P_B(\{\omega\}) = \begin{cases} \dfrac{P(\{\omega\})}{P(B)}, & \text{if } \omega \in B, \\ 0, & \text{if } \omega \in \Omega \sim B. \end{cases} \tag{1.3}$$

Consequently, we can conclude from (1.3) and Axiom 3 that[10]

$$P_B(E) = \frac{P(E \cap B)}{P(B)} \qquad \text{for each } E \subset \Omega. \tag{1.4}$$

To see this pictorially, consider the Venn diagram in Figure 1.1, and suppose that $P(\{\omega\}) = \frac{1}{16}$ for each of the 16 sample points in Ω. Then $P(B) = \frac{5}{16}$ and $P(E \cap B) = \frac{2}{16}$ so $P_B(E) = \frac{2}{5}$.

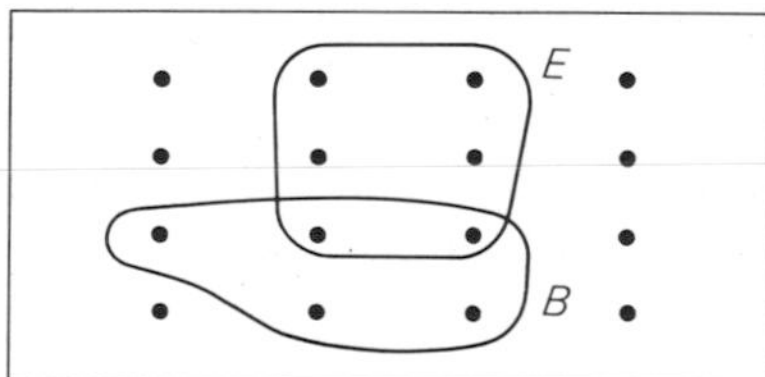

Figure 1.1 Venn Diagram of Conditional Probability

It is common to write $P(E \mid B)$, read "the conditional probability of E given B," instead of $P_B(E)$, even though the former notation is not consistent with the set notation we have been employing as "$E \mid B$" is not an event. Recapitulating, we have the following

[10] $$P_B(E) = \sum_{\omega \in E} P_B(\{\omega\}) = \sum_{\omega \in E \cap B} P_B(\{\omega\}) + \sum_{\omega \in E \cap \tilde{B}} P_B(\{\omega\}) = \sum_{\omega \in E \cap B} P_B(\{\omega\})$$
$$= \sum_{\omega \in E \cap B} \frac{P(\{\omega\})}{P(B)} = \frac{1}{P(B)} \sum_{\omega \in E \cap B} P(\{\omega\}) = \frac{P(E \cap B)}{P(B)}.$$

DEFINITION

If B is an event with positive probability, then for each event $E \subset \Omega$ we define the *conditional probability of E given B* by

$$P(E \mid B) = \frac{P(E \cap B)}{P(B)}.$$

In this case, we often refer to B as the reduced sample space.

Notice that Theorems 1 through 5 hold for the new probability model (B, P_B).

EXAMPLE 11

A card is chosen at random from an ordinary deck of cards, and we are told that it is at least as high as a 10. (An ace is considered high.) What is the probability that it is (a) an ace? (b) a five? (c) a heart?

SOLUTION: First define the following events:

$B \equiv$ the card is at least as high as a 10,
$A \equiv$ the card is an ace,
$5 \equiv$ the card is a five,
$H \equiv$ the card is a heart.

(a) Since the card is chosen at random, each of the 52 cards is as likely to be selected as any other, so $P(B) = \frac{20}{52}$ and $P(A) = \frac{4}{52}$. Hence, using the fact that $A \cap B = A$, we obtain

$$P(A \mid B) = \frac{P(A \cap B)}{P(B)} = \frac{P(A)}{P(B)} = \frac{\frac{4}{52}}{\frac{20}{52}} = \frac{4}{20} = \frac{1}{5}.$$

(b) Since $5 \cap B = \varnothing$, we have

$$P(5 \mid B) = \frac{P(5 \cap B)}{P(B)} = \frac{P(\varnothing)}{P(B)} = \frac{0}{P(B)} = 0.$$

(c) Finally, we note that $P(H \cap B) = \frac{5}{52}$, so

$$P(H \mid B) = \frac{P(H \cap B)}{P(B)} = \frac{\frac{5}{52}}{\frac{20}{52}} = \frac{5}{20} = \frac{1}{4}. \bullet$$

EXAMPLE 12

A certain family has exactly two children and we know that at least one of them is a girl. Assuming male and female births to be equally likely, and assuming that the sex of the older child in no way affects the sex of the younger child, we ask the question: What is the probability of the event $E \equiv$ this family has two girls?

SOLUTION: The sample space is $\Omega = \{(b,b), (b,g), (g,b), (g,g)\}$, and all outcomes are equally likely. [(b, g) means the older child is a boy and the younger child is a girl.] The added information tells us that the outcome is in $B = \{(b, g), (g, b), (g, g)\}$. Consequently, the probability that the family has two girls is $\frac{1}{3}$, as

$$P(E \mid B) = \frac{P(E \cap B)}{P(B)} = \frac{\frac{1}{4}}{\frac{3}{4}} = \frac{1}{3}. \bullet$$

EXAMPLE 13

We have discovered that your natural dislike for mathematics is such that the probability of your understanding the first section on sets is .2. If you understand it, however, you are obviously quite bright, so that .9 is the probability of your understanding the second section on functions. What is the probability that you will understand both sections?

SOLUTION: Let $A \equiv$ understand Section 1, and $B \equiv$ understand Section 2. We know that $P(B \mid A) = .9$ and $P(A) = .2$, so we can conclude from the definition of conditional probability that $P(A \cap B) = .9(.2) = .18$. Note that we haven't been given enough information to find $P(B)$ or $P(A \mid B)$. •

EXAMPLE 14

Three students—Alan, Barry, and Carl—all desire to borrow a book of which the lender has but one copy. The lender says: "I am thinking of a number from one to three inclusive. Alan, you guess first. If you guess right, you get the book. If you guess wrong, Barry guesses. If Barry guesses right, he gets the book. If Barry guesses wrong, Carl gets the book." All three boys will be present during the guessing and none of them is inordinately stupid. Do you think this method is fair?

SOLUTION: Clearly $P(A \text{ wins}) = \frac{1}{3}$. Using Theorem 5, the definition of conditional probability, and the fact that the two events "A wins" and "A loses" partition Ω, we have

$$\begin{aligned} P(B \text{ wins}) &= P(B \text{ wins and } A \text{ wins}) + P(B \text{ wins and } A \text{ loses}) \\ &= P(B \text{ wins and } A \text{ loses}) = P(B \text{ wins} \mid A \text{ loses}) \cdot P(A \text{ loses}) \\ &= \tfrac{1}{2} \cdot \tfrac{2}{3} \\ &= \tfrac{1}{3}. \end{aligned}$$

It now follows from Theorem 1 that $P(C \text{ wins}) = \frac{1}{3}$, so the method is fair.[11] •

[11] If we let $\Omega = \{(i, j, k): i, j, k = 1, 2, 3 \text{ and } k \neq j\}$, where i is the number chosen by the lender, j is Alan's guess, and k is Barry's guess, then each of the 18 points of Ω is seen to be equally likely. Moreover, the 6 points with $i = j$, with $i \neq j$ and $i = k$, and with $i \neq j$ and $i \neq k$ correspond to Alan winning, Barry winning, and Carl winning, respectively.

EXAMPLE 15

At a psychiatric clinic in San Francisco, the psychiatric social workers are so busy that, on the average, only 60 percent of (potential) new patients who call on the phone are able to talk immediately with a social worker when they call. The other 40 percent are always persuaded to leave their phone number. About 70 percent of the time, a social worker is able to return the call and contact the caller on the same day, and the other 30 percent of the time the caller is contacted the following day. Experience at the clinic indicates that the probability a caller will come to the clinic for consultation is .8 if the caller was able to talk to a social worker when he called or if he was contacted later on that day, whereas this probability is only .5 if the caller is not contacted until the following day. What is the probability that a caller comes to the clinic for consultation?

SOLUTION: First define the following events:

$C \equiv$ a caller comes to the clinic for consultation,
$I \equiv$ a caller is able to talk to a social worker immediately upon calling,
$S \equiv$ a caller is contacted later on in the same day,
$F \equiv$ a caller is not contacted until the following day.

Then,

$$\begin{aligned} P(S) &= P(S \cap I) + P(S \cap \tilde{I}) = P(S \cap \tilde{I}) \\ &= P(S \mid \tilde{I})P(\tilde{I}) = (.7)(.4) = .28. \end{aligned}$$

Similarly, $P(F) = (.3)(.4) = .12$, so

$$\begin{aligned} P(C) &= P(C \cap I) + P(C \cap S) + P(C \cap F) \\ &= P(C \mid I)P(I) + P(C \mid S)P(S) + P(C \mid F)P(F) \\ &= (.8)(.6) + (.8)(.28) + (.5)(.12) \\ &= .764. \bullet \end{aligned}$$

EXERCISES:

(The student is urged to *carefully* define the sample space and all events of interest when working the exercises.)

19. One of the campus sororities lists the following statistics of its membership: 15 blue-eyed blondes, 8 brown-eyed blondes, 9 blue-eyed brunettes, 12 brown-eyed brunettes, 1 blue-eyed redhead, and 0 brown-eyed redheads. Suppose you have arranged a blind date with one of the members and it is raining when you meet the girl. Her hair is completely covered, however, her sparkling blue eyes bid you welcome! What is the probability that she is a blonde?

20. Four keys, of which only one fits, are tried one after another until the door opens. (The same key is never tried more than once.) What is the probability that this procedure requires exactly i trials, $i = 1, 2, 3, 4$? What if two of the keys fit?

21. Suppose an urn contains 5 black balls (B) and 6 white balls (W). We draw two balls from the urn without replacement. What is the probability of getting two B's? one B and one W?

22. Urn A contains 1 white ball and 2 black balls, whereas urn B contains 3 white balls and 2 black balls. We select a ball from urn A if the outcome of rolling a fair die is less than 2; otherwise we select a ball from urn B. What is the probability of selecting a white ball?

23. Excluding our house, half of the other 20 houses on our block in the new housing development have been sold. What is the probability that the two houses adjacent to ours have also been sold? [*Hint:* Define $L \equiv$ adjacent house on the left has been sold and define R similarly. Then $P(R \cap L) = P(R \mid L)P(L)$.]

24. Suppose that the examination books of n students are randomly arranged in a pile and that no two scores are the same. Find the probability that (a) the best and second best scores (examination books) are adjacent in the pile [*Hint:* condition on the position of the best score]; (b) the best, second best, and third best scores are adjacent and are in the order named.

25. Show $P(A \cap B \cap C) = P(C \mid A \cap B)P(A \mid B)P(B)$.

2.2 Independence

Very often we encounter situations in which the occurrence of some event A in no way affects the occurrence of another event B. In this case we say that A and B are independent events. For example, if we flip one fair coin and then another, we would all agree that the probability of the second coin landing heads is $\frac{1}{2}$ without regard to the outcome of the first flip. Previously, we said that if we flip two fair coins then all of the points in $\Omega = \{(H, H), (H, T), (T, H), (T, T)\}$ are equally likely. Implicit was the idea that the two flips are independent of one another. The probability of (H, H) was reasonably assumed to be $\frac{1}{4}$, since half of the times the first coin came up heads and half of these times the second coin also came up heads, so that both coins come up heads $\frac{1}{4}$ of the time. This motivates the more general

DEFINITION

Two events A and B are said to be *independent* if

$$P(A \cap B) = P(A)P(B).$$

EXAMPLE 16

A card is chosen at random from an ordinary deck of cards. Consider the events B, A, and H defined in Example 11. Which pairs are independent?

SOLUTION: Since

$$P(A \cap B) = \tfrac{4}{52} \neq \tfrac{4}{52} \cdot \tfrac{20}{52} = P(A)P(B),$$

A and B are not independent; however, A and H as well as B and H are independent, since

$$P(A \cap H) = \tfrac{1}{52} = \tfrac{4}{52} \cdot \tfrac{13}{52} = P(A)P(H)$$

and

$$P(B \cap H) = \tfrac{5}{52} = \tfrac{20}{52} \cdot \tfrac{13}{52} = P(B)P(H). \bullet$$

EXAMPLE 17

Suppose we roll a fair die 3 times. Find the probability of the event $E \equiv$ no six appears.

SOLUTION: We choose to imagine that the three rolls in no way affect each other. Hence, the events $E_i \equiv$ "no six appears on the ith roll," $i = 1, 2, 3$, are independent,[12] so

$$P(E) = P(E_1 \cap E_2 \cap E_3) = P(E_1)P(E_2)P(E_3) = (\tfrac{5}{6})^3. \bullet$$

EXAMPLE 18

A marketing executive is about to place an advertisement in *Fortune* magazine, and is trying to determine the probability that the ad will be read. He has the following information from the *Fortune* circulation department:

Given a man has a yearly income of	*His probability of reading Fortune is*
less than \$5000	.004
\$5000 to \$10,000	.01
greater than \$10,000	.05

Furthermore, population statistics show:

[12] The events $E_1, E_2, \ldots, E_n$ are said to be *mutually independent* if for all combinations $1 \leq i < j < k < \cdots \leq n$ we have

$$\begin{aligned} P(E_i \cap E_j) &= P(E_i)P(E_j) \\ P(E_i \cap E_j \cap E_k) &= P(E_i)P(E_j)P(E_k) \\ &\cdots\cdots\cdots \\ P(E_1 \cap E_2 \cap E_3 \cap \cdots \cap E_n) &= P(E_1)P(E_2)P(E_3) \cdots P(E_n). \end{aligned}$$

35 percent of all men have incomes less than \$5000,
48 percent of all men have incomes between \$5000 and \$10,000,
17 percent of all men have incomes greater than \$10,000.

Finally, it has been found for *Fortune* that a man has a probability of .3 of reading an advertisement, given he is reading the magazine. Furthermore, this probability is independent of income. What is the probability this advertisement will be read by any particular man?

SOLUTION: First, define the following events

$R \equiv$ a man reads this ad,
$F \equiv$ a man reads *Fortune*,
$L \equiv$ a man's income is less than \$5000 (low),
$M \equiv$ a man's income is \$5000 to \$10,000 (middle),
$H \equiv$ a man's income is greater than \$10,000 (high).

Using Theorem 5 and the definition of conditional probability, we have

$$\begin{aligned} P(R) &= P(R \cap F) + P(R \cap \tilde{F}) = P(R \cap F) \\ &= P(R \mid F)P(F) = .3P(F). \end{aligned}$$

Similarly,

$$\begin{aligned} P(F) &= P(F \cap L) + P(F \cap M) + P(F \cap H) \\ &= P(F \mid L)P(L) + P(F \mid M)P(M) + P(F \mid H)P(H) \\ &= .004(.35) + .01(.48) + .05(.17) = .0147, \end{aligned}$$

so

$$P(R) = .00441.$$

(Although this probability is small, note that if there are 100 million men in the United States, then over 400,000 people will read the ad.) •

EXAMPLE 19[13]

To encourage Elmer's promising tennis career, his father offers him a groovy prize if he wins at least two tennis sets in a row in a three-set series to be played with his father and the club champion alternately: father–champion–father or champion–father–champion, according to Elmer's choice. The champion is a better player than Elmer's father. Which series should Elmer choose?

SOLUTION: Let W_1, W_2, and W_3 and L_1, L_2, and L_3 be the events that Elmer wins and loses the first, second, and third set, respectively, and let W be the event that Elmer wins two sets in a row. Then

$$W = [W_1 \cap W_2 \cap W_3] \cup [W_1 \cap W_2 \cap L_3] \cup [L_1 \cap W_2 \cap W_3].$$

[13] This example is due to Frederic Mosteller and appears as a problem (p. 1) in his book *Fifty Challenging Problems in Probability* (Reading, Mass.: Addison-Wesley Publishing Co., Inc., 1965).

(Note that the three sets in brackets are disjoint.) Finally, let f and c be the probability that Elmer wins a given set when playing father and the club champion, respectively. Now if we assume that the sets are independent of one another and Elmer chose the series $F - C - F$, then

$$\begin{aligned} P(W) &= P(W_1 \cap W_2 \cap W_3) + P(W_1 \cap W_2 \cap L_3) + (L_1 \cap W_2 \cap W_3) \\ &= fcf + fc(1 - f) + (1 - f)cf \\ &= fc(2 - f). \end{aligned}$$

Similarly, if Elmer chose the other series, then $P(W) = cf(2 - c)$. Since $f > c$, it now follows that Elmer should choose the series C–F–C. Surprisingly, the importance of winning the middle game outweighs the disadvantage of playing the champion twice. •

EXAMPLE 20. (Craps)

One of the most exciting and best-known gambling games is craps. The rules are as follows: the player rolls a pair of fair dice and observes the total. If the total is 7 or 11, the player wins immediately, while he loses immediately if the total is 2, 3, or 12. Any other total is called his point. If the game is not ended on the first toss, the player continues to roll the dice until either he rolls his point again, in which case he wins, or he rolls a 7, in which case he loses. We desire to know the probability that the player wins (W).

SOLUTION: We shall find this probability using two distinct approaches. First, note that the game is completely determined by the first and the last rolls, so that one reasonable sample space is given by

$$\begin{aligned} \Omega = &\{2, 3, 7, 11, 12\} \\ &\cup \{(4, 4), (5, 5), (6, 6), (8, 8), (9, 9), (10, 10), (4, 7), (5, 7), (6, 7), \\ &\qquad (8, 7), (9, 7), (10, 7)\}, \end{aligned}$$

or more simply,

$$\Omega = \{2, 3, 7, 11, 12\} \cup \{(i, j) : i \in \{4, 5, 6, 8, 9, 10\} \text{ and } j \in \{i, 7\}\}.$$

Here, the outcome of say 3 represents obtaining a 3 on the first roll. Of course, if a 3 is obtained on the first roll, then the first roll is also the last role. The pair (i, j) means that the first roll resulted in i and the last (but not necessarily second) roll resulted in j. Now, it only remains to find the probabilities of the points in Ω that correspond to winning. This can be done using conditional probabilities. For example, let's find the probability of $\omega = (4, 4)$. First note that there are three ways to roll 4 and six ways to roll 7. Thus, given that our first roll is a 4, the probability of winning is $3/(3 + 6)$, since the only rolls that matter are those which result in either 4 or 7. Thus,

$$P(\{(4, 4)\}) = \tfrac{3}{9} \cdot \tfrac{3}{36}.$$

We find the probabilities for the other favorable points of Ω in a similar way except for 7 and 11. Of course, $P(\{7\}) = \frac{6}{36}$ and $P(\{11\}) = \frac{2}{36}$. We now find that

$$\begin{aligned} P(W) &= P(\{7, 11, (4, 4), (5, 5), (6, 6), (8, 8), (9, 9), (10, 10)\}) \\ &= \tfrac{6}{36} + \tfrac{2}{36} + \tfrac{3}{9} \cdot \tfrac{3}{36} + \tfrac{4}{10} \cdot \tfrac{4}{36} + \tfrac{5}{11} \cdot \tfrac{5}{36} + \tfrac{5}{11} \cdot \tfrac{5}{36} \\ &\qquad + \tfrac{4}{10} \cdot \tfrac{4}{36} + \tfrac{3}{9} \cdot \tfrac{3}{36} \approx .493. \end{aligned}$$

▶▶

Next we can view each game as a sequence, say $(x_1, x_2, \ldots, x_n)$, of rolls where n is the length of the particular game and x_i is the outcome of the ith roll. In this case we write

$$\Omega = \bigcup_{n=1}^{\infty} \Omega_n,$$

where $\Omega_1 = \{2, 3, 7, 11, 12\}$ and $\Omega_n = \big\{(x_1, x_2, \ldots, x_n): x_i \notin \{x_1, 7\}$ for $1 < i < n$, and $x_n \in \{x_1, 7\}\big\}$ for $n \geq 2$. Of course, Ω_n is the set of games of length n. We now use Theorem 5 and condition upon the outcome of the first roll to obtain

$$P(W) = \sum_{i=2}^{12} P(W \mid i)P(i), \tag{1.5}$$

where "i" represents the event "i on the first roll." All of the quantities on the right-hand side of the equation are quite easily obtained except for $P(W \mid i)$ for $i = 4, 5, 6, 8, 9, 10$. To obtain these probabilities we let E_n denote the event that the length of the game is exactly n. Since $E_1, E_2, \ldots$ partition Ω, we can conclude from Theorem 5 [and $P(W \cap E_1 \mid i) = 0$]

$$P(W \mid i) = \sum_{n=2}^{\infty} P(W \cap E_n \mid i).$$

Using the fact that the n rolls are independent of one another, we obtain

$$P(W \cap E_n \mid i) = [1 - P(i) - P(7)]^{n-2} P(i).$$

Then by summing the infinite series we can conclude that

$$P(W \mid i) = \frac{P(i)}{P(i) + P(7)}, \qquad \text{for } i = 4, 5, 6, 8, 9, 10.$$

Substituting these values into (1.5), we see that the probability of winning is approximately .493.

One small but important point we have overlooked is the possibility that the game never ends. The rules of the game do allow such outcomes [for example, (4, 8, 8, 8, ...)], but the probability that the game never ends is 0. To see this, suppose the first roll resulted in a 4. Then for each n we have from Theorem 3 that

$$P(\text{never ending} \mid 4) \leq P(\text{length of game is at least } n \mid 4) = (1 - \tfrac{9}{36})^{n-2}$$

which decreases (to zero) as n increases (to infinity). •

▶

EXERCISES:

(In solving the exercises below, state explicitly the assumptions you need to make —for example, independence and equally likely events.)

26. An urn contains 5 white, 4 red, and 3 black balls. Another urn contains 5 white, 6 red, and 7 black balls. One ball is selected from each urn. What is the probability that they will be the same color?

27. An urn contains 3 black balls and 5 white balls. A second urn contains 6 white balls and 8 black balls. One ball is drawn from urn one and placed in urn two. If we now draw a ball from urn two, what is the probability that the ball is black?

28. Ten dice are thrown. Find the probability of getting at least one 3. [*Hint:* Use Theorem 1.]

29. A die is thrown until a 3 appears. What is the probability that it must be thrown more than ten times?

30. In a town of $n + 1$ inhabitants, a person tells a rumor to a second person, who in turn repeats it to a third person, and so forth. At each step the recipient of the rumor is chosen at random from the n people available. Find the probability that the rumor will be told r times without: (a) returning to the originator, (b) being repeated by any person. (This problem is to be found on page 56 of Feller's book; see Reference 4.)

31. In Example 15, find the probability that a caller spoke to a social worker on the same day he called $(I \cup S)$ if the caller came to the clinic for consultation.

32. Consider the system depicted below. The system is said to be operative if each stage is operative. We assume that the components within each stage work in parallel, that is, they are supplied with switching circuits that have the property of shunting a new component into the circuit when the old one fails. Of course, if all components in the stage fail, then the stage

is *not* operative. Suppose that the components within a stage are identical. In particular, each of the components in stages 1, 2, 3, and 4 have probabilities $\frac{1}{5}$, $\frac{1}{3}$, $\frac{1}{2}$, and $\frac{3}{4}$ of failing, respectively. Find the probability that the system operates. (This probability is called the *reliability* of the system.)

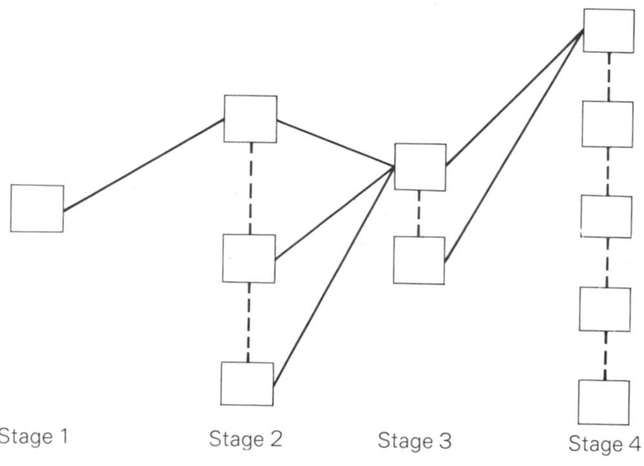

Figure 1.2

33. Someone has devised a system for winning at roulette. It is surprisingly simple and works as follows. When he bets, he only bets on odd, and he places a bet if and only if the ten previous spins of the roulette wheel have resulted in even. (Recall, the outcome is said to be odd if one of the numbers 1, 3, 5, . . . , 35 appears, and it is said to be even if one of the numbers 2, 4, 6, . . . , 36 appears. If 0 or 00 appears, the result is neither odd nor even.) He reasons that by using his system the chances of winning are greatly enhanced beyond $\frac{18}{38}$—as the probability of eleven consecutive spins resulting in "not odd" is terribly small. What do you think of his system? Where has he erred?

34. Three players I, II, and III take turns at playing a game. At the start I and II play while III is out. The loser is replaced by III, so the winner and III play the second game while the loser of the first game is out. The game continues in this way until a player wins twice in succession. Assume that (i) there is never a tie in one of the individual games, (ii) the three players are equally skillful, and (iii) the individual games are independent of one another.
    (a) Display both sample points of "the grand game" such that play never terminates. Show that each of these two points has probability zero.

(b) Show that I and II each have probability $\frac{5}{14}$ of winning while III has probability $\frac{4}{14}$ of winning. [*Hint:* Use the fact that $\sum_{i=0}^{\infty} (\frac{1}{8})^i = \frac{8}{7}$.]

(c) Show that the conditional probability of winning "the grand game" given that you have won the last game played (but not the last two) is $\frac{4}{7}$.

35. Consider a random experiment in which $A \subset B \subset \Omega$, $P(A) > 0$ and $P(B) < 1$. Are A and B independent events? Are A and Ω independent events?

36. Prove that if $A \cap B = \emptyset$, $P(A) > 0$, and $P(B) > 0$, then A and B are not independent.

37. Show that if A and B are independent, then A and $\tilde{B}$, $\tilde{A}$ and $\tilde{B}$, and $\tilde{A}$ and B are also independent.

38. Suppose that $P(A) > 0$ and $P(B) > 0$.
(a) Show that $P(A \mid B) = P(A)$ and $P(B \mid A) = P(B)$ if A and B are independent events.
(b) Show that $P(A \mid B) = P(A)$ implies that A and B are independent events.

In light of (a) and (b), we see that we could have defined A and B to be independent events if $P(A \mid B) = P(A)$.

39. Suppose that a red and a green die are thrown, and define the events A, B, and C as follows: A = the red die shows 1, 3, or 5, B = the green die shows 1, 3, or 5, and C = the sum on the two dice is 3, 5, 7, 9, or 11. Show that A and C, A and B, and B and C are independent. But also show that A, B, and C are not independent. Hence, pairwise independence is not as strong a property as mutual independence.

2.3 Bayes' Theorem

We have shown how to find the conditional probabilities of events when we are given additional information concerning the outcome of our random experiment. We now consider a special case of this general problem wherein our random experiment can often be thought of as being performed in two successive stages. In this case, we seek the conditional probabilities of the outcomes of the first stage of our experiment when we are given the outcome of the second stage. For example, suppose we first flip a fair coin and then draw a ball from urn H or from urn T depending whether the outcome was heads or tails, respectively. (Urn H contains 2 white and 1 black balls, and urn T contains 3 white and 3 black balls.) In order to find the conditional probability that the ball came from urn H

given that it is white $[P(H \mid W)]$, we employ Theorem 5 and the definition of conditional probability as shown below:

$$P(H \mid W) = \frac{P(H \cap W)}{P(W)} = \frac{P(W \mid H)P(H)}{P(W \cap H) + P(W \cap T)}$$
$$= \frac{P(W \mid H) \cdot P(H)}{P(W \mid H) \cdot P(H) + P(W \mid T) \cdot P(T)} = \frac{\frac{2}{3} \cdot \frac{1}{2}}{\frac{2}{3} \cdot \frac{1}{2} + \frac{1}{2} \cdot \frac{1}{2}} = \frac{4}{7}.$$

Also, $P(T \mid W) = \frac{3}{7}$. More generally, we have

BAYES' THEOREM

If $E_1, E_2, \ldots$ partition Ω, then for any event A with $P(A) > 0$ we have for each fixed i $(1 \le i < \infty)$,

$$P(E_i \mid A) = \frac{P(A \mid E_i)P(E_i)}{\sum_{j=1}^{\infty} P(A \mid E_j)P(E_j)}.$$

The sets $\{P(E_i): i = 1, 2, \ldots\}$ and $\{P(E_i \mid A): i = 1, 2, \ldots\}$ are called the *prior* and the *posterior* probabilities, respectively, since the former are the appropriate probabilities *before* we know the outcome of our experiment, while the latter are the appropriate probabilities *after* we know the outcome of the second stage of our experiment. The power of this method (theorem) is made evident in the following examples and exercises.

EXAMPLE 21. (Pólya's Urn Scheme)

An urn contains b black balls and r red balls. One of the balls is drawn at random, but when it is put back in the urn an additional $c > 0$ balls of the same color are put in with it. Now suppose that we draw another ball. What is the probability that the first ball drawn was black given that the second ball drawn was red?

SOLUTION: Using Bayes' theorem and assuming that each ball is drawn with equal probability, we have

$$P(\text{1st } B \mid \text{2nd } R) = \frac{P(\text{2nd } R \mid \text{1st } B) \cdot P(\text{1st } B)}{P(\text{2nd } R \mid \text{1st } B) \cdot P(\text{1st } B) + P(\text{2nd } R \mid \text{1st } R) \cdot P(\text{1st } R)}$$
$$= \frac{\frac{r}{b+r+c} \cdot \frac{b}{b+r}}{\frac{r}{b+r+c} \cdot \frac{b}{b+r} + \frac{r+c}{b+r+c} \cdot \frac{r}{b+r}} = \frac{b}{b+r+c}. \bullet$$

Bayes' theorem is also helpful even when the experiment is not performed in two successive stages. This is illustrated in our next example. (Also see Exercises 42, 43, 44, and 45.)

EXAMPLE 22

George has a motorcycle that is extremely fast but not too dependable. At any rate, it quit suddenly on him one day. He called Mac, the mechanic, who asked him whether it quit suddenly, or coughed and gasped before dying. When told it quit suddenly, Mac made a telephone diagnosis of distributor trouble (which turned out to be correct). How did he accomplish this? Well, Mac knows that with this kind of cycle (essentially 100 percent of) the trouble is due either to the distributor, the carburetor, or the fuel pump, with these accounting for (approximately) 40 percent, 50 percent, and 10 percent of the breakdowns, respectively. If the distributor malfunctions, the cycle quits suddenly 75 percent of the time, while the chances that the cycle quits suddenly when either the carburetor or the fuel pump malfunctions are 10 percent and 20 percent. Mac used this information to make his telephone diagnosis even though he wasn't certain of his diagnosis, since when he is correct it makes a good impression on the customer, and he is willing to risk an occasional incorrect diagnosis to achieve this. Nevertheless, he would like to know the probability that the trouble was in the distributor (D) given the information—namely, the cycle quit suddenly (QS). Find this probability.

SOLUTION: We use Bayes' theorem to obtain

$$P(D \mid QS) = \frac{P(QS \mid D) \cdot P(D)}{P(QS \mid D) \cdot P(D) + P(QS \mid C) \cdot P(C) + P(QS \mid FP) \cdot P(FP)}$$

$$= \frac{30}{37}. \bullet$$

EXERCISES:

40. In Exercise 22, find the probability that urn A was selected when the outcome of the experiment was a white ball.

41. In Exercise 27, find the probability that the first ball drawn was black given that the second ball drawn was white.

42. Suppose that 1 male student out of 5, and 1 female student out of 20, is a science major. A science major is chosen at random from a student population that has twice as many males as females. What is the probability that this student is a male?

43. In Exercise 13, find the probability that a student picked at random and found to own a car is an upperclassman.

44. Wearing beads is very stylish at the university this year. In fact, 20 percent of the men and 60 percent of the women wear them. Taking into account

the fact that 70 percent of the student body is male, find the probability that the next person you meet on campus wearing beads is a male.

45. On a certain multiple-choice examination there is but one question. If a student knows the subject material on the test, he should be able to give the correct answer with probability $\frac{3}{4}$. If he does not know the material, he will have a probability of $\frac{1}{5}$ of answering the question. In the past, the instructor has found that 30 percent of the students know the material and 70 percent do not. If a student takes the test and answers the question correctly, then what is the probability he knows the material?

46. A ball is withdrawn from either urn I or II with probabilities $\frac{3}{4}$ and $\frac{1}{4}$, respectively. This ball is placed in urn III. If a ball is then drawn at random from urn III, find the probability that it came from urn I if the urns contain 3, 4, and 5 white balls and 7, 6, and 5 black balls, respectively, and if the ball drawn from urn III was white.

47. Consider a family of three—mother, father, son. On any particular day, the probability of each deciding (individually) to go to the beach is $\frac{1}{3}$, $\frac{1}{4}$, and $\frac{4}{5}$, respectively. However, if the father goes, the mother will go with probability $\frac{3}{5}$. Also, if both mother and father go, the son's probability of going is $\frac{2}{5}$. If the father decides to go to the beach, what is the probability that all three will go?

48. Our geologist tells us that .8 is the probability that there is oil on our land; moreover, if there is oil on our land, the probability of hitting oil with the first well is .5. What would you say about the probability that there is oil on our land if we drilled a dry well?

49. Joe is known to be less than honest; in fact, 10 percent of the time Joe uses 2-headed coins (90 percent of the time he uses a fair coin). On one particular occasion, the first 5 flips resulted in heads. What is the probability that Joe was using a 2-headed coin?

50. The secretary has carelessly placed your letter in one of 6 folders. Let q_i be the probability that she will find your letter upon making one examination of folder i if it is, in fact, in folder i ($i = 1, 2, \ldots, 6$). (We may have $q_i < 1$.) Suppose she looks in folder i and doesn't find the letter. What is the probability of the letter being in folder i? What if she looks a second time and doesn't find it?

51. The x-group hires Mr. X to observe the outcomes of experiments performed by members of the y-group. The outcomes of those experiments are either favorable (F) to the y-group or unfavorable (U), with $P(F)$ being the probability of a favorable outcome. Moreover, the experiments are independent of one another. Since the two groups are in competition and Mr. X is loyal to the x-group, he finds it hard to report favorable outcomes, so there is a bias in his reports. This is reflected in the fact that given the outcome is actually favorable, the probability he reports it as favorable (F') is only $\frac{1}{2}$. When the outcome is unfavorable, he delightedly reports

it accurately so that $P(U' \mid U) = 1$, where U' stands for the event Mr. X reports outcome as unfavorable. You, as a bystander, have noted a long series of Mr. X's reports and thereby have been able to estimate $P(F')$ directly from your own experience. What you don't know is $P(F)$, which reflects how well the y-group is really doing in its experiments. What can you conclude about $P(F)$? Express your answer in terms of $P(F')$. (This exercise and Example 22 are due to James B. MacQueen.)

52. By Axiom 2, $P(\Omega) = 1$. However, in Example 22 about Mac the mechanic, we don't require that the probabilities .75, .1, and .2 add up to 1. Why?

3. COUNTING POINTS IN A SAMPLE SPACE

Counting, like factoring an algebraic expression, is an art that one learns through practice. And, as in factoring, there are a few basic methods that, when employed properly, enable us to solve a great variety of problems. It is the purpose of this section to present some of these methods.[14]

3.1 Sampling Without Replacement

Fortunately, Nancy, a pretty young housewife and brilliant psychologist, is available to help us motivate and explain our perusal of counting—or what is more commonly called combinatorial analysis. Nancy has just completed a course in psychology wherein she learned of the many advantages inherent in exposing her young child Laura to as many different environmental stimuli as possible. Today, Nancy noticed Laura focusing her attention upon the 5 library books lined up on the bookshelf. Being only six months old, Laura can't read the titles, but she can distinguish colors. Each of the books is a different color. Noting this, Nancy gasped gleefully, for she quickly realized that she could present Laura with 120 different environmental stimuli using only the 5 books on the shelf. Nancy reasoned as follows: "I can place any one of the 5 books at the far left (position one). Having done this, I am now free to arrange the four remaining books in the four remaining positions. Consequently, there are 5 times as many environments available with 5 books as there are with 4 books." (Think about this for a few minutes!) Continuing in this manner, she finally concluded that there were $5 \cdot 4 \cdot 3 \cdot 2 \cdot 1 = 120$ different environments.

The following week Nancy returned home from the library with 5 different colored books, but this time the books were so thick that she could only put 2 on the shelf at any one time. Laura didn't seem to mind, but Nancy sighed sadly, for she realized that this week Laura would have

[14] The reader is cautioned to proceed slowly through this section. "Good things come to those who wait."

but 20 different environmental stimuli. Nancy figured this out in a manner similar to that above: Any one of the 5 books can be placed in position one. Having done this, I can choose any one of the 4 remaining books to place in position two, so there are $5 \cdot 4 = 20$ possibilities in all.

In the two examples above, we have illustrated what is called sampling without replacement. Presently we shall see how this ties in with probability theory, but first we generalize the ideas of the preceding two examples.

DEFINITION

Let $a_1, a_2, \ldots, a_n$ be a population (set) of n distinguishable elements (such as colored books). Any ordered arrangement of s distinct elements of the population is called an *ordered sample of size s without replacement;* that is, if $(j_1, j_2, \ldots, j_s)$ is a sequence of s distinct integers between 1 and n, then $(a_{j_1}, a_{j_2}, \ldots, a_{j_s})$ is an ordered sample of size s without replacement from our population. Less formally, we often say that such a sample is a *sample of size s without replacement.*

Define $n! \equiv n(n-1)(n-2)\cdots 3\cdot 2\cdot 1$ for each positive integer n.[15] Then, employing the reasoning in the example above, we have

THEOREM 6

For a given population of size n, there are $n!/(n-s)!$ distinct samples of size s without replacement.

Next, consider the random experiment wherein Nancy places the 5 books on the shelf blindfolded. Then our sample space Ω consists of the 120 distinct environments, and each point (environment) in Ω is equally likely so that $P(\{\omega\}) = \frac{1}{120}$ for each point ω in Ω. Similarly, we have $P(\{\omega\}) = \frac{1}{20}$ for each of the 20 points ω in Ω in the second example. Of course, a great many random experiments are abstractly equivalent to placing books on a shelf. In this case, we attach probability $1/[n!/(n-s)!]$ to each (*ordered*) *random sample of size s without replacement* from a population of size n.

Evidently, it is often the case that finding the probability of a particular event is equivalent to counting both the number of points in the sample space and the number of points that comprise the event itself. That is,

$$P(E) = \frac{\text{number of points in } E}{\text{number of points in } \Omega}.$$

We make this point more obvious in the following example and in Exercises 55 through 69.

[15] We define $0! = 1$, and $n!$ is read "n factorial."

EXAMPLE 23. (Blackjack)

A hand of blackjack contains two cards. If one of them is an ace and the other is a card of value 10 (a ten, jack, queen, or king), then the hand is called a blackjack. What is the probability of being dealt a blackjack from a single well-shuffled deck of cards?

SOLUTION: The sample space Ω consists of all ordered pairs of cards, so there are $52 \cdot 51$ members of Ω, since the number of ordered samples of size 2 from a population of size 52 is $52 \cdot 51$ by Theorem 6. Moreover, each of these sample points has probability $1/(52 \cdot 51)$, as the cards are well shuffled. Next, observe that there are $4 \cdot 16 + 16 \cdot 4 = 128$ ways (sample points) of being dealt a blackjack; $4 \cdot 16$ is the number of ways of first being dealt an ace and then a card of value 10, whereas $16 \cdot 4$ is the number of ways of first being dealt a card of value 10 and then an ace. Hence, the probability of being dealt a blackjack is $128/(52 \cdot 51)$. •

Continuing with our original example, the third week Nancy returned home with 4 books thin enough to be placed on the shelf simultaneously; however, this time they weren't all different colors. Two of the books were red. Laura smiled unperturbed while Nancy murmured mournfully, for this time she figured that there were but 12 distinct environments (samples of size 4). Nancy arrived at the figure 12 by observing that Laura would have seen twice as many environments as she did see had she been able to differentiate between what to her were 2 indistinguishable books. It follows from Theorem 6 that Laura would have seen 4! environments had she been able to differentiate between the two red books. Consequently, Laura could see but $4!/2 = 4!/2!1!1! = 12$ environments. More generally, we have

THEOREM 7

If a population of n elements can be divided into k parts or subpopulations so that elements from different parts are distinguishable and elements from the same part are indistinguishable, then the number of distinguishable ordered samples of size n without replacement is

$$\frac{n!}{r_1!r_2!\cdots r_k!},$$

where r_i is the number of elements in part i, $i = 1, 2, \ldots, k$.

A very important special case of Theorem 7 occurs when $k = 2$. In this case we have $r_2 = n - r_1$, and we write

$$\binom{n}{r_1} = \frac{n!}{r_1!(n - r_1)!}.$$

The symbol $\binom{n}{r_1}$ is read "n objects taken r_1 at a time." The reason for this terminology is that $\binom{n}{r_1}$ is the number of ways of selecting r_1 objects from a population of size n. That is, $\binom{n}{r_1}$ is the number of sets (so order is not relevant) of r_1 objects we can obtain from a population of n distinguishable objects. By selecting r_1 objects, we have in effect partitioned the population into two subpopulations of size r_1 and $n - r_1$, where we think of the elements within a subpopulation as being indistinguishable. The subpopulation of size r_1 contains those objects we have selected, while the subpopulation of size $n - r_1$ contains those objects we have not selected.

The inquisitive student might ask why the number of ways of selecting r_1 objects from a population of size n is not $n!/(n - r_1)!$, as indicated by Theorem 6. The reason is partly one of semantics. In this case, when we say "selecting r_1 objects," we mean "forming a subpopulation of size r_1," and two subpopulations are different only if they do not contain the same elements—that is, order does not count. Thus, the (inappropriate) use of Theorem 6 results in counting each distinct subpopulation of size r_1 $r_1!$ times. Consequently, $n!/[(n - r_1)!r_1!]$ is the correct number.

When order is not important, many authors use the word *combinations*, while they use the word *permutations* when order is important. We do not employ these two words, as their use often results in a disfunctional rigidity of thought via the students' dichotomization of combinatorial problems.

EXAMPLE 24. (Beat the Dealer)

In mid-1969, the Union Oil Company introduced a popular promotional game called "Beat the Dealer." Every time you visited a Union Oil service station you would get one ticket, a sample of which is shown in the accompanying illustration. (Note that the numbers in the shaded areas are not visible.) Here's how to play: Carefully use the edge of a coin to rub off the spot below the words "Dealer's Hand." Underneath is the score you must beat to win and the cash amount you can win by beating that score. Carefully rub off any 3—*but only* 3—of the spots on the "playing card." If the numbers revealed under the 3 spots you rub off add up to *more* than the score of the "Dealer's Hand," you win the prize specified. As in the sample ticket shown, there is always one, but only one, winning combination. What is the probability that a person wins on any given ticket?

Figure 1.3 Sample Ticket

SOLUTION: Since there are $\binom{16}{3} = 560$ possible choices of 3 spots to rub off and since exactly one of these choices leads to winning, the probability of winning is $\frac{1}{560}$. •

EXAMPLE 25. (Poker[16])

A poker hand contains 5 cards. When a well-shuffled deck is used, what is the probability of getting a flush (all five cards of the same suit)? a full house (a pair and three of a kind)?

SOLUTION: Since the deck is well shuffled, it follows from Theorem 7 that each of the $\binom{52}{5}$ possible hands is equally likely. Next, note that there are $4\binom{13}{5}$ ways of getting a flush because there are 4 different suits of 13 cards each. Thus, the probability of a flush is

$$4\binom{13}{5}\Big/\binom{52}{5} \approx .002.$$

Similarly, there are $\binom{4}{2}\binom{4}{3}$ ways of getting, say, 2 aces and 3 kings. But there are 13 different choices for the kind of pair, and, having chosen a

[16] An ordinary deck of playing cards contains 52 cards arranged in four suits of thirteen each. There are thirteen face values (2, 3, . . ., 10, jack, queen, king, ace) in each suit. The four suits are called spades, clubs, diamonds, and hearts. The last two are red, and the first two black. Cards of the same face value are called of the same kind. For our purposes, playing poker means selecting five cards from the deck.

pair, there are 12 other choices for the kind of three of a kind. Thus, the probability of a full house is

$$13 \cdot 12 \binom{4}{2} \binom{4}{3} \Big/ \binom{52}{5} \approx .0014. \bullet$$

EXERCISES:

53. Suppose that all foods eaten affect the taste of subsequently eaten foods. (a) How many different 7-course dinners can be served if the 7 individual courses themselves are unchanged? (b) Suppose that we are late so that there is time enough for only 5 courses? (c) Suppose that, in addition to our being late, the meal must include the meat course?

54. How many "environments" can Nancy construct with Laura's 11 blocks, which spell Mississippi when properly aligned? What if Nancy uses only 10 blocks instead of 11? (Does it matter which 10 she uses?)

55. What is the probability of being dealt a blackjack from two well-shuffled decks of cards? Compare your answer with Example 23.

56. Solve Exercises 23 and 24 using counting techniques.

57. Suppose you purchased 3 lottery tickets, 100 tickets were sold in all, and 5 prizes were awarded (fairly). What is the probability that you won at least one prize?

58. A ship carrying 12 men, 10 women, and 4 children is sinking. In order to determine who will board the lifeboat, six names are to be drawn from a hat containing all 26 names. (a) What is the probability that the lifeboat will contain only women and children? (b) What is the conditional probability of this event given that the names of all children are drawn?

59. The merry-go-round has 22 seats, and there are 22 people in all. If everyone simply "scrambles" for a seat, what is the probability that your girlfriend will (a) sit in front of you? (b) sit next to you—that is, in front or in back? Compare this with Exercise 24.

60. If two people randomly choose a seat at a lunch counter that has n seats, what is the probability that they will sit next to each other? Compare this with Exercise 24.

61. (Snowball) In order to get the party rolling, the host and hostess each asked someone to dance. At the end of the dance, these four people (including the host and hostess) each asked someone else (not yet on the dance floor) to dance, and so on until everyone was dancing. If there were $2n + 2$ people at the party (including the host and hostess), if $n + 1$ people were

women, and if men ask only women to dance and vice versa, then what is the probability that you will be asked to dance the third dance?

62. (Musical chairs) The game of musical chairs is usually played as follows: n people are seated on chairs; when the music starts, they all get up and walk in a circle around the chairs. While they are walking, one chair is removed so that one person cannot find a seat when the music stops. This person is eliminated from the game. The music starts again, another chair is removed, and so on until there is but one person left. This person is declared the winner. What is the probability that (a) you win? and (b) you are removed the third time the music stops?

63. Find (*a*) and (*b*) in Exercise 62 when $2n + 1$ people start and two people are removed each time the music stops.

64. In a deal of bridge, 13 cards are dealt to each of the four players. Find the probability that in a deal of bridge, player one gets (a) all 4 aces; (b) exactly 3 aces; (c) exactly 2 aces. [*Hint:* Use Theorem 5.]

65. Find the probability that in a poker hand you are dealt (a) two pairs; (b) three of a kind; (c) a straight;[17] (d) four of a kind; (e) one pair.

66. Suppose you have been dealt one pair in a poker hand. Find the probability of getting exactly two pairs if you now draw (a) 2 more cards; (b) 3 more cards. Of course, you "discard" the number of cards that you draw.

67. In order to get the job done, there must be at least 2, 4, 6, and 7 men working on subjobs 1, 2, 3, and 4, respectively. When no one is sick, there are 3, 5, 7, and 8 men on the subjobs. Looking at the number of time cards, the foreman discovered that there were 4 men absent. What is the probability that the job can be done? What if 5 men had been absent and the usual number of men on the subjobs was 4, 6, 8, and 9?

68. How many environments could Laura distinguish if Nancy brought home 5 books so thick that only 3 could be placed on the shelf simultaneously and two of them were the same color? [*Hint:* None of the theorems presented are appropriate here. Develop an ad hoc procedure.]

69. Show that the probability that some one particular element is part of an ordered random sample of size r without replacement is r/n if the population size is n. [*Hint:* Use Theorem 1.]

3.2 Sampling with Replacement

After spring cleaning, Nancy had 4 shelves upon which to place her 5 different colored library books. Each shelf had room to accommodate all 5 books. After a moment's reflection, Nancy trilled triumphantly for she

[17] A straight is 5 consecutive numbers where jack, queen, and king are 11, 12, and 13, respectively, while ace is either 1 or 14.

knew that this week there were $4^5 = 1024$ different environments.[18] She reasoned as follows: I can place book number one on any one of the 4 shelves. Having done this, I am still free to place the 4 remaining books on any of the 4 shelves (because each shelf is wide enough to accommodate all 5 books). Consequently, there are 4 times as many environments available with 5 books and 4 shelves as there are with 4 books and 4 shelves. Continuing in the indicated manner, she finally concluded that there were $4 \cdot 4 \cdot 4 \cdot 4 \cdot 4 = 4^5$ different environments in all. (Go back and compare this with the first counting method presented!)

In his book on probability (see Reference 4), W. Feller provides another interpretation of the general situation described above by considering each book to be a ball and each shelf to be a cell or box. [On pages 10–11, Feller also gives 16 distinct examples (with applications) which are abstractly equivalent to placing books on shelves.] Then there are 4^5 ways of placing 5 balls in 4 cells. That is, there are 4^5 ways of choosing one cell for each ball—*not* one ball for each cell. We now make all this more precise by introducing the following

DEFINITION

Let $a_1, a_2, \ldots, a_n$ be a population of n distinguishable elements (such as shelves). Any ordered arrangement of s (not necessarily distinct) elements of the population is called an *ordered sample of size s with replacement;* that is, if $(j_1, j_2, \ldots, j_s)$ is a sequence of (not necessarily distinct) integers between 1 and n, then $(a_{j_1}, a_{j_2}, \ldots, a_{j_s})$ is an ordered sample of size s with replacement from our population. Less formally, we often say that such a sample is a *sample of size s with replacement.*

In the example above, the books themselves were also distinguishable, so we had to take into account which book was on which shelf. (Ponder this point.) Also, note that it is possible to have $s > n$ when sampling with replacement, but this is not possible if we are sampling without replacement. Employing the reasoning above, we can now establish

THEOREM 8

For a given population of size n there are n^s distinct ordered samples of size s with replacement.

Thus, there are n^s ways of placing s balls on n shelves. If we attach probability $1/n^s$ to each ordered sample of size s with replacement, then we refer to the samples as (*ordered*) *random samples with replacement.*

[18] Owing to the added complexity of this week's configurations, Laura noticed only which book was on which shelf and not, in addition, the order of the books on each shelf.

EXAMPLE 26. (Three Sisters)[19]

In a certain family, three sisters take turns at washing the dinner dishes. During the year, 4 dishes were broken, 3 of these by the youngest. Thereafter she was called clumsy. Was she justified in attributing this to chance?

SOLUTION: First identify each dish with a book and each girl with a shelf. We will find the probability of the event $E \equiv$ exactly 3 books on shelf number 1 under the assumptions that the girls are equally clumsy and that the breakages occur independently. Under these assumptions, each of the 3^4 points in the sample space has probability $1/3^4$. Since there are 3 books on shelf number one, the remaining book can be placed on either shelf two or three. Moreover, the remaining book could have been any one of the 4 books. Thus, there are $4 \cdot 2 = 8$ points in E, so that $P(E) = 8/3^4 \approx .099$. Apparently this is evidence that the youngest sister is indeed clumsy; however, we shall pursue this point further in Chapter 4. •

EXAMPLE 27

Suppose we rolled a fair die 3 times. Find the probability that $A_0 \equiv$ no six appears, $A_1 \equiv$ one six appears, $A_2 \equiv$ two sixes appear, and $A_3 \equiv$ three sixes appear.

SOLUTION: Identify each number between 1 and 6 with a shelf and each roll of the die with a book. Then there are 6 shelves and 3 books. Hence, there are $6^3 = 216$ sample points in Ω—that is,

$$\Omega = \{(i, j, k) : i, j, k \in \{1, 2, \ldots, 6\}\}.$$

If there are no books on the sixth shelf, then each of them can be put on any one of the 5 remaining shelves. By Theorem 8, there are 5^3 ways of doing this. Thus, $P(A_0) = 5^3/6^3$. There $\binom{3}{1} = 3$ ways of choosing one book to be placed on the sixth shelf, and the remaining 2 books can be placed on the other 5 shelves in 5^2 ways. Hence, $P(A_1) = 3 \cdot 5^2/6^3$. Similarly, $P(A_2) = 3 \cdot 5/6^3$, as there are $\binom{3}{2} = 3$ ways of choosing two books to be placed on the sixth shelf and 5^1 ways of placing the one remaining book on the other 5 shelves. Finally, $P(A_3) = 1/6^3$. •

EXAMPLE 28. (The Birthday Problem)

Suppose that there are s people in a room. What is the probability that $E \equiv$ at least two people have a common birthday?

[19] This example is due to William Feller and appears as an exercise (p. 56) in his renowned book (Reference 4).

Solution: We assume that the probability of being born on any given day of the year is $\frac{1}{365}$ and that births are independent of one another. Consequently, there are 365^s equally probable points in our sample space. (Here the cells represent days of the year and the balls represent people.) Specifying that no two persons have the same birthday is easily seen to be equivalent to sampling (from the set of 365 days in the year) without replacement so the number of points in $\tilde{E}$ is $365!/(365 - s)!$ Thus,

$$P(E) = 1 - \frac{365!}{365^s(365 - s)!}$$

by Theorem 1. For $s = 23$, we have $P(E) > \frac{1}{2}$, a surprising result indeed. •

More generally, we have the very useful

THEOREM 9

The probability that an ordered random sample of size $s(\leq n)$ with replacement consists entirely of distinct elements from a population of size n is $n!/[n^s(n - s)!]$.

Shortly after spring cleaning, Nancy again returned from the library with 5 books. This time, however, all 5 books were red; that is, to Laura they were indistinguishable. She gulped gloomily, for she ascertained that there were only $\binom{4 + 5 - 1}{5} = 56$ different ways of placing the 5 indistinguishable books on the 4 shelves. She figured this out only after having seen the following nonobvious theorem.

THEOREM 10

There are $\binom{n - 1 + r}{r}$ ways of placing r indistinguishable balls in n cells.

Proof.[20] Represent each ball by the symbol ∘ (circle), and let the n spaces between the $n + 1$ symbols | (bar) represent the n cells. (For example, the arrangement |∘∘∘| |∘| means that the first cell contains 3 balls, the second cell contains no balls, and the third cell contains one ball.) For each way of placing r indistinguishable balls in n cells there corresponds precisely one arrangement of the r circles and the $n - 1$ interior bars. Conversely, each such arrangement corresponds to precisely one way of placing r indistinguishable balls in n cells. Now it easily follows from Theorem 7 with $k = 2, n_1 = r$, and $n_2 = n - 1$ that there are $\binom{n - 1 + r}{r}$ distinguishable arrangements of the r circles and the $n - 1$ interior bars. ◆

[20] This well-known proof is due to Feller.

Applications and extensions of this theorem are given in Exercises 81 through 84.

▶

Our perusal of counting is now over. We have presented four basic methods for counting. There are, however, many other methods with application to a plethora of fascinating problems, and we refer the reader to Feller's excellent book for additional material and references.

Before concluding this section, we remark that just as "there is more than one way to skin a cat," there is often more than one (good) way to solve a counting problem. Furthermore, Theorems 6 through 10 are not sufficient to solve all the exercises, let alone all counting problems. We must be flexible and develop our own theorems or ad hoc procedures when necessary.

EXERCISES:

70. California license plates have 3 letters and 3 numerals with the 3 letters appearing first. How many different license plates can the state make?

71. Solve Exercise 30 using counting techniques.

72. Suppose that a bag contains 100 tickets, 3 of which are yours. Five tickets are drawn at random with replacement. What is the probability that at least one of your tickets was drawn at least once? Compare this with Exercise 57.

73. Each day substitute teachers are chosen by randomly drawing their names from a list. The list has 10 names and 3, 1, 3, 5, and 2 people were called each day this week. If your name were on the list and you weren't called this week, would you have reason to believe the names weren't being drawn at random?

74. Use Theorem 1 to find the probability of throwing at least one six with 9 dice.

75. Generalize Example 26 by finding the probability that a particular cell contains exactly k of the s balls when the balls are placed randomly in the n cells. $\left[\textit{Answer: } \binom{s}{k}(n-1)^{s-k}/n^s.\right]$

76. Suppose that in Example 26 the first two dishes were broken by different girls. Find the probability that exactly one of the girls broke exactly 2 dishes; 3 dishes.

77. Find the appropriate probability in Exercise 75 given that the first j balls have been placed in different cells.

78. If s balls are randomly placed in n cells, what is the probability that cell number 1 is occupied? Notice the similarity to Exercise 69, and give a heuristic explanation of why this probability is smaller than the one found in Exercise 69.

79. If s balls are randomly placed in n cells and no two balls are in the same cell, what is the probability that cell number 1 is occupied?

80. Suppose that cell number 1 is occupied. What is the probability that cell number 1 is multiply occupied when s balls have been randomly placed in the n cells?

81. An elevator starts at floor 1 with 7 passengers (including Laura) and discharges them all by the time it reaches the top floor, number 10. If Laura got off at the top floor, then how many different ways of discharging the passengers could she have perceived if the passengers all looked alike to her? What if there were 4 men and 3 women and she can tell a man from a woman?

82. An art collection that was on auction included 2 Dalis, 3 Picassos, and 4 Kandinskys, and at the auction were 5 art collectors. The society page reporter only observed how many Dalis, Picassos, and Kandinskys each collector acquired. How many different results could she have reported?

83. How many ways are there of placing r indistinguishable balls in n cells so that no cell is unoccupied? (This question arises in connection with the theory of runs. See Section 3 of Chapter 9.)

84. How many solutions in nonnegative integers are there to the inequality $x_1 + x_2 + \cdots + x_n \leq r$ where r is an integer? (This problem arises in connection with integer programming.)

2

RANDOM VARIABLES AND EXPECTATION

1. RANDOM VARIABLES

In this section, we undertake the study of real-valued functions[1] whose domain is the sample space Ω of our probability model (Ω, P).

Often we are given a particular probability model, but our interest centers only on a class of (disjoint) subsets of Ω (whose union is Ω) and not on all the subsets of Ω. Consider, for example, the outcome of each individual roll in the game of craps (see Example 20 of Chapter 1). Here we are interested in the various totals of the two dice rather than the points showing on each die; that is, we are only interested in knowing $P(E_k)$, where

$$E_k = \{(i, j): i + j = k,\ i, j = 1, 2, \ldots, 6\},\ k = 2, 3, \ldots, 12.$$

In this instance, we identify the event E_k with the real number k. Implicitly, we have defined a map X of the sample space Ω into the real numbers by the rule (function) $X(\omega) = k$ for each $\omega \in E_k$, $k = 2, 3, \ldots, 12$. The function X is called a random variable. Of course, X is simply the total showing on the two dice. Other examples of random variables we have encountered include

(1) the number of people having the same birthday in a group of 23,
(2) the number of dishes broken by the youngest of three sisters,
(3) the number of pairs in a poker hand,
(4) the number of sixes that appears when a fair die is tossed 3 times,

[1] Recall that a function is real-valued if its range is a subset of the real numbers.

(5) the number of white balls drawn from an urn with 2 white balls and 3 black balls when two balls are drawn without replacement,
(6) the number of heads that appears when a coin is tossed n times,
(7) the number of flips of a coin required until a head appears, and
(8) the number of girls in a family with two children.

This motivates the following

DEFINITION

A real-valued function X whose domain is the sample space Ω is called a *random variable.*

Thus, a random variable attaches a value to each sample point. The term random variable is slightly misleading. Given a point ω in our sample space, $X(\omega)$ is simply a single real number which is by no means random. What is random, however, is the particular value X will assume, since the outcome ω of the random experiment is not known with certainty.

Let X be a random variable that assumes the values $x_1, x_2, \ldots$; that is, the range of X is $\{x_1, x_2, \ldots\}$. For each i, we shall find it convenient to write $X = x_i$ for the event consisting of those sample points mapped into x_i by X; that is,

$$X = x_i \equiv \{\omega \in \Omega: X(\omega) = x_i\}.$$

Usually, we shall denote the probability of the event $X = x_i$ by $f_X(x_i)$ instead of $P(X = x_i)$.

DEFINITION

Given a random variable X, the function f_X defined by

$$f_X(x_i) = P(X = x_i), \qquad i = 1, 2, \ldots,$$

is called the *probability mass function* (p.m.f.) of X.

We remark that if Ω is a set of real numbers, then $Y(\omega) \equiv \omega$ is a random variable with p.m.f. f_Y defined by $f_Y(\omega) = P(\{\omega\})$. Sometimes we shall simply write f instead of f_X or f_Y if there is no ambiguity as to the random variable corresponding to the p.m.f. f.

EXAMPLE 1

Let X be the number that appears when a single fair die is rolled. Then $\Omega = \{1, 2, 3, 4, 5, 6,\}$ and $X(\omega) = \omega$. Clearly the p.m.f. of X is given by

$$f_X(1) = \tfrac{1}{6}, \quad f_X(2) = \tfrac{1}{6}, \quad f_X(3) = \tfrac{1}{6}, \quad f_X(4) = \tfrac{1}{6}, \quad f_X(5) = \tfrac{1}{6}, \quad f_X(6) = \tfrac{1}{6}. \bullet$$

EXAMPLE 2

Let X be the total number that appears when a pair of fair dice is rolled. Then $\Omega = \{(i, j): i, j = 1, 2, 3, 4, 5, 6\}$ and $P(\{\omega\}) = \frac{1}{36}$ for each $\omega \in \Omega$. Thus, $X((i, j)) = i + j$ for each $(i, j) \in \Omega$. For example, $X((2, 2)) = 4$, $X((3, 1)) = 4$, and $X((1, 3)) = 4$. Note too that the event

$$X = 4 = \{(i, j) \in \Omega: i + j = 4\} = \{(2, 2), (1, 3), (3,1)\}$$

($= E_4$ as described above), so $f_X(4) = P(X = 4) = \frac{3}{36}$. Similarly, we obtain

$$f_X(2) = \tfrac{1}{36},\quad f_X(3) = \tfrac{2}{36},\quad f_X(4) = \tfrac{3}{36},\quad f_X(5) = \tfrac{4}{36},\quad f_X(6) = \tfrac{5}{36},$$
$$f_X(7) = \tfrac{6}{36},\quad f_X(8) = \tfrac{5}{36},\quad f_X(9) = \tfrac{4}{36},\quad f_X(10) = \tfrac{3}{36},$$
$$f_X(11) = \tfrac{2}{36},\quad f_X(12) = \tfrac{1}{36}. \bullet$$

EXAMPLE 3. (Three Sisters)

Let X be the number of dishes broken by the youngest sister in Example 26 of Chapter 1. The sample space Ω consists of 81 points (distributions of the 4 dishes among the 3 sisters), each having probability $\frac{1}{81}$. Using our counting techniques, we see that the p.m.f. of X is given by

$$f_X(0) = \frac{2^4}{81},\quad f_X(1) = \frac{4\cdot 2^3}{81},\quad f_X(2) = \frac{\binom{4}{2}2^2}{81},\quad f_X(3) = \frac{4\cdot 2}{81},\quad f_X(4) = \frac{1}{81}. \bullet$$

EXAMPLE 4

Let X be the number of flips of a fair coin required until the first head appears. Then the sample space Ω is seen to be

$$\Omega = \{H, TH, TTH, TTTH, \ldots\},$$

and, for example, $X(TTH) = 3$. Of course, the range of X is $\mathfrak{N}$, the set of positive integers. Since the flips are independent and the coin is fair, it follows that

$$f_X(1) = \tfrac{1}{2}, \qquad f_X(2) = (\tfrac{1}{2})^2, \quad \text{and} \quad f_X(3) = (\tfrac{1}{2})^3.$$

More generally, the p.m.f. of X is given by $f_X(i) = (\frac{1}{2})^i$, $i = 1, 2, 3, \ldots$. •

Since the events $X = x_i$, $i = 1, 2, \ldots$, are disjoint and their union is Ω, we can use Axioms 1, 2, and 3 to obtain

$$f(x_i) \geq 0 \quad \text{and} \quad \sum_{i=1}^{\infty} f(x_i) = 1. \tag{2.1}$$

Now, in light of Exercise 17 of Chapter 1, it is evident that specifying a probability mass function f is equivalent to specifying a probability model

(Ω', P'), where $\Omega' = \{x_1, x_2, \ldots\}$ and $P'(\{x_i\}) = f(x_i)$, $i = 1, 2, \ldots$. Consequently, we shall unabashedly speak of the random variable X which takes on the value x_i with probability $f(x_i)$ ($i = 1, 2, \ldots$) without further reference to our old probability model (Ω, P).

1.1 Joint Distributions

Of course, it is possible to define more than one random variable for a given probability model (Ω, P). For example, let (Ω, P) represent the random experiment consisting in one toss of a pair of fair dice, let X be the number of dice showing odd scores, let Y be the total points showing on the two dice, and let f_X and f_Y be their respective p.m.f.'s. As we shall soon see, f_X and f_Y don't contain all the relevant probabilistic information (see Table 2.1). For instance, we cannot obtain the joint and conditional probabilities from them [for example, $P(X = 2 \text{ and } Y = 12)$]. This motivates the need for a new probability mass function.

Let f_X and f_Y be the p.m.f.'s of two random variables X and Y defined on (Ω, P) which assume the values $x_1, x_2, \ldots$ and $y_1, y_2, \ldots$, respectively. We define the event $X = x_i$, $Y = y_j$ to be the intersection of the events $X = x_i$ and $Y = y_j$, and we denote the probability of this event by $f_{X,Y}(x_i, y_j)$ instead of $P(X = x_i, Y = y_j)$.

DEFINITION

The function $f_{X,Y}$ defined by

$$f_{X,Y}(x_i, y_j) = P(X = x_i, Y = y_j), \qquad i, j = 1, 2, \ldots,$$

is called the *joint probability mass function* of X and Y.

Let (Ω, P) be the probability model for the random experiment that consists in tossing a pair of fair dice once, let X be the number of dice showing odd scores, and let Y be the total points showing on the two dice. Then the joint probability mass function of X and Y is given within the heavy lines in Table 2.1. As shown in Table 2.1, we see, for example, that $f_{X,Y}(1, 7) = \frac{6}{36}$ and $f_{X,Y}(0, 8) = \frac{3}{36}$.

Now if X and Y are any pair of random variables, then clearly

$$f_{X,Y}(x_i, y_j) \geq 0 \quad \text{and} \quad \sum_{i=1}^{\infty} \sum_{j=1}^{\infty} f_{X,Y}(x_i, y_j) = 1. \tag{2.2}$$

Furthermore, a straightforward application of Theorem 5 of Chapter 1 yields

$$f_X(x_i) = \sum_{j=1}^{\infty} f_{X,Y}(x_i, y_j), \qquad i = 1, 2, \ldots, \tag{2.3}$$

and

$$f_Y(y_j) = \sum_{i=1}^{\infty} f_{X,Y}(x_i, y_j), \qquad j = 1, 2, \ldots. \tag{2.4}$$

In view of (2.3), (2.4), and Table 2.1, f_X and f_Y are often called the marginal distributions. Note that $f_{X,Y}$ determines f_X and f_Y, but we cannot determine $f_{X,Y}$ from f_X and f_Y.

Table 2.1 Joint Distribution of X and Y

X \ Y	2	3	4	5	6	7	8	9	10	11	12	f_X
0	0	0	$\frac{1}{36}$	0	$\frac{2}{36}$	0	$\frac{3}{36}$	0	$\frac{2}{36}$	0	$\frac{1}{36}$	$\frac{9}{36}$
1	0	$\frac{2}{36}$	0	$\frac{4}{36}$	0	$\frac{6}{36}$	0	$\frac{4}{36}$	0	$\frac{2}{36}$	0	$\frac{18}{36}$
2	$\frac{1}{36}$	0	$\frac{2}{36}$	0	$\frac{3}{36}$	0	$\frac{2}{36}$	0	$\frac{1}{36}$	0	0	$\frac{9}{36}$
f_Y	$\frac{1}{36}$	$\frac{2}{36}$	$\frac{3}{36}$	$\frac{4}{36}$	$\frac{5}{36}$	$\frac{6}{36}$	$\frac{5}{36}$	$\frac{4}{36}$	$\frac{3}{36}$	$\frac{2}{36}$	$\frac{1}{36}$	

X = number of dice showing odd scores
Y = total points showing of the dice

Extending our notation in the obvious manner, the conditional probability of the event $X = x_i$ given that $Y = y_j$ [with $f_Y(y_j) > 0$] becomes

$$f_{X|Y}(x_i \mid y_j) = \frac{f_{X,Y}(x_i, y_j)}{f_Y(y_j)}, \qquad i,j = 1, 2, \ldots,$$

and we obtain the

DEFINITION

If $f_{X,Y}(x_i, y_j) = f_X(x_i) \cdot f_Y(y_j)$ for each pair (x_i, y_j), then X and Y are said to be *independent random variables.*

Of course, all of the above generalizes to any finite set of random variables.

EXERCISES:

1. Find the probability mass function f_X of the random variables X defined in items (3), (4), (5), and (8) of the list on pp. 41–42, but first define these random variables as mappings whose domains are the appropriate sample spaces.

2. Suppose that N balls in an urn are numbered $1, 2, \ldots, N$, and that n $(\leq N)$ of these balls are drawn without replacement. Let Y be the largest number

on the n balls that were withdrawn. Find the probability mass function of Y.

$$\left[\textit{Answer:}\quad P(Y = k) = n\binom{k-1}{n-1}(n-1)!\bigg/\binom{N}{n}n!\right.$$
$$\left. = \binom{k-1}{n-1}\bigg/\binom{N}{n},\ n \le k \le N.\right]$$

3. A certain UCLA professor always draws 2 of the 5 exam questions from the last two exams (6 questions each). You have been able to work 6 of these, whereas the probability of your being able to work any given one of the new questions is only $\frac{1}{3}$. Let X and Y be the number of new and old exam questions you will be able to work, respectively, and let $Z = X + Y$ be the total number you will be able to work. (i) Find f_X and f_Y. (ii) Find f_Z. [*Hint:* Use Theorem 5 of Chapter 1 and independence to show

$$f_Z(i) = \sum_{j=0}^{i} f_X(j)\cdot f_Y(i-j).]$$

4. Suppose two dice are rolled and the usual 36 outcomes are taken to be equally likely. Find the joint p.m.f. of X and Y, where X is the number of dice showing even scores and Y is the number of dice showing scores of 2. Find f_X and f_Y and make a table similar to Table 2.1. Are X and Y independent? What is the probability that Y is smaller than X? [In other words, find $P(S)$ where $S = \{\omega \in \Omega\colon Y(\omega) < X(\omega)\}$.]

5. A box contains 3 red, 2 green, and 1 black ball. Three balls are to be drawn without replacement. Let X and Y be the number of red and green balls drawn, respectively. Perform all tasks described in Exercise 4. Also, find $f_{X|Y}$.

6. Let Ω be the set of all distinguishable arrangements of the six letters a, a, a, b, b, b, and suppose that all of the elements of Ω are equally likely.

   A sequence of one or more identical letters that is preceded and followed by a different letter (or preceded by no letter if it is at the beginning or followed by no letter if it is at the end of the entire sequence) is called a *run*. For example, the sequence a a b a b b contains 4 runs of lengths 2, 1, 1, and 2, respectively.

   For each element ω in Ω, let $Y(\omega)$ be the number of runs in the element ω, and let $X(\omega)$ be the number of runs formed by the first 4 letters in the element ω.

   (a) Find the joint p.m.f. of X and Y, the p.m.f. of X, and the p.m.f. of Y.
   (b) Are X and Y independent?

7. Suppose 3 distinguishable balls are randomly placed in 3 cells, let X denote the number of balls in cell 1, and let Y denote the number of unoccupied cells. Find f_X, f_Y, and $f_{X,Y}$. Are X and Y independent?

8. Verify equation (2.1).

9. Verify equations (2.3) and (2.4).

10. Let X and Y be two random variables defined with respect to the same probability model such that $f_X = f_Y$. Does it follow that $X = Y$? [*Hint:* Toss a fair coin; let $\Omega = \{H, T\}$, $X(H) = Y(T) = 1$, $X(T) = Y(H) = 0$.]

11. Suppose you are given f_X and $f_{X,Y}$ but not f_Y. How would you determine whether or not X and Y are independent?

2. EXPECTATION

In order to deal effectively with large quantities of information, we must somehow reduce them to manageable proportions. We did this, for instance, when we constructed a probability model for a real (random) experiment, and then condensed all the relevant information into the probability mass function f_X of some random variable X. For some purposes, especially those of comparison and decision making, a single number would be much easier to deal with than the function f_X. In condensing the information, however, we invariably lose some of it—but we gain greatly in simplicity. The question, then, is how to condense all the information contained in f_X in some "good" way into a single number.

In many instances, we are primarily interested in the central tendency of our random variable X. One important measure of central tendency is the average (or center of gravity) or what we call the expectation of X, denoted by $E(X)$. For example, if you roll a fair die once, the average number of points to appear is $3\frac{1}{2}$, since each of the six numbers appears about one sixth of the time, so the average is $(1 + 2 + \cdots + 6)/6 = 3\frac{1}{2}$. Or consider the following example:[2] In a certain town, n_k families have exactly k children ($k = 0, 1, 2, \ldots, 10$), so the total number of children is $c = n_1 + 2n_2 + \cdots + 10n_{10}$, while the total number of families is $f = n_0 + n_1 + \cdots + n_{10}$. Then the average number of children per family in this town is c/f. The analogy between probabilities and frequencies suggests the following

DEFINITION

Let X be a random variable with probability mass function f. Then the *expected value* of X is defined by

$$E(X) = \sum_{i=1}^{\infty} x_i f(x_i).$$

[2] This example appears on page 221 of Feller's book.

It is common in many texts to write μ_X in place of $E(X)$ or more simply μ instead of μ_X if no possible ambiguity can result. The symbol "μ" is the lower-case Greek letter mu.

Note that $E(X)$ need not be in the range of X—that is, a possible value of X. For example, let X be the number of points that appear when we roll a fair die; then $E(X) = 3\frac{1}{2}$, yet $E(X) \notin \{1, 2, 3, 4, 5, 6\}$, which is the range of X.

To further illustrate the idea of the expectation as a number that locates the center of the distribution of a random variable, consider the (weightless) teeter-totter shown in Fig 2.1, upon which are placed various

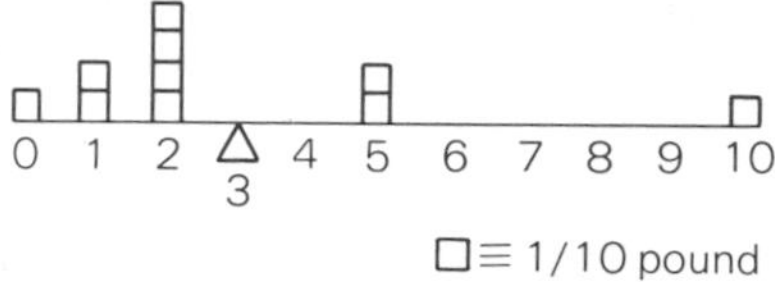

Figure 2.1 Expectation as the Center of Gravity

weights at various distances from the end. Clearly the teeter-totter will balance only if the fulcrum (△) is placed at the center of gravity, which is seen to be 3, since

$$(3 - 0)\tfrac{1}{10} + (3 - 1)\tfrac{2}{10} + (3 - 2)\tfrac{4}{10} = (5 - 3)\tfrac{2}{10} + (10 - 3)\tfrac{1}{10}.$$

The analogy between expectation and center of gravity becomes clear when we interpret the distances as the values assumed by our random variable and the weights as the probabilities of these values.

EXAMPLE 5

Find $E(X)$ where X is the number of sixes that appear when a fair die is rolled 3 times.

Solution: By Example 27 of Chapter 1, we have

$$f_X(0) = \tfrac{125}{216},\quad f_X(1) = \tfrac{75}{216},\quad f_X(2) = \tfrac{15}{216},\quad \text{and}\quad f_X(3) = \tfrac{1}{216},$$

so

$$E(X) = 0\cdot\tfrac{125}{216} + 1\cdot\tfrac{75}{216} + 2\cdot\tfrac{15}{216} + 3\cdot\tfrac{1}{216} = \tfrac{1}{2}. \bullet$$

EXAMPLE 6

Consider the following game: you pick a number between 1 and 6, and then roll a pair of fair dice. If your number appears on 0, 1, or 2 dice, then you pay 0, 1, or 2 dollars. How much should you be paid to play this game in order to make it fair?

SOLUTION: Let X be the number of dice that turn up with your number showing. Then $f_X(0) = \frac{25}{36}$, $f_X(1) = \frac{10}{36}$, and $f_X(2) = \frac{1}{36}$, so the amount you will lose on the average is

$$0 \cdot \tfrac{25}{36} + 1 \cdot \tfrac{10}{36} + 2 \cdot \tfrac{1}{36} = \tfrac{1}{3} = E(X).$$

Thus, $\$\frac{1}{3}$ is a fair inducement for you to play. •

EXAMPLE 7

Let X be the number of tosses of a fair coin required until the first head appears. Then as shown in Example 4, $f_X(i) = (\frac{1}{2})^i$, $i = 1, 2, \ldots$. Consequently,

$$E(X) = 1(\tfrac{1}{2})^1 + 2(\tfrac{1}{2})^2 + 3(\tfrac{1}{2})^3 + \cdots.$$

Using calculus, it can now be shown that $E(X) = 2$ (see the proof of Theorem 12). •

For further illustrations of how to compute the expectation of a random variable, the reader should consult Exercises 11, 20, 21, 22, and 36 of Chapter 1 (see Exercise 17).

Just as we constructed new sets and new probability models from old ones, we can construct new random variables from old ones. Let g be a real-valued function whose domain is the real numbers and let X be a random variable, then the function $Z = g(X)$ defined by $Z(\omega) = g(X(\omega))$ for each $\omega \in \Omega$ is a random variable. Keep in mind the fact that $g(X)$ is a function but $g(X(\omega))$, the value of $g(X)$ at ω, is a number. Since we already know f_X, we can find $f_{g(X)}$ (see Exercise 27). This, however, is often quite laborious. The following theorem shows that we can find the expectation of $g(X)$ without finding the probability mass function of $g(X)$.

THEOREM 1

If g is a real-valued function whose domain is the real numbers and X is a random variable,[3] then

$$E(g(X)) = \sum_{i=1}^{\infty} g(x_i) f_X(x_i).$$

[3] This theorem is also known as "the law of the unconscious statistician" because it is used "unconsciously" as if it were the definition of expectation. When X is a continuous random variable (see Chapter 3), the proof becomes quite difficult; it can be found in H. L., Royden, *Real Analysis* (New York: The Macmillan Company 1963), pages 260–261, or in K. L. Chung, *A Course in Probability Theory* (New York: Harcourt, Brace & World, Inc., 1968), page 43.

PROOF. Let $Z = g(X)$; then

$$E(Z) = \sum_{j=1}^{\infty} z_j f_Z(z_j) = \sum_{j=1}^{\infty} \left\{ z_j \sum_{\{i: g(x_i) = z_j\}} f_X(x_i) \right\}$$
$$= \sum_{i=1}^{\infty} g(x_i) f_X(x_i). \ \blacklozenge$$

In the next section we will want to find not only $E(X)$ but also $E(X^2)$. If we let g be that function defined by $g(t) = t^2$ for each real number t, we see that $g(X) = X^2$, so it follows from Theorem 1 that

$$E(X^2) = E(g(X)) = \sum_{i=1}^{\infty} g(x_i) f_X(x_i)$$
$$= \sum_{i=1}^{\infty} x_i^2 f_X(x_i).$$

An important special case of Theorem 1 is the following corollary, whose proof is left as an exercise.

COROLLARY 1

Let a and b be constants, then

$$E(a + bX) = a + bE(X).$$

If $X(\omega) = a$ for each $\omega \in \Omega$, then X is called a constant (random variable) and $E(X) = a$—that is, the expected value of a constant is the same constant. Thus, an immediate consequence of this corollary is the fact that

$$E(X - E(X)) = 0.$$

The reader should ponder the intuitive meaning of this last equality and judge for himself whether it is obvious or incredible.

The next theorem, which is also quite useful, shows that the expectation of a sum of random variables is the sum of their expectations. Of course, the sum of random variables is itself a random variable, and the random variable $Z \equiv X + Y$ satisfies $Z(\omega) = X(\omega) + Y(\omega)$ for each $\omega \in \Omega$.

THEOREM 2

Let $X_1, X_2, \ldots, X_n$ be random variables, then

$$E(X_1 + X_2 + \cdots + X_n) = E(X_1) + E(X_2) + \cdots + E(X_n).$$

PROOF. Making use of induction, it suffices to prove the theorem for two random variables X and Y.

$$\begin{aligned} E(X+Y) &= \sum_{i,j} (x_i + y_j) f_{X,Y}(x_i, y_j) \\ &= \sum_{i,j} x_i f_{X,Y}(x_i, y_j) + \sum_{i,j} y_j f_{X,Y}(x_i, y_j) \\ &= \sum_i x_i f_X(x_i) + \sum_j y_j f_Y(y_j) \\ &= E(X) + E(Y). \blacklozenge \end{aligned}$$

The power of this theorem is well illustrated in the next two examples.

EXAMPLE 8

Let Y be the total number of points that appear when a fair die is rolled 3 times. Find $E(Y)$.

SOLUTION: We could present an ad hoc procedure for finding the p.m.f. of Y (see Exercise 21). But why go to all that trouble if we are only interested in finding $E(Y)$? First, observe that $Y = X_1 + X_2 + X_3$, where X_i is the number of points that appear on the ith roll, $i = 1, 2, 3$. Second, note that for $i = 1, 2$, and 3, $f_{X_i}(k) = \frac{1}{6}$ for $k = 1, 2, \ldots, 6$, so that

$$E(X_i) = 1\cdot\tfrac{1}{6} + 2\cdot\tfrac{1}{6} + \cdots + 6\cdot\tfrac{1}{6} = 3\tfrac{1}{2}.$$

We can now use Theorem 2 to obtain

$$E(Y) = E(X_1 + X_2 + X_3) = E(X_1) + E(X_2) + E(X_3) = 10\tfrac{1}{2}. \bullet$$

EXAMPLE 9

Let Z be the number of tosses of a fair coin required until 2 heads appear. Then the sample space Ω is seen to be

$$\Omega = \{HH; HTH, THH; HTTH, THTH, TTHH; \\ HTTTH, THTTH, TTHTH, TTTHH; \ldots\}.$$

Although it is easily seen that $f_Z(2) = \frac{1}{4}$, $f_Z(3) = \frac{2}{8}$, $f_Z(4) = \frac{3}{16}$, and $f_Z(5) = \frac{4}{32}$, the values of $f_Z(k)$ for $k > 5$ are not so obvious. They can, of course, be found (see Exercise 69), but only after careful thought. Observe, however, that if we define X_1 to be the number of flips required until the first head appears, and if we define X_2 to be the number of additional flips required until the second head appears, then $Z = X_1 + X_2$. (Incidentally, X_1 and X_2 are independent.) Now using Theorem 2 and the fact that $E(X_1) = E(X_2) = 2$ (see Example 7), it follows that

$$E(Z) = E(X_1 + X_2) = E(X_1) + E(X_2) = 4. \bullet$$

If X and Y are random variables, then their product XY is also a random variabe. Of course, $Z \equiv XY$ satisfies $Z(\omega) = X(\omega)\cdot Y(\omega)$ for each $\omega \in \Omega$. It is clear that the analog of Theorem 2 does not hold for products (see Exercise 29); however, if the random variables are independent, we can obtain

THEOREM 3

If X and Y are independent random variables, then

$$E(XY) = E(X)E(Y).$$

PROOF. Since X and Y are independent,

$$f_{X,Y}(x_i, y_j) = f_X(x_i)\cdot f_Y(y_j).$$

Consequently, we can use Theorem 1 to obtain

$$E(XY) = \sum_{i=1}^{\infty}\sum_{j=1}^{\infty} x_i y_j f_{X,Y}(x_i, y_j) = \sum_{i=1}^{\infty}\sum_{j=1}^{\infty} x_i y_j f_X(x_i) f_Y(y_j)$$

$$= \sum_{i=1}^{\infty} x_i f_X(x_i) \sum_{j=1}^{\infty} y_j f_Y(y_j) = E(X)E(Y). \blacklozenge$$

Note that independence is required in Theorem 3 but not in Theorem 2 (see Exercises 29 and 30). For an application of this theorem, see Exercises 22 and 23.

2.1 Other Measures of Central Tendency

As the reader has noticed, the expected value $E(X)$ need not be a possible value of X. Moreover, it need not even be representative of X, for it is strongly affected by extreme values. Our intention is illustrated in the following example: $f_X(0) = \frac{3}{11}$, $f_X(i) = \frac{1}{11}$, $i = 1, 2, \ldots, 7$, and $f_X(93) = \frac{1}{11}$. Here, $E(X) = 11$, yet there is no value of X is near 11 and $P(X \leq 11) = \frac{10}{11}$. Evidently $E(X)$ does not reflect the central tendency of X. Perhaps a more typical value of X or at least a more reasonable measure of central tendency for this random variable is the number 3, which satisfies $P(X \leq 3) \geq \frac{1}{2}$ and $P(X \geq 3) \geq \frac{1}{2}$. Such a number is called the *median*, and we shall denote it by $M(X)$. That is, if $P(X \geq M(X)) \geq \frac{1}{2}$ and $P(X \leq M(X)) \geq \frac{1}{2}$, then $M(X)$ is the median of X. The median always exists, but it need not be unique. When studying the distribution of personal income of families, especially in underdeveloped countries, it becomes readily apparent that the median contains much more information than does the expected value. This is because relatively few families receive most of the income. Thus, sometimes the median is a better measure of central tendency than the expectation.

A third (and less important) measure of central tendency is the mode. The *mode*, denoted by $m(X)$, is the most probable value of X. That is, if $f_X(m(X)) \geq f_X(x_i)$, $i = 1, 2, \ldots$, then $m(X)$ is said to be the mode of X. Note that the mode will always exist (and be in the range of X) although it need not be unique. In the example above, $m(X) = 0$. Of course, we should use the mode as our measure of central tendency when we are interested in the single most probable outcome. For example, what we have in mind when we speak of the typical number of children in a family is the mode rather than the expectation or the median.

Our intention in discussing the median and the mode was merely to point out that expectation is neither the only measure of central tendency nor always the most reasonable one. But, because it is often a reasonable measure of central tendency, because it has desirable statistical properties, and because it is easy to work with (manipulate), we shall adopt it for use as our measure of central tendency.

EXERCISES:

12. Suppose that you received the grades of A, B, A, D, and C in a 3, 4, 3, 3, and $\frac{1}{2}$ unit course, respectively. Also suppose that your school uses the 4-point system; that is, an A is worth four grade points, a B three, a C two, and a D one. Compute your grade point average (G.P.A.). Make a comparison between expectation and G.P.A.

13. Making use of Exercise 1, find $E(X)$ for items (5) and (8) of the list on pp. 41–42.

14. Define Y as in Exercise 2 and show that

$$E(Y) = \frac{n}{n+1}(N+1).$$

[*Hint:* You will need to use the following identity, which is given as equation (12.8a) on page 64 in Feller:

$$\sum_{k=0}^{M} \binom{k}{m} = \binom{M+1}{m+1}.\Big]$$

15. Find $E(X)$ and $E(Y)$ where X and Y are defined in Exercises 4 and 5.

16. Find $E(X)$ and $E(Y)$ where X and Y are defined in Exercise 6.

17. Find $E(X)$ after defining X in the most reasonable way in the following exercises of Chapter 1: (a) 11; (b) 20; (c) 21; (d) 22; (e) 36.

18. Find $E(Z)$ where Z is given in Exercise 3.

19. Use Theorem 2 to solve Exercise 18.

20. Use Theorem 2 to find $E(Z)$, where Z is the total points showing when 10 fair dice are rolled.

21. Find $E(Y)$ where Y is defined in Example 8. Do not use Theorem 2.

22. Consider the following game called a parlay: You roll a fair die and receive X dollars where X is the number of points showing on the die. Next, you bet the X dollars you won on a horse called Charlie. If Charlie comes in first, you win (in all) 5 times your bet; if he comes in second, you win 3 times your bet; otherwise, you lose. Let Z be the total amount you win. Find $E(Z)$ when the probabilities that Charlie wins and comes in second are $\frac{1}{5}$ and $\frac{2}{5}$, respectively.

23. Use Theorem 3 to solve Exercise 22.

24. A random variable which we shall find very useful is the *indicator random variable* I_A, defined by $I_A(\omega) = 1$ if $\omega \in A$ and $I_A(\omega) = 0$ if $\omega \notin A$, where A is some specified subset of Ω. Show that $E(I_A) = P(A)$.

25. Use Exercise 24 and Theorem 2 to find the expected value of the appropriate random variables defined in Examples 5 and 6, in item (5) of the list on pp. 41–42, and in Exercise 64 of Chapter 1.

26. Consider the problem of the 3 sisters breaking 4 dishes. As before, let X be the number of dishes broken by the youngest. Use indicator random variables to find $E(X)$. [*Hint:* Let X_i be 1 or 0 if the youngest sister does or does not break the ith dish, $i = 1, 2, 3, 4$.]

27. Show how to construct $f_{g(X)}$ from f_X.

28. Prove Corollary 1.

29. Give an example where $E(XY) \neq E(X)E(Y)$. [*Hint:* Consider Exercise 4.]

30. Give an example to show that $E(XY) = E(X)E(Y)$ does not necessarily imply that X and Y are independent. [*Hint:* Let $f_{X,Y}(1, 1) = f_{X,Y}(1, -1) = \frac{1}{4}$ and $f_{X,Y}(-1, 0) = \frac{1}{2}$.]

31. (Conditional Expectation) Let X and Y be random variables. Then the *conditional expectation* of X given Y, denoted by $E(X \mid Y)$, is also a random variable which assumes the value $E(X \mid Y = y_j)$ with probability $f_Y(y_j)$, where

$$E(X \mid Y = y_j) = \sum_{i=1}^{\infty} x_i f_{X|Y}(x_i \mid y_j).$$

Show that $E(X) = E(E(X \mid Y))$; that is, show

$$E(X) = \sum_{j=1}^{\infty} E(X \mid Y = y_j) f_Y(y_j).$$

32. (Continuation) Let X be the length (that is, number of rolls) of a game of craps (see Example 20 of Chapter 1). Show that $E(X) = 3.375$. [*Hint:* Make use of Exercise 31 and equation (2.10).]

33. (Continuation) A large class of probability problems can be formulated as follows: Let $Y, X_1, X_2, X_3, \ldots$ be independent random variables, let the range of Y be contained in the set of nonnegative integers, let the X_i's have the common p.m.f. f_X, and let S be defined by

$$S = \sum_{i=1}^{Y} X_i.$$

Thus, S is the sum of a random number of random variables. Among the more obvious questions we could ask might be, what is f_S and what is $E(S)$? Exercise 31 is helpful in finding $E(S)$, while finding f_S is generally beyond our means. (See Exercise 35 for an exception to this.) Show that $E(S) = E(X)\,E(Y)$.

34. (Continuation) Let X_i be the number of children your ith child will have, $i = 1, 2, \ldots$, and let Y be the number of children you (will) have. What is the expected number of grandchildren you will have if $f_Y = f_{X_i} = f_X$, $i = 1, 2, \ldots$, and $f_X(0) = \frac{2}{15}$, $f_X(1) = \frac{1}{5}$, $f_X(2) = \frac{1}{3}$, $f_X(3) = \frac{1}{5}$, and $f_X(4) = \frac{2}{15}$?

35. (Continuation) A random variable that arises frequently in geographical and biological statistics is the Poisson. We say that X is a *Poisson random variable* with parameter λ (> 0) if $P(X = k) = e^{-\lambda}\lambda^k/k!$, $k = 0, 1, 2, \ldots$. Suppose, as is reasonable, that the number N of patients arriving each day at the hospital emergency room is a Poisson random variable with parameter λ and that the probability that a patient will need immediate surgery is p. Assuming that the patients' need for immediate surgery is independent, show that the number S of patients needing immediate surgery is a Poisson random variable with parameter λp. [*Hint:* Note that $S = \sum_{i=1}^{N} X_i$, where X_i equals 1 or 0 according to the ith patient's need for surgery, and use Theorem 5 of Chapter 1.]

36. Suppose that you toss a fair coin until a head appears and you receive $X = \$2^n$ if exactly n tosses are required. Then $f_X(2^n) = 1/2^n$ for $n = 1, 2, \ldots$. How much should you pay to make this a fair game? What if you receive $\$2^n$ if n is even and $-\$2^n$ if n is odd? (For a thorough discussion of this problem see Feller, pp. 251–253.)

3. VARIANCE

In the last section we saw that the expectation of a random variable was a good measure, but not the only measure, of central tendency. In addition to central tendency, we are also interested in the degree to which our random variable is spread out or scattered. To understand this more fully, place yourself in the position of having to choose between the two investments whose returns are depicted in Figure 2.2 and have the

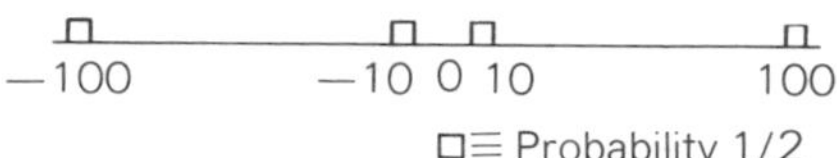

Figure 2.2 Risk as Measured by the Spread

probability mass functions defined by $f_X(10) = f_X(-10) = f_Y(100) = f_Y(-100) = \frac{1}{2}$. Clearly $E(X) = E(Y) = 0$, but investment Y has much more variability or spread than investment X. The larger variability is to be interpreted as greater risk, so naturally most of us would choose investment X.[4]

In the example above, it was quite obvious which variable had the larger spread, but generally the comparison cannot be made so easily. Consequently, we must find a good measure of spread. It seems reasonable that our measure should be the expected value of some (symmetric) function of the difference between X and $E(X)$. We will use $E[(X - E(X))^2]$, denoted by $\text{Var}(X)$, as our measure. It is common in many texts to write σ_X^2 in place of $\text{Var}(X)$ or more simply σ^2 if no possible ambiguity can result. The symbol "σ" is the lower-case Greek letter sigma. Three other reasonable measures are $E(|X - E(X)|)$, $E(|X - M(X)|)$, and $E[(X - M(X))^2]$, but they are more difficult to manipulate and do not possess as nice mathematical properties as $\text{Var}(X)$.

DEFINITION

Let X be a random variable with probability mass function f. Then the *variance* of X is defined by

$$\text{Var}(X) = E[(X - E(X))^2].$$

[4] If we had $E(Y) > E(X)$ and risk $Y >$ risk X, then it would be quite difficult to choose between the two investments [see H. Markowitz, *Portfolio Selection, Efficient Diversification of Investments* (New York: John Wiley & Sons, Inc., 1959)].

Hence, it follows from Theorem 1 that

$$\operatorname{Var}(X) = \sum_{i=1}^{\infty} (x_i - E(X))^2 f_X(x_i).$$

Its nonnegative square root is called the *standard deviation* of X. One important feature of the standard deviation is the fact that it is expressed in the original units of measurement and not in squared units like the variance.

EXAMPLE 10

Let X be the random variable given in Example 6. Then

$$\begin{aligned}\operatorname{Var}(X) &= \sum_{i=0}^{2} (x_i - E(X))^2 f_X(x_i) = \sum_{i=0}^{2} (i - \tfrac{1}{3})^2 f_X(i) \\ &= \tfrac{1}{9}\cdot\tfrac{25}{36} + \tfrac{4}{9}\cdot\tfrac{10}{36} + \tfrac{25}{9}\cdot\tfrac{1}{36} = \tfrac{90}{324}. \bullet\end{aligned}$$

Noting that $E(X)$ is a constant, writing $(X - E(X))^2 = X^2 - 2E(X)X + E(X)^2$, and applying Theorem 2 to the definition of variance, we can obtain the very useful computing formula

$$\operatorname{Var}(X) = E(X^2) - E(X)^2. \tag{2.5}$$

Generally, it is easier to compute the variance using equation (2.5) rather than directly applying the definition.

Analogous to Corollary 1 in Section 2 we have the following theorem, whose proof is left as an exercise.

THEOREM 4

Let a and b be constants, then

$$\operatorname{Var}(a + bX) = b^2 \operatorname{Var}(X).$$

In particular, notice that choosing $a = 0$ and $b = 1/\sqrt{\operatorname{Var}(X)} = 1/\sigma_X$, we have

$$\operatorname{Var}\left(\frac{X}{\sigma_X}\right) = 1.$$

Combining this result with our remark following Corollary 1, we have $E(X^*) = 0$ and $\operatorname{Var}(X^*) = 1$, where

$$X^* \equiv \frac{X - E(X)}{\sqrt{\operatorname{Var}(X)}}.$$

The random variable X^* is called the standardized form of X. We will use this result extensively in our study of statistics.

We showed in Theorem 2 that the expectation of a sum of random variables is the sum of their expectations. The next theorem shows that a comparable result is true for variances with the proviso that the variables are independent.

THEOREM 5

If X and Y are independent random variables, then

$$\operatorname{Var}(X + Y) = \operatorname{Var}(X) + \operatorname{Var}(Y).$$

PROOF. Using Theorem 2, Theorem 3, and equation (2.5), we have

$$\begin{aligned}\operatorname{Var}(X + Y) &= E[(X + Y - E(X) - E(Y))^2] \\ &= E(X^2) + E(Y^2) + E(X)^2 + E(Y)^2 + 2E(XY) \\ &\qquad + 2E(X)E(Y) - 2(E(X) + E(Y))E(X + Y) \\ &= E(X^2) + E(Y^2) + E(X)^2 + E(Y)^2 + 2E(X)E(Y) \\ &\qquad + 2E(X)E(Y) - 2E(X)^2 - 2E(Y)^2 \\ &\qquad - 4E(X)E(Y) \\ &= E(X^2) - E(X)^2 + E(Y)^2 - E(Y)^2 \\ &= \operatorname{Var}(X) + \operatorname{Var}(Y). \ \blacklozenge\end{aligned}$$

The next theorem is just a special case of Theorems 2 and 5. However, we shall use it so frequently in Chapters 4 through 9 as to justify its formal presentation.

THEOREM 6

Let $X_1, X_2, \ldots, X_n$ be a sequence of independent random variables each having the same probability mass function f_X. Then

$$E\left(\frac{1}{n}\sum_{i=1}^{n} X_i\right) = E(X),$$

and

$$\operatorname{Var}\left(\frac{1}{n}\sum_{i=1}^{n} X_i\right) = \frac{1}{n}\operatorname{Var}(X).$$

The proof is left as an exercise.

One important case where the random variables of interest are not independent is when they form a random sample without replacement. Fortunately, we can still find the variance of the sum even though the variables are not independent.

THEOREM 7

Let $X_1, X_2, \ldots, X_n$ be a random sample of size n without replacement from a finite population with mean μ and variance σ^2 whose N members have the values $v_1, v_2, \ldots, v_N$. Then

$$\operatorname{Var}\left(\frac{1}{n}\sum_{i=1}^{n} X_i\right) = \frac{N-n}{N-1}\cdot\frac{\sigma^2}{n}.$$

▶▶

Proof. Applying equation (2.5) to the random variable $\sum_{i=1}^{n} X_i$ yields

$$\begin{aligned}
\operatorname{Var}\left(\sum_{i=1}^{n} X_i\right) &= E\left[\left(\sum_{i=1}^{n} X_i\right)^2\right] - (n\mu)^2 \\
&= E\left(\sum_{i=1}^{n} X_i^2\right) + E\left(\sum_{i\neq j} X_iX_j\right) - n^2\mu^2 \\
&= \sum_{i=1}^{n} E(X_i^2) - n\mu^2 + \sum_{i\neq j} E(X_iX_j) - n(n-1)\mu^2 \\
&= n\sigma^2 + \sum_{i\neq j} E(X_iX_j) - n(n-1)\mu^2.
\end{aligned}$$

In order to determine $E(X_iX_j)$, we fix i and j with $i \neq j$. Then, since $P(X_i = v_s \text{ and } X_j = v_t) = 1/[N(N-1)]$ for any pair of values from our population of size N, it follows that

$$\begin{aligned}
E(X_iX_j) &= \sum_{s\neq t} v_sv_t \frac{1}{N(N-1)} = \frac{1}{N-1}\sum_{s\neq t} v_sv_t\frac{1}{N} \\
&= \frac{1}{N-1}\left[\sum_{s=1}^{N}\sum_{t=1}^{N} v_sv_t\frac{1}{N} - \sum_{s=1}^{N} v_s^2\frac{1}{N}\right] \\
&= \frac{1}{N-1}\left[\sum_{s=1}^{N} v_s \sum_{t=1}^{N} v_t\frac{1}{N} - (\sigma^2+\mu^2)\right] \\
&= \frac{1}{N-1}\left[N\mu^2 - (\sigma^2+\mu^2)\right] = \mu^2 - \frac{\sigma^2}{N-1}
\end{aligned}$$

so

$$\operatorname{Var}\left(\sum_{i=1}^{n} X_i\right) = n\sigma^2 + \sum_{i \neq j}\left(\mu^2 - \frac{\sigma^2}{N-1}\right) - n(n-1)\mu^2$$

$$= n\sigma^2 + n(n-1)\left(\mu^2 - \frac{\sigma^2}{N-1}\right) - n(n-1)\mu^2$$

$$= n\sigma^2 - n(n-1)\frac{\sigma^2}{N-1} = \frac{N-n}{N-1}n\sigma^2.$$

The desired result now follows from Theorem 4. ◆

Since $(N - n)/(N - 1) < 1$, it follows that sampling without replacement results in less variance (or equivalently, yields more information) than sampling with replacement.

EXERCISES:

[In Exercises 37–39, find the variance of the indicated random variables first by using the definition directly and then by making use of equation (2.5). Exercise 37 is designed to further illustrate computation of the variance.]

37(a) Let X be the points showing when a single fair die is rolled.
(b) Let X be the random variable in Examples 2, 3, and 5.
(c) Let X be the random variable in items (3) and (8) listed on pp. 41–42 (see Exercises 1 and 13).
(d) Let Z be the random variable in Exercise 3 (see Exercise 18).
(e) Let X and Y be the random variables in Exercise 4 (see Exercise 15).
(f) Let X and Y be the random variables in Exercise 5 (see Exercise 15).
(g) Let X and Y be the random variables in Exercise 6 (see Exercise 16).
(h) Let Z be the random variable in Exercise 20. Use Theorem 5, and do not apply the definition.
(i) Let X be the random variable in Example 8.
(j) Let Z be the random variable in Exercise 22.

38. Let I_A be the indicator random variable for an event A and let $P(A) = p$. Show that $\operatorname{Var}(I_A) = p(1 - p)$.

39. Let X be the random variable in Exercise 26. Find $\operatorname{Var}(X)$ a third way by using Theorem 5 and indicator random variables (see Exercise 24).

40. Let X be the random variable in Exercise 32.

41. Sometimes it is easier to find $E(Y^2)$ by using the identity $E(Y^2) = E[Y(Y + 1)] - E(Y)$. Use this device and equation (2.5) to show that

$$\text{Var}(Y) = \frac{n(N+1)(N-n)}{(n+2)(n+1)^2},$$

where Y is defined in Exercise 2. [*Hint:* Use the hint in Exercise 14.]

42. Suppose that 4 balls in an urn are numbered 1, 2, 3, and 4 and that 2 of these balls are drawn without replacement. Let X be the larger of the numbers on the two balls that were withdrawn. Find: (a) f_X, (b) $E(X)$, (c) $\text{Var}(X)$. (d) Let Y be the larger of the numbers on the two balls that were not withdrawn. Then by symmetry, $f_X = f_Y$, so $E(X) = E(Y)$ and $\text{Var}(X) = \text{Var}(Y)$. Are there any theorems presented in this chapter that are helpful in finding $E(X + Y)$ and/or $\text{Var}(X + Y)$? If so, use these theorems to find these quantities. If not, explain why.

43. Let $X_1, X_2, \ldots, X_n$ be independent random variables each having p.m.f. f_X. Use Theorems 4 and 5 to show that

$$\text{Var}\left(\frac{1}{n}\sum_{i=1}^{n} X_i\right) = \frac{\text{Var}(X)}{n}.$$

44. Show that $E(X^2) \geq E(X)^2$ and that equality holds if and only if the range of X is a single number.

45. Prove Theorem 4.

46. Show that $\text{Var}(X - Y) = \text{Var}(X) + \text{Var}(Y)$ if X and Y are independent random variables. (Convince yourself in an intuitive way that this should be true.)

47. Let X and Y be independent random variables. Find $\text{Var}(2 + 3X - 4Y)$.

48. Suppose X is a random variable with mean μ and variance σ^2. What is the mean and variance of the random variable $Y = (X - \mu)/\sigma$?

49. (Covariance) The *covariance* of two random variables X and Y, denoted by $\text{Cov}(X, Y)$, is defined by

$$\text{Cov}(X, Y) \equiv E([X - E(X)][Y - E(Y)]).$$

Show that:

(a) $\text{Cov}(X, Y) = E(XY) - E(X)E(Y)$.

(b) $\text{Cov}(X, Y) = 0$ if X and Y are independent, but $\text{Cov}(X, Y) = 0$ does not imply that X and Y are independent. [*Hint:* Suppose 3 distinguishable balls are randomly placed in 3 cells, let X denote the number of balls in cell 1, and let Y denote the number of occupied cells.]

(c) $\text{Var}(X + Y) = \text{Var}(X) + \text{Var}(Y) + 2\,\text{Cov}(X, Y)$.

(d) $\text{Cov}(a + bX, c + dY) = bd\,\text{Cov}(X, Y)$.
(e) $\text{Cov}(X + Y, Z) = \text{Cov}(X, Z) + \text{Cov}(Y, Z)$.
(f) $-\sigma_X\sigma_Y \leq \text{Cov}(X, Y) \leq \sigma_X\sigma_Y$, where $\sigma_X = \sqrt{\sigma_X^2}$ and $\sigma_X^2 = \text{Var}(X)$. [*Hint:* Use the fact that $\text{Var}(X/\sigma_X \pm Y/\sigma_Y) \geq 0$.] This shows that $\rho(X, Y)$, the *correlation coefficient* of X and Y, satisfies $-1 \leq \rho(X, Y) \leq 1$, where $\rho(X, Y) \equiv \text{Cov}(X, Y)/\sigma_X\sigma_Y$.
(g) Let X be the number of ones and let Y be the number of twos in n rolls of a fair die. Find $\text{Cov}(X, Y)$ using indicator random variables.

4. THE CHEBYCHEV INEQUALITY AND THE WEAK LAW OF LARGE NUMBERS

Often we are not given f_X but only $E(X)$ and perhaps $\text{Var}(X)$. Nevertheless, we would like to make probabilistic statements about X even though we don't know f_X. The next two theorems enable us to do this, but first we make the point clearer with an example.

EXAMPLE 11

The author was working with a random variable X such that $E(X) = 1470$ and $\text{Var}(X) = 35^2$, but he forgot f_X. Still, he wanted to know $P(X \geq 1470)$, $P(X \geq 2520)$, and $P(1400 \leq X \leq 1540)$. A little later he recalled that X was the sum of 420 independent random variables, each having the probability mass function f_Y with $E(Y) = 3\frac{1}{2}$, and $\text{Var}(Y) = \frac{35}{12}$. Of course, he forgot f_Y. (In Chapter 3 he recalled that Y corresponded to rolling a fair die.) This time he wanted to know $P(3 \leq W \leq 4)$, where $W \equiv X/420$. Later, using our knowledge of f_Y, we shall show that these probabilities are (about) $\frac{1}{2}$, 0, .9544, and .999; however, without knowing f_Y, we can only give upper and lower bounds on these probabilities. We do so after presenting Theorems 8 and 9. •

THEOREM 8. (Markov Inequality)

If X is a random variable that takes only nonnegative values, then for any number $t > 0$,

$$P(X \geq t) \leq \frac{E(X)}{t}.$$

PROOF.

$$E(X) = \sum_{x\geq 0} xf_X(x) = \sum_{0\leq x<t} xf_X(x) + \sum_{x\geq} xf_X(x)$$

$$\geq \sum_{x\geq t} xf_X(x) \geq t\sum_{x\geq t} f_X(x) = tP(X \geq t). \ \blacklozenge$$

THEOREM 9. (Chebychev Inequality)

Let X be a random variable, then for any number $t > 0$,

$$P(|X - E(X)| \geq t) \leq \frac{\text{Var}(X)}{t^2}.$$

Proof. Since $(X - E(X))^2$ is a nonnegative random variable, we can apply Theorem 8 to obtain

$$P((X - E(X))^2 \geq t^2) \leq \frac{E[(X - E(X))^2]}{t^2} = \frac{\text{Var}(X)}{t^2}.$$

But also, $(X(\omega) - E(X))^2 \geq t^2$ if and only if $|X(\omega) - E(X)| \geq t$, establishing the desired result. ◆

Referring to Example 11, Theorem 8 enables us to assert that $P(X \geq 1470) \leq 1$ and $P(X \geq 2520) \leq .583$, while we can use Theorem 9 to conclude that

$$P(1400 \leq X \leq 1540) = 1 - P(|X - E(X)| \geq 2 \cdot 35) \geq 1 - \frac{35^2}{4 \cdot 35^2} = \frac{3}{4}$$

and $P(3 \leq W \leq 4) \geq \frac{35}{36} \approx .972$. The reason why the bounds for X are not very sharp (that is, close to the true probabilities) is that they work for *all* random variables with expectation 1470 and variance 35^2, and not just for rolling 420 fair dice. Thus, while our inequalities aren't always terribly useful as a means of estimating probabilities, they are often useful; moreover, they have useful theoretical implications as we shall see in the proof of Theorem 10. We also hasten to add that the situation described in Example 11 is not as artificial as it may seem. You need only read today's newspaper to find examples wherein all the relevant information has been condensed into one real number, namely, the average W.

When properly interpreted, the following theorem (which is often called the "law of averages") states that our intuitive notion of probability has been properly axiomatized. For example, let X_i be 1 if the ith toss of a fair coin results in tails and 0 otherwise ($i = 1, 2, \ldots$), and define $T(n)$, the total number of tails observed on the first n tosses, by $T(n) = \sum_{i=1}^{n} X_i$. If our intuitive notion of probability as discussed in Section 1 of Chapter 1 is to comply with our axioms, then the relative frequency of tails, $T(n)/n$, should get closer and closer to $\frac{1}{2}$ as the number n of tosses increases. That is, the probability that the relative frequency of tails will differ from $\frac{1}{2}$ by more than some arbitrary but fixed number $\epsilon > 0$ tends to zero as the number of tosses tends to infinity.[5] In symbols: given any $\epsilon > 0$,

$$P\left(\left|\frac{T(n)}{n} - \frac{1}{2}\right| > \epsilon\right) \to 0 \qquad \text{as } n \to \infty.$$

We establish a more general version of this statement in the following

THEOREM 10. (Weak Law of Large Numbers)

Let $X_1, X_2, \ldots$ be a sequence of independent random variables with the same probability mass function f_X. Then given $\epsilon > 0$,

$$P\left(\left|\frac{X_1 + \cdots + X_n}{n} - E(X)\right| > \epsilon\right) \to 0 \qquad \text{as } n \to \infty.$$

PROOF. Using Theorem 6, we can apply the Chebychev inequality to yield (for each fixed $\epsilon > 0$)

$$P\left(\left|\frac{X_1 + \cdots + X_n}{n} - E(X)\right| > \epsilon\right) \leq \frac{\operatorname{Var}(X)}{n\epsilon^2} \to 0 \qquad \text{as } n \to \infty. \ ♦$$

EXERCISES:

50. Verify footnote 5 by using the fact that $\binom{2n}{n} \approx 2^{2n}(\pi n)^{-1/2}$.

51. Let $X_1, X_2, \ldots$ be a sequence of independent but not necessarily identically distributed random variables, and suppose that $\operatorname{Var}(X_i) \leq M < \infty$ for each i. Show that for each $\epsilon > 0$,

$$P\left(\left|\frac{X_1 + \cdots + X_n}{n} - \frac{E(X_1) + \cdots + E(X_n)}{n}\right| > \epsilon\right) \to 0 \qquad \text{as } n \to \infty.$$

52. (Continuation) Use part (c) of Exercise 49 to extablish the same result when we assume that $\operatorname{Cov}(X_i, X_j) \leq 0$ for $i \neq j$ instead of independence.

A One-Sided Inequality. The following inequality is of interest in that it often provides much stronger bounds than do the Chebychev or Markov inequalities. The proof depends upon the Hölder inequality and can be found on page 256 in H. Cramér, *Mathematical Methods of Statistics* (Princeton, N.J.: Princeton University Press, 1946). Let X be a random

[5] This is not the same as saying that the number $T(n)$ of tails will be close to the expected number $n/2$ of tails; in fact, given $\epsilon > 0$, $P\left(\left|T(n) - \frac{n}{2}\right| > \epsilon\right) \to 1$ as $n \to \infty$.

variable such that $E(X) = 0$, then

$$P(X \leq t) \leq \frac{\text{Var}(X)}{\text{Var}(X) + t^2}, \qquad \text{for } t < 0,$$

and

$$P(X \geq t) \leq \frac{\text{Var}(X)}{\text{Var}(X) + t^2}, \qquad \text{for } t \geq 0.$$

Thus, in Example 11, we have $P(X \geq 2520) \leq 35^2/(35^2 + 1050^2) \approx .01$, a much sharper bound than the one obtained using the Markov inequality. Of course, this is to be expected, since we now assume knowledge of $\text{Var}(X)$, whereas for the Markov inequality we only require knowledge of $E(X)$. This suggests that if we have more information we should be able to obtain a better inequality—and, indeed, this is usually the case.

The next observation applies to Exercises 53 through 57. Over the years, a certain professor has found that X, the test scores of students taking the first test,[6] satisfy $E(X) = 1000$ and $\text{Var}(X) = 2 \cdot 100^2$.

53. This year the average score in his class of 100 was 1100. Was this good performance due to improved lectures or to pure chance? (Use both the Markov and the one-sided inequalities.)

54. What can you assert about the probability that none of the 100 students gets a score above 1400? Assume that the scores are independent.

55. What can you assert about the probability that each of the 100 scores will lie between 900 and 1100? What about the average score?

56. How many students would have to be enrolled to ensure with probability .9 that the class average is within 20 of 1000?

57. The average time required for the students to complete the test was found to be 70 minutes with a variance of 144 minutes. When should the professor terminate the test if he wishes to allow sufficient time for 90 percent of the students to complete the test?

5. SPECIAL DISCRETE RANDOM VARIABLES

Certain random variables that we have already encountered arise so frequently as to merit special consideration. In this section we present four of these special random variables.

[6] Here Ω is the (hypothetical) infinite set of all students who could ever take his test and $X(\omega)$ is the score that student ω would receive.

5.1 Binomial

The simplest and most important of the special discrete random variables is the binomial. A binomial random variable simply counts the number of "successes" in a sequence of independent, identical trials (experiments), each of which results in either "success" or "failure." Our most familiar example of a binomial random variable arises in connection with flipping a coin several times. In this experiment, the total number X of heads that appears is a binomial random variable, and each flip is a trial with heads being success and tails being failure. The number X of dishes broken by the youngest sister in Example 26 of Chapter 1 is also seen to be a binomial random variable if we identify success at trial i with the youngest sister breaking the ith dish, $i = 1, 2, 3, 4$.

DEFINITION

We say that a random experiment is a *binomial experiment* if

(i) there are a fixed number n of trials,
(ii) the outcome of each trial can be classified into one of two categories: success and failure,
(iii) each trial has the same probability p of success, and
(iv) the trials are mutually independent.

If X is the number of successes in a binomial experiment with n trials and probability p of success, then X is said to be a *binomial random variable* with parameters n and p, written "X is $B(n, p)$" and read "X is binomial n, p." (When $n = 1$, X is often called a *Bernoulli random variable.*)

EXAMPLE 12

A fair coin is tossed three times and X is the total number of heads. Here $n = 3$ and $p = \frac{1}{2}$, so X is $B(3, \frac{1}{2})$. •

EXAMPLE 13

If X is the number of dishes broken by the youngest sister in Example 26 of Chapter 1, then X is $B(4, \frac{1}{3})$, since, according to the youngest sister, the three sisters are equally clumsy. •

EXAMPLE 14

Six balls are drawn at random with replacement from an urn that contains 3 white and 5 red balls. The number X of white balls drawn is $B(6, \frac{3}{8})$, since there are six trials and the probability of success on each trial is $\frac{3}{8}$ as the drawing is done with replacement. This also implies the trials are independent. •

EXAMPLE 15

If the balls were not replaced in Example 14, then the experiment would not be a binomial experiment, as (iv) would be violated. •

EXAMPLE 16

Fred, an avid fisherman, has vowed to go fishing every day this week or until he catches a fish. The probability that he catches at least one fish on any given day is $\frac{2}{5}$, and we say he had a success on the ith trial if he caught at least one fish on the ith day, $1 \leq i \leq 7$. This is not a binomial experiment, for if he caught a fish the first day, then he would not fish again that week. This violates (i). •

EXAMPLE 17

A certain family of 5 has agreed to toss a fair coin. If a head appears, they will all go to the beach; otherwise they all stay home. Identify success on the ith trial with the ith member of the family going to the beach, $i = 1, 2, 3, 4, 5$. This is not a binomial experiment, since if one member of the family goes to the beach they all do; that is, (iv) is violated. •

Next, we find the relevant probabilistic information associated with a binomial random variable.

THEOREM 11

If X is $B(n, p)$, then

$$P(X = i) = \binom{n}{i} p^i(1 - p)^{n-i}, \qquad i = 0, 1, 2, \ldots, n, \tag{2.6}$$

$$E(X) = np, \tag{2.7}$$

and

$$\operatorname{Var}(X) = np(1 - p). \tag{2.8}$$

PROOF. Fix i, $0 \leq i \leq n$. The probability of observing any particular sequence of i successes (S's) and $n - i$ failures (F's) is $p^i(1 - p)^{n-i}$, since the trials are independent by (iv) and the probability of success and failure on each trial is p and $1 - p$ by (iii). Moreover, it follows from Theorem 7 of Chapter 1 that there are $\binom{n}{i}$ sequences of length n containing i S's and $(n - i)$ F's. Equation (2.6) is now established upon observing that these $\binom{n}{i}$ elementary events partition the event $X = i$.

To show (2.7) and (2.8) we make use of our knowledge of indicator random variables developed in Exercises 24 and 38.[7] Let X_j be 1 if trial j results in success and 0 otherwise, $1 \leq j \leq n$. Then $X = X_1 + X_2 + \cdots + X_n$. Hence, we can conclude from Theorem 2 and Exercise 24 that

$$E(X) = E\left(\sum_{j=1}^{n} X_j\right) = \sum_{j=1}^{n} E(X_j) = \sum_{j=1}^{n} p = np,$$

while we can employ (iv), Theorem 5, and Exercise 38 to obtain

$$\text{Var}(X) = \text{Var}\left(\sum_{j=1}^{n} X_j\right) = \sum_{j=1}^{n} \text{Var}(X_j) = \sum_{j=1}^{n} p(1-p) = np(1-p). \blacklozenge$$

5.2 Geometric

The geometric random variable can be thought of as simply counting the number of independent, identical trials until a success occurs. For example, if a fair coin is tossed repeatedly until the first head appears and if X is the number of flips required, then X is a geometric random variable.

DEFINITION

We say that a random experiment is a *geometric experiment* if

(i) the outcome of each trial can be classified into one of two categories: success and failure,
(ii) each trial has the same probability p of success,
(iii) the trials are mutually independent, and
(iv) the trials are repeated until one success is obtained.

If X is the number of trials required until a success is obtained, then X is said to be a *geometric random variable* with parameter p, written "X is $G(p)$" and read "X is geometric p."

THEOREM 12

If X is geometric random variable with probability p of success, then

$$P(X = n) = (1-p)^{n-1}p, \qquad n = 1, 2, \ldots, \tag{2.9}$$

$$E(X) = \frac{1}{p}, \tag{2.10}$$

and

$$\text{Var}(X) = \frac{1-p}{p^2}. \tag{2.11}$$

[7] Direct proofs can be given using only (2.6) and the binomial theorem.

PROOF. Fix n. Then the event $X = n$ can occur if and only if there are $n - 1$ failures followed by a single success. Since the trials are independent, this happens with probability $(1 - p)^{n-1}p$. Verifying (2.10) and (2.11) is more difficult, and many readers will want to skip over it. Using the definition of expectation and the fact that a power series can be differentiated term by term within its radius of convergence, we have

$$E(X) = \sum_{n=1}^{\infty} n(1 - p)^{n-1}p = -p \sum_{n=1}^{\infty} \frac{d}{dp} (1 - p)^n$$

$$= -p \frac{d}{dp} \sum_{n=1}^{\infty} (1 - p)^n = -p \frac{d}{dp}\left[\frac{1 - p}{1 - (1 - p)}\right] = \frac{p}{p^2} = \frac{1}{p}.$$

Similarly, we obtain

$$E(X^2) + E(X) = \sum_{n=1}^{\infty} n(n + 1)(1 - p)^{n-1}p = \frac{2}{p^2},$$

so

$$\text{Var}(X) = \frac{1 - p}{p^2} \qquad \text{by (2.5).} \ \blacklozenge$$

A very simple argument can be given to verify that $E(X) = 1/p$ if X is $G(p)$. Define $E(X \mid S)$ and $E(X \mid F)$ to be, respectively, the expected number of *additional* trials needed to obtain the first success given that the first trial resulted in a success or in a failure. Then

$$E(X) = [1 + E(X \mid S)]p + [1 + E(X \mid F)](1 - p).$$

Clearly, $E(X \mid S) = 0$, and $E(X \mid F) = E(X)$, since the trials are independent. Thus,

$$E(X) = [1 + 0]p + [1 + E(X)](1 - p) = 1 + (1 - p)E(X),$$

so $E(X) = 1/p$ as desired. We hasten to point out that this simple argument arises frequently (see Exercises 31 and 32 and Exercise 25 of Chapter 3), particularly in the study of the important field of stochastic processes.

EXAMPLE 18. (Craps)

Suppose that the very first roll in a game of craps results in the number 4. Find the expectation and the distribution of the number X of additional rolls needed to conclude the game.

SOLUTION: The game will end when either a 4 or a 7 appears, and the probability of a 4 or a 7 appearing on any given roll is $(3 + 6)/36 = \frac{1}{4}$. Since the rolls are mutually independent, X is a geometric random variable with parameter $\frac{1}{4}$. Hence, $E(X) = 4$. Making use of this type of reasoning, we can show (see Exercise 32) that the expected number of rolls in a game of craps is roughly 3.375. Results such as these are useful in the successful operation of gambling casinos. (The rate of winning must cover the overhead expenses in addition to the return on invested capital.) •

EXAMPLE 19

There is but one person left in a spelling bee, and the teacher is asking him to spell equally difficult words. If he misses with probability $\frac{1}{10}$ and each trial takes $\frac{1}{2}$ minute, on the average, how long will the rest of the class have to wait until he misses? What is the probability they will have to wait no more than 1 minute?

SOLUTION: Clearly, the number X of words that need to be asked until he misses (success) is a geometric random variable with $p = \frac{1}{10}$. Hence, the average waiting time is 5 minutes, since $E(X) = 10$. The probability that the class waits no longer than on minute is

$$P(X \leq 2) = P(X = 1) + P(X = 2) = \tfrac{1}{10} + \tfrac{9}{10} \cdot \tfrac{1}{10} = \tfrac{19}{100}. \bullet$$

More generally, rather than counting the number of trials until the first success, we could let X be the number of trials until the rth success. In this case, X is called a *negative binomial* random variable (see Exercises 69 through 71).

5.3 Hypergeometric

DEFINITION

If a random sample of size n is drawn without replacement from a population of size N in which there are r special items, then the number X of special items drawn (successes) is said to be a *hypergeometric random variable* with parameters n, r, and N, and we write "X is $H(n, r, N)$."

Examples of these special items include the r red beads in a jar of N beads, the r defective items in a lot of N items (such as light bulbs), the r registered Republican voters among the N registered voters, the 7 linemen on the 11-man football team, the 8 men out of our sample of 1697 who have an XYY chromosome pattern (see Example 7 of Chapter 5), and the 13 no-load funds included in our study of 115 mutual funds (see Example 1 of Chapter 9).

EXAMPLE 20. (Mutual Funds)

There has been considerable disagreement among investors concerning the relative performance of two types of mutual funds called load and no-load funds. They differ in that the former charges investors an 8 percent commission whereas the latter charges only a nominal fee. Recently, M. C. Jensen[8] made a study evaluating the performance of 115 mutual funds in the twenty-year period 1945–1964. There were 102 load and 13 no-load funds in Jensen's sample. Using a sophisticated technique for evaluating performance (which does not take the commission into account), Jensen found that only 39 of the 115 funds received a rating of at least zero. Consequently, *if* there is no difference in the performance ratings of load and no-load funds, then the number X of no-load funds receiving a performance rating of at least zero is a hypergeometric random variable with $n = 39$, $r = 13$, and $N = 115$. The question of whether or not there is a difference is considered in Example 1 of Chapter 9. (Incidentally, the observed value X was 9. Would this lead you to believe that there is, in fact, a difference?) •

THEOREM 13

If X is $H(n, r, N)$, then

$$P(X = i) = \frac{\binom{r}{i}\binom{N-r}{n-i}}{\binom{N}{n}}, \qquad i = 0, 1, 2, \ldots, \min(n, r), \tag{2.12}$$

$$E(X) = n\frac{r}{N}, \tag{2.13}$$

and

$$\text{Var}(X) = \frac{N-n}{N-1}\,n\frac{r}{N}\left(1 - \frac{r}{N}\right). \tag{2.14}$$

PROOF. Fix i, $0 \leq i \leq \min(n, r)$. Then our sample of size n contains i out of the r special items and $n - i$ out of the $N - r$ other items. Since the i special items can be chosen in $\binom{r}{i}$ ways and the $n - i$ others in $\binom{N-r}{n-i}$ ways by Theorem 7 of Chapter 1, and all such choices of special

[8] M. C. Jensen, "Problems in Selection of Security Portfolios—The Performance of Mutual Funds in the Period 1945–1964," *Journal of Finance,* 1968, pp. 389–416.

items are compatible with the choices of other items, it follows that in all there are $\binom{r}{i}\binom{N-r}{n-i}$ ways of choosing i special and $n - i$ other items. Since the n drawings form a random sample without replacement, there are $\binom{N}{n}$ equally likely points in our sample space. Combining these two facts establishes (2.12).

Again we use indicator random variables to find $E(X)$ and $\text{Var}(X)$. Let X_i be 1 if the ith drawing results in a special item and 0 otherwise. Then $X = X_1 + X_2 + \cdots + X_n$, so by Theorem 2 it suffices to show that $E(X_i) = r/N$ for each i, $1 \leq i \leq n$. To show $E(X_i) = r/N$, note that each of the N items in our population has probability $1/N$ of being drawn at the ith trial. Since there are r special items, it follows that $P(X_i = 1) = r/N$. Hence,

$$E(X_i) = P(X_i = 1) = \frac{r}{N}.$$

Equation (2.14) follows from Theorem 7 and the fact that

$$\text{Var}(X_i) = \frac{r}{N}\left(1 - \frac{r}{N}\right). \quad \blacklozenge$$

EXAMPLE 21

Suppose that r fish are caught, tagged, and released back into a lake that contains N fish, $r = 100$, and $N = 2000$. After a while a new catch of $n = 60$ fish is made, and the number X of tagged fish is observed to be 5. Assuming that the number of fish in the lake didn't change between the two catches and that the second catch constituted a random sample of the fish in the lake, then X is $H(60, 100, 2000)$. Hence we can conclude from (2.12) that

$$P(X = 5) = \binom{60}{5}\binom{1940}{95}\Big/\binom{2000}{60} = .1019429. \tag{2.15}$$

An interesting problem would have been to make an estimate of N on the basis of our observation (of $X = 5$) if we did not know the value of N. We shall do this in Chapter 7.

If we had been sampling with replacement, then X would have been $B(n, r/N)$ instead of $H(n, r, N)$. In this case, we would have found

$$P(X = 5) = \binom{60}{5}\left(\frac{100}{2000}\right)^5\left(1 - \frac{100}{2000}\right)^{95} = .1016158. \bullet \tag{2.16}$$

Notice how close the two probabilities in (2.15) and (2.16) are. This gives us an idea of how similar $B(n, r/N)$ and $H(n, r, N)$ are when the sampling fraction n/N is small. Moreover, notice that these two random variables have the same expected value regardless of the size of n/N, and their variances are almost the same if n/N is small. Consequently, we are led to believe that if n/N is small, say less than .05, then we can rather accurately find probabilities for a hypergeometric random variable by finding the appropriate probabilities for a binomial random variable. This is, in fact, true, and we verify this for the more advanced reader in the next theorem. We also illustrate the approximation for the cases where X is $B(6, \frac{1}{3})$ and Y is $H(6, r, 3r)$ for $r = 5$, 25, and 100 in what is known as a *histogram* (see Figures 2.3 and 2.4). In the histogram, the area of the ith rectangle in solid lines represents the probability that $Y = i - 1$ and the area of the ith rectangle in dotted lines represents the probability

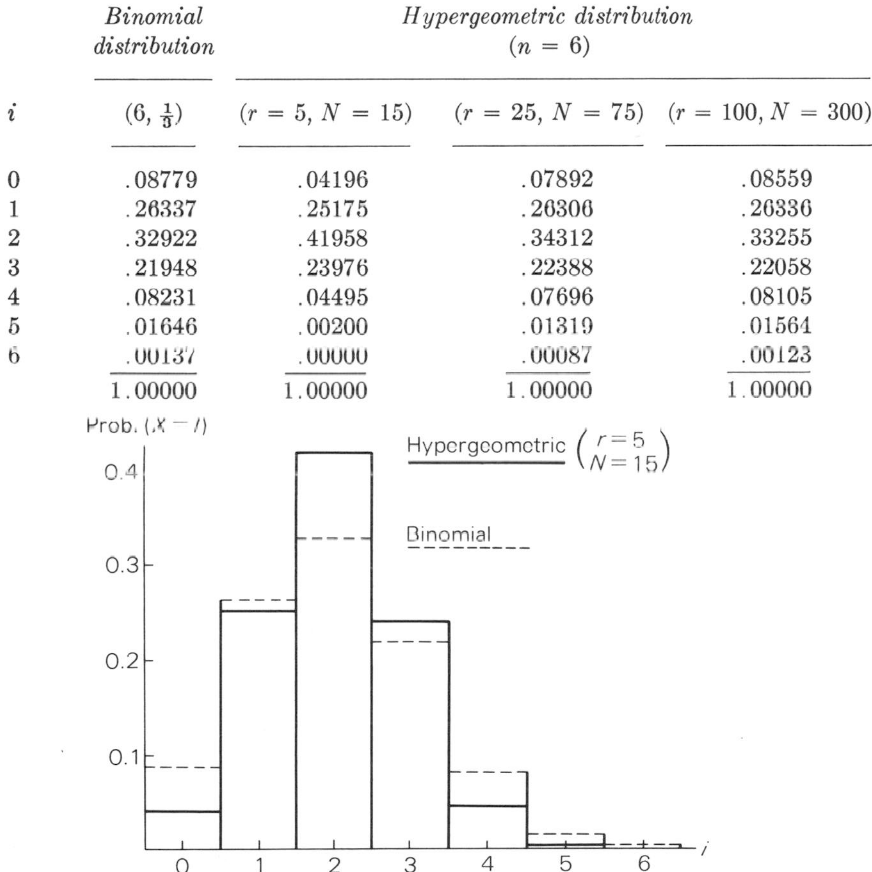

	Binomial distribution	*Hypergeometric distribution* ($n = 6$)		
i	$(6, \frac{1}{3})$	$(r = 5, N = 15)$	$(r = 25, N = 75)$	$(r = 100, N = 300)$
0	.08779	.04196	.07892	.08559
1	.26337	.25175	.26306	.26336
2	.32922	.41958	.34312	.33255
3	.21948	.23976	.22388	.22058
4	.08231	.04495	.07696	.08105
5	.01646	.00200	.01319	.01564
6	.00137	.00000	.00087	.00123
	1.00000	1.00000	1.00000	1.00000

Figure 2.3

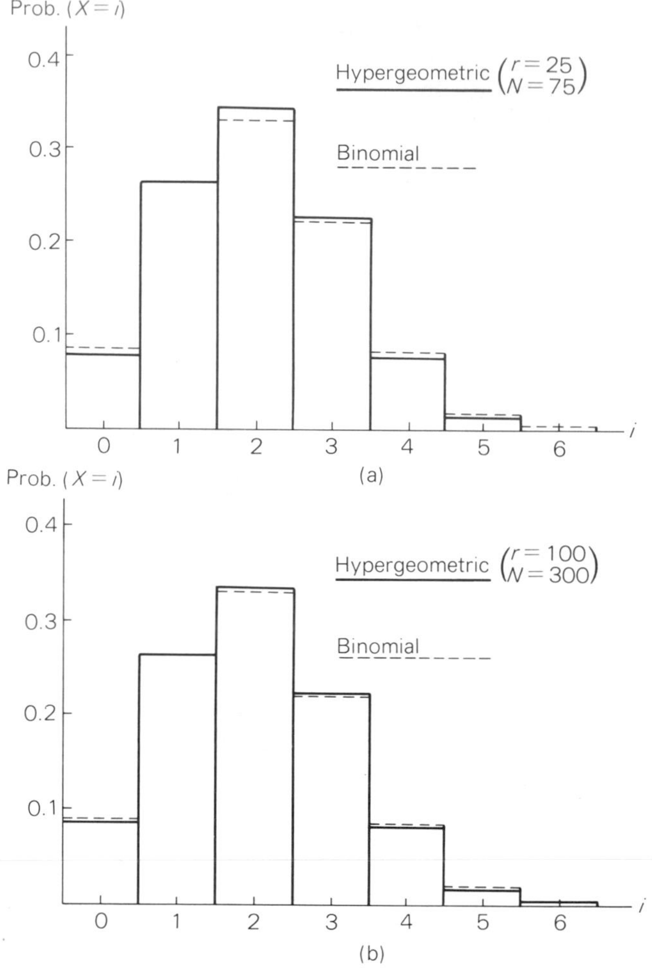

Figure 2.4

that $X = i - 1$. Of course, the small loss in accuracy concomitant with using a binomial in place of a hypergeometric is more than offset by the ease with which we can compute probabilities for a binomial random variable. (In fact, probabilities for binomial random variables can be found in tables.)

THEOREM 14

Given $\epsilon > 0$, n, and p, there is an integer M so large that if X is $H(n, r, N)$ with $N > M$, if $p = r/N$, and if Y is $B(n, p)$, then

$$|P(X = k) - P(Y = k)| < \epsilon, \qquad k = 0, 1, 2, \ldots, n.$$

PROOF.

$$P(X = k) = \frac{\binom{r}{k}\binom{N-r}{n-k}}{\binom{N}{n}} = \binom{n}{k}\frac{r!}{(r-k)!}\cdot\frac{(N-n)!}{N!}\cdot\frac{(N-r)!}{(N-r-n+k)!}$$

$$= \binom{n}{k}\frac{r(r-1)\cdots(r-k+1)}{N(N-1)\cdots(N-k+1)}$$

$$\cdot\frac{(N-r)(N-r-1)\cdots(N-r-n+k+1)}{(N-k)(N-k-1)\cdots(N-n+1)}$$

$$\geq \binom{n}{k}\left(\frac{r}{N}\right)^k\left(\frac{N-r}{N}\right)^{n-k} - \epsilon = P(Y = k) - \epsilon,$$

since the ratios $(r-1)/(N-1)$, $(N-r-1)/(N-k-1)$, and so on, are close to r/N and $(N-r)/N$ when N is large. Similarly, we can show that $P(X = k) \leq P(Y = k) + \epsilon$. ◆

▶

5.4 Poisson

A random variable that arises frequently in a wide variety of disciplines is the Poisson random variable. It arises in the physical sciences, natural sciences, social sciences, and in engineering. Examples of Poisson random variables include (i) the number of machine breakdowns in a day, (ii) the number of customers who arrive during a time period of length t, (iii) the number of calls placed through a given switchboard in a minute, (iv) the number of α particles emitted by a radioactive substance in a second, and (v) the number of deaths on a given day among the policyholders of a life insurance company—to name just a few.

DEFINITION

We say that X is a *Poisson random variable* with parameter λ (> 0), written "X is Po(λ)," if

$$P(X = k) = \frac{e^{-\lambda}\lambda^k}{k!} \qquad \text{for } k = 0, 1, 2, \ldots.$$

It is easily shown (see Exercise 80) that if X is Po(λ), then $E(X) = \lambda$ and Var(X) $= \lambda$.

EXAMPLE 22. (Emission of α Particles)

An interesting application of the Poisson distribution to an important problem in physics was made by H. Bateman.[9] He used it to explain experimental data collected by two of the great pioneers of atomic physics, E. Rutherford and H. Geiger (after whom the geiger counter was named).

In counting the α particles emitted from radioactive substances, it had been observed that although the average number of α particles emitted is approximately constant over long time intervals, the number emitted is subject to wide fluctuations over short time intervals. Rurtherford and Geiger[10] performed an experiment to find out whether these wide fluctuations are in agreement with the laws of probability. Specifically, they sought to determine whether or not α particles are expelled independently and at random with regard to both time and space. This question is of extreme importance in physics. For instance, it might be conceived that the emission of an α particle precipitates the disintegration of neighboring nuclei or that radioactive nuclei are subject to "aging" (that is, the emission of the first α particle from a given nucleus becomes more and more likely with the passage of time).

The apparatus used by Rutherford and Geiger is shown in Figure 2.5. As schematically shown there, the radioactive source (a small disk coated with polonium), which is contained in a sealed vacuum tube, emits α particles in all directions. The particles hitting the zinc sulphide coating produce tiny flashes of light called "scintillations," which the experimenter can see through the microscope. Thus, counting the scintillations is equivalent to counting the particles that reach the portion of the window in the experimenter's field of vision.

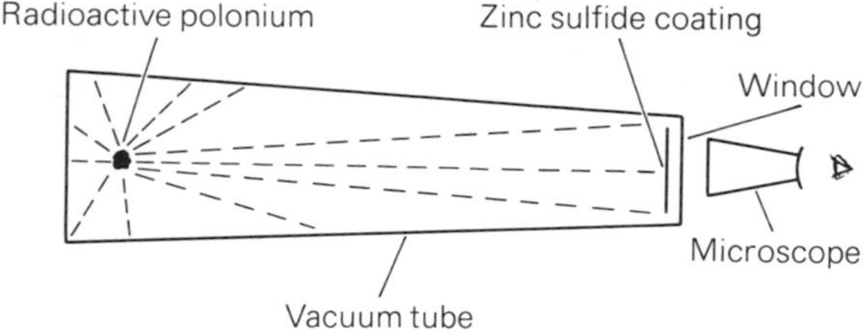

Figure 2.5 Experiment Apparatus

[9] H. Bateman, "On the Probability Distribution of α Particles," *Philosophical Magazine*, pages 704–707, Ser. 6, 20 (1910).

[10] E. Rutherford, and H. Geiger, "The Probability Variations in the Distribution of α Particles," *Philosophical Magazine*, pages 698–704, Ser. 6, 20 (1910).

Rutherford and Geiger recorded the number of scintillations in 2608 successive intervals of length 7.5 seconds over a 5-day period. In all they counted 10,097 scintillations, so that on the average there were 3.87 scintillations per time interval. Bateman showed that if their atomic theory were correct and if λ is the average number of scintillations for a time interval of given length, then the number X of scintillations in this time interval is a Poisson random variable with parameter λ. Thus, using 3.87 as our best guess of λ, their atomic theory predicted that X is a Poisson random variable with parameter 3.87. The actual results and the theoretical results are shown in Table 2.2. Using the methods in Section 1 of Chapter 6, we can show the remarkable agreement between their theory and the experimental results. •

Table 2.2 Results of the Experiment

	Number of time intervals (of length 7.5 seconds) during which n particles were recorded	
n	*Observed*	*Predicted (rounded off)*
0	57	54
1	203	210
2	383	407
3	525	525
4	532	508
5	408	394
6	273	254
7	139	140
8	45	68
9	27	29
10	10	11
11	4	4
12	0	1
$\geqq 13$	2	1

For further examples of Poisson random variables, see Exercise 35 and Exercise 36 of Chapter 4.

The importance of the Poisson random variable is, perhaps, best explained by the next theorem which states that if X is $B(n, p)$, with n quite large and p quite small, then the distribution of X is very nearly the same as the distribution of a Poisson random variable with parameter $\lambda = np$.

THEOREM 15

Given $\epsilon > 0$ and $\lambda > 0$, there is an integer M so large that if X is $B(n, \lambda/n)$ with $n > M$ and if Y is $\text{Po}(\lambda)$, then

$$|P(X = k) - P(Y = k)| < \epsilon, \qquad \text{for } k = 0, 1, 2, \ldots.$$

The proof of this theorem is similar to that of Theorem 14 (see Exercise 81).

EXAMPLE 23. (Geographical Location)

In a recent issue of *Geographical Analysis*,[11] Dacey describes a single model that subsumes many of the location models that occur in geographical analysis. Specifically, his model supposes that a region consists of J disjoint subregions and that n objects—such as people, buildings, businesses, or other facilities—will locate in the region. The actual locating of these n objects takes place one by one. Naturally, the location of the kth object depends upon the location of the first $k - 1$ objects in addition to the attributes of the J individual subregions. Of particular interest is the number X_j of these n objects that will locate in subregion $j (1 \le j \le J)$ and Y_i, the location of the ith object.

Now the location of the first object depends only on the attributes a_j of the individual subregions themselves. So if, for example, a_j is the area of the jth subregion, $A = \sum_{j=1}^{J} a_j$, and the first object locates at random in the region, then

$$P(Y_1 = j) = \frac{a_j}{A}, \qquad j = 1, 2, \ldots, J.$$

While it might be true that the random variables Y_i are identically distributed, it would usually not be the case that the Y_i are independent. In fact, it is often the case that the objects have a tendency to cluster. This tendency is quite evident in the growth of cities. Pólya's urn scheme (see Example 21 of Chapter 1) provides one nice way of modeling this tendency of the objects to cluster in particular subregions. This is seen as follows.

Suppose that an urn contains a_j balls of type j $(1 \le j \le n)$ and that

[11] Michael F. Dacey, "A Hypergeometric Family of Discrete Probability Distributions: Properties and Applications to Location Models," *Geographical Analysis*, Vol. 1, pp. 283–317 (1969).

each ball has probability $1/A$ of being drawn. If the first ball drawn is of type j, then we say that object 1 located in subregion j—that is, $Y_1 = j$. To incorporate the concentration bias in the location process, the ball withdrawn and $h \geq 0$ more of the same type are placed in the urn. Then a second ball is withdrawn from the urn. If it is of type k, then we say that the second object located in subregion k so $Y_2 = k$. In this case,

$$P(Y_2 = k) = \begin{cases} \dfrac{a_k}{A + h}, & \text{if } k \neq Y_1, \\[2ex] \dfrac{a_k + h}{A + h}, & \text{if } k = Y_1. \end{cases}$$

The location process continues in this manner, so that in general we have

$$P(Y_i = k) = \frac{a_k + th}{A + (i - 1)h} \qquad \text{if } Y_j = k, \text{ for } t \text{ values of } j < i.$$

It can also happen that the objects have a tendency to disperse, as, for example, hospitals do. In this case, we can model the location process as above, but choose $h < 0$ so that the ball withdrawn is replaced but h balls of like kind are removed.

If $h = 0$, then X_j is a binomial random variable with parameters n and a_j/A, whereas X_j is hypergeometric with parameters n, a_j, and A if $h = -1$. ●

EXERCISES:

58. On the average, John hits 70 percent bull's-eyes. Let X be the number of bull's-eyes John hits in his first 10 shots today. Is it clear that X is $B(10, .7)$?

59. On a multiple-choice exam with 5 possible answers for each of the 20 questions, what can you say about the number of correct answers given by a person who knew absolutely nothing about the material being tested?

60. Approximately 10 percent of the population in the United States is Negro. Assuming there is no racial prejudice, and so forth, what are the chances of having at least 3 Negroes enrolled in a class of 10?

61. To determine the effectiveness of a certain diet to reduce the amount of cholesterol in the blood stream, 10 people were put on a new diet. After they had been on the new diet for a sufficient length of time, their cholesterol count was taken. The person running this experiment will endorse this diet only if at least 7 out of the 10 people lowered their cholesterol count after going on the diet. What is the probability that he endorses the new diet if, in fact, it has no effect on the cholesterol level?

62. As part of a continuing study of parapsychology, volunteers are regularly tested for clairvoyant capabilities. The test procedure uses a special card deck consisting of five identical cards in each of five "suits": $\approx$, $*$, $\bigcirc$, $\square$, and $+$. (There are five cards in all!) The test is given as follows: The staff member giving the test sits in one room and shuffles the deck. Then he turns over one card at a time, writes down its value, concentrates on it for one minute, and goes on to the next card. The subject is synchronized by a buzzer that sounds in both rooms. At the end of the test the number of matched cards is determined. Assuming that the person being tested is not clairvoyant and is not unwilling to write, for example, $*$, even if he has already written it once, what is the probability that he will name at least 4 cards correctly? What if the same value can not be written more than once?

63. On a certain multiple-choice examination there are 5 questions. If a student knows the subject material on the test, he should be able to give the correct answer to any question on the test with probability $\frac{3}{4}$. If he does not know the material, he will have a probability of $\frac{1}{5}$ of answering any question correctly. In the past the instructor has found that 30 percent of the students know the material and 70 percent do not. If a student takes the test and answers 3 questions correctly, what is the probability he knows the material? Compare this with Exercise 45 of Chapter 1.

64. Let X, Y, and Z be independent binomial random variables, X is $B(n, p)$, Y is $B(m, p)$, and Z is $B(n, p')$, $p \neq p'$. Are $X + Y$ and $X + Z$ binomial random variables? If so, what are their probability mass functions? Explain why.

65. If X is $B(n, p)$ and Y is $B(n, 1 - p)$, what is the relationship between $P(X = j)$ and $P(Y = n - j)$?

66. A certain well-known textbook states that if X is $B(n, p)$, then $X = np$ is the most likely event. Comment on this statement.

67. Suppose 365 identical blue capsules are placed in an urn and thoroughly mixed. Each capsule contains a slip of paper with a different day of the year written on it. Next, 120 of the capsules are drawn without replacement. Denote by X the number of these 120 capsules that contain one of the days from the month of January. What is the distribution of the random variable X?

68. Suppose X is $G(p)$, Y is $G(p)$, and X and Y are independent. Show that for each positive integer n,

$$P(X = k \mid X + Y = n) = \frac{1}{n + 1}, \qquad k = 0, 1, \ldots, n.$$

69. (Negative Binomial) If X is the number of trials needed to obtain exactly r successes when the probability of success on each of the independent trials

is p, then X is said to be a *negative binomial random variable* with parameters r and p, and we write "X is $NB(r, p)$."

(a) Find the probability mass function of X if X is $NB(r, p)$.

(b) Suppose $X_1, X_2, \ldots, X_r$ are independent random variables and X_i is $G(p)$, $1 \leq i \leq r$. Find the probability mass function of X if $X = \sum_{i=1}^{r} X_i$.

(c) Using part (b), find $E(X)$ and $\text{Var}(X)$ if X is $NB(r, p)$.

70. No pitcher since Dizzy Dean in 1934 had won 30 games in a single season. With 6 starts left to go in the 1969 season, the Detroit Tiger's Denny McLain had already scored 27 victories. Assuming that over his lifetime he wins 75 percent of the games he starts, find the probability that he will equal Dizzy Dean's record. What is the probability that he will achieve his thirtieth victory on his last start?

71. A marketing research firm is interested in obtaining information only from families with 3 or more children. If 10 percent, 15 percent, and 35 percent of all families have 0, 1, and 2 children, respectively, then what is the probability that the firm will need to contact 1000 or more families if it needs a sample of size 100? (Do not attempt to perform the calculations.)

72. A student of probability and statistics has divided the course subject matter into twenty separate problem topics that could appear on his final exam. The student feels the professor will choose his problems from among these topics with equal likelihood and only one from each topic. The final will have five problems. This student has only prepared sixteen of these topics and feels he will flunk the final if *two or more* unprepared topics are on the final. Find the probability the student will flunk.

73. Solve Exercises 23 and 57 of Chapter 1 using hypergeometric random variables.

74. Let X be the number of aces in a poker hand. Find the probability mass function of X.

75. In a recent basketball league expansion, the new team was allowed to pick any 10 players from a list of 20. By definition, a balanced team consists of 4 guards and 6 frontcourt men (forwards and centers). If the team picked the players by ability without regard to whether or not they were guards and if 8 out of the 20 players were guards, what is the probability that a balanced team was chosen?

76. (Keno) The popular game of Keno is played in Las Vegas as follows. The player selects 8 of the numbers 1 through 80. Then 20 balls are withdrawn from a tumbler which contains 80 balls numbered 1 through 80. The player pays \$.60 to play the game, and he receives \$5, \$50, \$1100, or \$12,500 if 5, 6, 7, or 8 of his numbers are drawn.

(a) What is the probability that he will win?

(b) What is the expected value of this game? [*Hint:* $P(X = 5) = .0183026$, $P(X = 50) = .0023727$, $P(X = 1100) = .0001605$, $P(X = 12{,}500) = .0000043$, where X is the amount won.]

77. (Acceptance Sampling) A purchaser of intricate electrical components will reject lots of size 15 if there are more than 5 defective components in the lot. Owing to the high cost of inspection, he inspects only 3 components in each lot, and he accepts the lot only if there are no defectives in his sample of size 3. If 50, 40, and 10 percent of the lots have 4, 6, and 7 defectives, respectively, what is his probability of rejecting a good lot or accepting a bad lot?

78. A hand of seven cards is drawn at random from an ordinary deck of playing cards. What is the probability that it contains exactly 2 hearts? Compare this with the corresponding probability if the cards had been drawn with replacement. [*Answer:* .336 and .311.]

79. When purchasing entire coin collections, the dealer classifies each coin into one of three categories: rare, uncommon, and common. Any collection with at least 1 rare and 2 uncommon coins is worth buying. My collection has 20 coins of which 2 are rare and 7 are uncommon. What is the probability that the dealer would be willing to buy the collection if he examined only 3 of the coins? 5 of the coins?

80. Suppose X is $\text{Po}(\lambda)$. Show that:

(a) $E(X) = \lambda$. [*Hint:* $\sum_{n=0}^{\infty} \frac{x^n}{n!} = e^x$.]

(b) $\text{Var}(X) = \lambda$. [*Hint:* Use equation (2.5) and the above hint, and consider the hint from Exercise 41.]

81. Prove Theorem 15.

3

CONTINUOUS RANDOM VARIABLES

So far, we have dealt exclusively with discrete sample spaces and discrete random variables. In this chapter, and in much of Part II, we shall be dealing with continuous random variables. The mathematics required for a completely rigorous treatment of this subject is far beyond our means. Nevertheless, we will be able to give a thorough and intelligible presentation of this subject without even having recourse to calculus.

1 CONTINUOUS RANDOM VARIABLES

Consider the random experiment wherein a perfectly balanced spinner is spun around the dial shown in Figure 3.1. Then $\Omega = \{x: 0 \leq x \leq 1\}$.

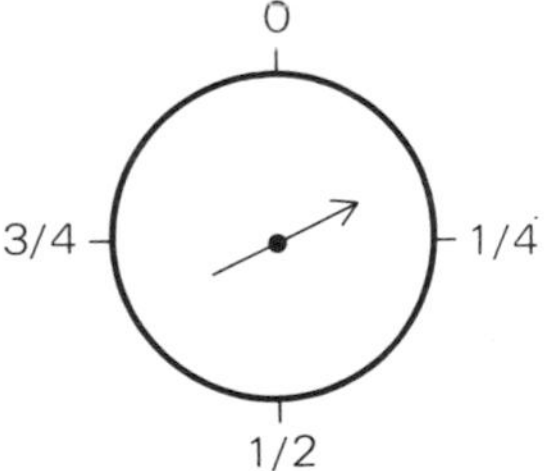

Figure 3.1 A Perfectly Balanced Spinner

Now, our probability model [or equivalently our random variable $X(x) \equiv x$] will be completely specified if we write down the probability function P defined on the subsets of Ω. Since the spinner is perfectly balanced, all of the points of Ω are equally likely, so that there is some nonnegative number p such that $P(\{x\}) = p$ for all $x \in \Omega$. We will now show that p cannot be positive. Suppose p were positive, then we could choose n so large that $np > 1$. Let

$$E = \left\{\frac{1}{2}, \frac{1}{3}, \frac{1}{4}, \ldots, \frac{1}{n+1}\right\}.$$

Then

$$1 = P(\Omega) \geq P(E) = \sum_{i=2}^{n+1} P\left(\left\{\frac{1}{i}\right\}\right) = \sum_{i=2}^{n+1} p = np > 1,$$

which is impossible. Consequently, we must have $p = 0$.

This is a situation we have not previously encountered, and at first glance it seems ridiculous: the probability of each of the possible outcomes is zero, yet one of them will indeed occur. Therefore, let us reassess the statement, "a perfectly balanced spinner is spun." Certainly this statement means that the probability of X lying in some interval $I \equiv \{x: a \leq x \leq b\}$, $0 \leq a < b \leq 1$, is simply the length of that interval. Thus, if $0 \leq a < b \leq 1$, then

$$P(a \leq X \leq b) = b - a. \tag{3.1}$$

Using (3.1), we can show again that $P(\{x\}) = 0$ for all $x \in \Omega$. Here is how. Choose an interval I_x such that $\{x\} \subset I_x \subset \Omega$ and the length of I_x is less than $P(\{x\})$—say, $\frac{1}{2}P(\{x\})$. Then by Theorem 3 of Chapter 1 and equation (3.1) we have

$$P(\{x\}) \leq P(I_x) = \tfrac{1}{2}P(\{x\}),$$

which implies $P(\{x\}) = 0$.

Thus, the two interpretations are consistent with each other, and both are correct. It is the latter interpretation, however, that we shall find more useful. That is, we shall specify the probability function P by assigning probabilities to intervals in the sample space and *not* to individual points.

The former interpretation of each point having the same probability or relative weight is depicted in Figure 3.2, where $f_X(x)$ is the relative weight of the point x. Also, notice that the area under the curve f_X, above the x-axis, and between a and b, is $b - a$ [since the area of a rectangle equals the length of its base times its height $((b - a) \cdot 1 = b - a)$].

Next, suppose that there is a magnet under the dial, and its strength at any given spot x on the dial is (directly) proportional to x. Thus, twice the force is exerted at $2x$ as is exerted at x so $P(\{2x\}) = 2P(\{x\})$. But as in the example above, it is clear that $P(\{x\}) = 0$ for all $x \in \Omega$. Consequently, we must reassess the meaning of the statement that the strength of the magnet at x is proportional to x. We do this graphically in Figure 3.3, where the value $f_X(x)$ of the function f_X at x gives the relative strength or weight of the magnet at x.

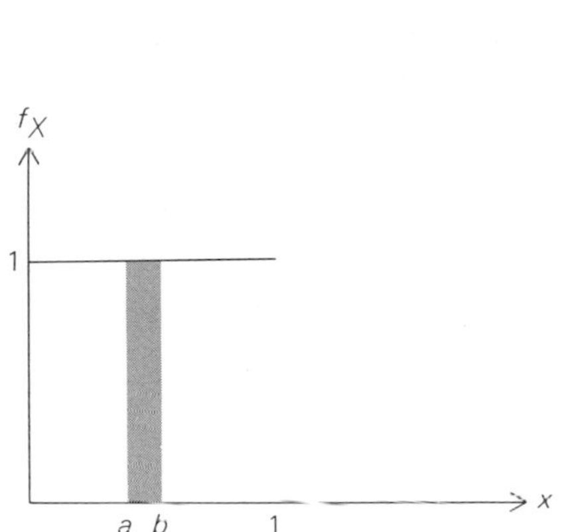

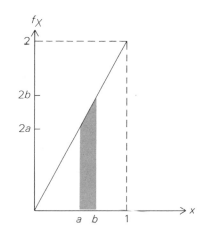

Figure 3.2 The Relative Weights of the Points on the Dial

Figure 3.3 Representation of the Magnet's Strength

Now if we identify probabilities with areas, then in light of Figure 3.3 it follows that if $0 \leq a < b \leq 1$, then[1]

$$P(a \leq X \leq b) = b^2 - a^2.$$

These two examples are suggestive. When the range of a random variable X is an interval, say $\{x: u \leq x \leq v\}$, then in order to find the probability that the random variable X is between a and b, we need to

(1) know the relative weights $f_X(x)$ with which each of the points $x \in \{x: u \leq x < v\}$ occur, and
(2) identify the area below f_X, above the x-axis, and between a and b with $P(a \leq X \leq b)$.

Of course, the weighting function or the probability density function f_X for a continuous random variable is analogous to the probability mass

[1] The area of the triangle formed by the points (0, 0), $(b, 0)$, and $(b, f(b))$ is $\frac{1}{2}(b - 0)(f(b) - 0) = \frac{1}{2}b(2b) = b^2$, since the area of a triangle is one half its height times the length of its base. Similarly, the area of the triangle formed by (0, 0) $(a, 0)$, and $(a, f(a))$ is a^2. Thus, the area under the curve f_X, above the x-axis, and between a and b is just the difference between the areas of the two triangles.

function for a discrete random variable, and we shall require f_X to satisfy analogous properties, namely $f_X(x) \geq 0$ and $\Sigma f_X(x) = 1$.

We now make all of this precise by introducing two definitions.

DEFINITION

Let f be a function such that $f(x) \geq 0$ for all real numbers x. Then we call the area below the graph of f, above the x-axis, and between a and b $(-\infty \leq a \leq b \leq \infty)$ *the integral of f from a to b*, and we denote it by $\int_a^b f(x)\,dx$, or simply $\int_a^b f$.

The reader unfamiliar with integrals should interpret the symbol $\int_a^b f(x)\,dx$ as meaning $\sum_{a<x<b} f(x)$. Integrals possess many of the properties of sums. In particular, let f and g be two functions and let $a \leq b \leq c$ and k be constants; then

(i) $\int_a^b kf(x)\,dx = k\int_a^b f(x)\,dx$,

(ii) $\int_a^b (f(x) + g(x))\,dx = \int_a^b f(x)\,dx + \int_a^b g(x)\,dx$,

(iii) $\int_a^c f(x)\,dx = \int_a^b f(x)\,dx + \int_b^c f(x)\,dx$.

DEFINITION

Let X be a random variable and let f_X be a function such that

(i) $f_X(x) \geq 0$, for all real numbers x,

(ii) $P(a \leq X \leq b) = \int_a^b f_X(t)\,dt$, all $a \leq b$,

(iii) $\int_{-\infty}^{\infty} f_X(t)\,dt = 1$;

then X is a *continuous random variable* and f_X is called the *probability density function* (p.d.f.) of X. We associate with f_X the *cumulative distribution function* (c.d.f.) F_X defined by

$$F_X(x) = \int_{-\infty}^{x} f_X(t)\,dt, \qquad \text{for all numbers } x.$$

Obviously, $F_X(x) = P(X \leq x)$. These definitions are illustrated by letting X be the spot where the spinner stops when it is perfectly balanced and letting Y be the spot where it stops when the magnet is beneath the dial. Then we have

$$f_X(x) = \begin{cases} 1, & \text{if } 0 < x \leq 1, \\ 0, & \text{otherwise,} \end{cases} \quad \text{and} \quad f_Y(x) = \begin{cases} 2x, & \text{if } 0 < x \leq 1, \\ 0, & \text{otherwise,} \end{cases}$$

and

$$F_X(x) = \begin{cases} 0, & \text{if } x < 0 \\ x, & \text{if } 0 \le x \le 1 \\ 1, & \text{if } x > 1 \end{cases} \quad \text{and} \quad F_Y(x) = \begin{cases} 0, & \text{if } x < 0, \\ x^2, & \text{if } 0 \le x \le 1, \\ 1, & \text{if } x > 1. \end{cases}$$

REMARKS:

1. The definitions in Chapter 2 can be extended in the obvious manner so that all the theorems in Chapter 2 are true for continuous random variables. For example, the expectation of a continuous random variable X is defined by

$$E(X) \equiv \int_{-\infty}^{\infty} x f_X(x)\, dx.$$

2. Henceforth, we will be dealing with both discrete and continuous random variables. Thus, instead of referring to the probability mass function or the probability density function of the random variable X, we simply refer to the *distribution* of X.

EXERCISES:[2]

1. Using either elementary calculus or simple facts concerning areas of rectangles and triangles, show that $E(X) = \frac{1}{2}$, $\text{Var}(X) = \frac{1}{12}$, $E(Y) = \frac{2}{3}$, and $\text{Var}(Y) = \frac{1}{18}$, where X and Y are defined above. [*Hint:* $\int_0^1 x = \frac{1}{2}$, $\int_0^1 x^2 = \frac{1}{3}$, and $\int_0^1 x^3 = \frac{1}{4}$.]

*2. Suppose that the magnet is such that the points between 0 and $\frac{1}{4}$ were equally likely, whereas the points between $\frac{1}{4}$ and 1 were also equally likely but twice as likely as the point $\frac{1}{8}$. Find f_X and show that $E(X) = \frac{31}{56}$.

*3. Suppose that the force exerted by the magnet is proportional to x for x between 0 and $\frac{1}{2}$, while inversely proportional to x for x between $\frac{1}{2}$ and 1, so that $f_X(x) = f_X(1 - x)$ for $0 \le x \le \frac{1}{2}$. Find f_X and $E(X)$. [*Hint:* $\int_0^{1/2} x^n = (\frac{1}{2})^{n+1}/(n+1)$.]

4. (Uniform) Let $U(a, b)$ be a random variable such that

$$f_{U(a,b)}(x) = \begin{cases} \dfrac{1}{b-a}, & \text{if } a \le x < b, \\ 0, & \text{otherwise.} \end{cases}$$

Show that f is a bona fide p.d.f. and that $E(U(a, b)) = \frac{1}{2}(a + b)$. Also, find its c.d.f. Such a random variable is called a uniformly distributed random

[2] Exercises marked with an asterisk require calculus.

variable. When $a = 0$ and $b = 1$, we refer to our random variable as the *uniform random variable*, and we denote it by U instead of $U(0, 1)$.

5. (Continuation) Let $X_1, X_2, \ldots, X_n$ be independent random variables each having the distribution of $U(0, b)$, and let $M_n \equiv \max(X_1, X_2, \ldots, X_n)$. Show that

   (a) $F_{M_n}(t) = \left(\frac{t}{b}\right)^n$, for $0 \le t \le b$.

   *(b) $f_{M_n}(t) = \frac{n}{b^n} t^{n-1}$, for $0 < t < b$.

   *(c) $E(M_n) = \frac{n}{n+1} b$.

6. (Exponential) A random variable of considerable importance is the exponential. It is the continuous analogue of the geometric random variable. For $\lambda > 0$, we say that X is *exponential* with parameter λ ,written "X is Exp (λ)," if

$$f_X(t) = \lambda e^{-\lambda t} \qquad \text{for } t > 0.$$

   It can be shown that (i) $E(X) = 1/\lambda$, (ii) $\mathrm{Var}(X) = 1/\lambda^2$, and (iii) $F_X(t) = \int_0^t \lambda e^{-\lambda s}\, ds = 1 - e^{-\lambda t}$ for $t > 0$. Often X represents the time at which a component breaks or fails. An interesting property is that for such components there is no "wearout" effect—only random breakdowns. That is, for any $t > 0$ and $s > 0$,

   (iv) $P(X > t + s \mid X > s) = P(X > t)$.

   Property (iv) is sometimes referred to as the memoryless property of the exponential random variable.
   *(a) Verify (i), (ii), and (iii). [*Hint:* Use integration by parts.]
   (b) Verify (iv). [*Hint:* Use (iii)—this does not involve calculus.]

7. What can you say about that probability model (Ω, P) wherein $\Omega = \{x : x \ge 0\}$ and each point in Ω is equally likely—that is, $f(x) = f(y)$ for all $x, y \ge 0$?

2. THE NORMAL DISTRIBUTION AND THE CENTRAL LIMIT THEOREM

Without doubt, the most important and the most widely used (and misused!) random variable is the so-called "normal." In this section we define it, show how to use it, and explain why it is so important.

DEFINITION

We say that X is a *normal random variable* (or simply that X is normally distributed) with mean[3] μ and variance σ^2, written "X is $N(\mu, \sigma^2)$," if the probability density function of X is given by[4]

$$f_X(x) = \frac{1}{\sqrt{2\pi}\,\sigma} \exp\{-(x - \mu)^2/2\sigma^2\}, \qquad \text{for all } x. \tag{3.2}$$

At this point the name normal seems to be a misnomer, for there appears to be nothing common or normal about the function defined in (3.2). The name normal originated in connection with the theory of errors of measurement in the eighteenth century, when it was found that errors "normally" had the distribution given by equation (3.2).

The graph of f_X with $\mu = 1$ and $\sigma^2 = \frac{1}{4}$, 1, and 4 is given in Figure 3.4.

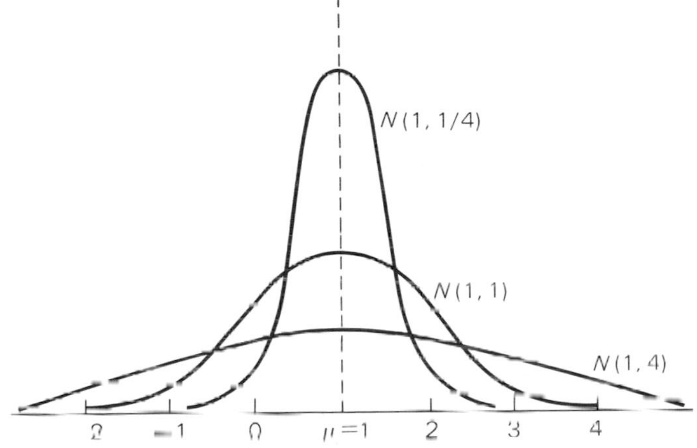

Figure 3.4 Normal Probability Density Functions

Notice that

(i) $f_X(x) > 0$, for all x,
(ii) $f_X(x)$ decreases (to zero) as the distance between x and μ increases (to infinity),
(iii) f_X is symmetric about μ; that is, $f_X(\mu + x) = f_X(\mu - x)$ for all x,
(iv) the mean, the median, and the mode are all equal to μ, and

[3] Henceforth, we shall use the words *expectation* and *mean* interchangeably.

[4] For readability, we sometimes write $\exp\{t\}$ in place of e^t if t is a complicated expression. Both e and π are well-known numbers; their values are approximately 2.718 and 3.1416.

(v) for any fixed value of μ, the curve f_X "heightens and narrows" as σ^2 decreases; in particular, given any $x > 0$, the area under the curve, above the x-axis, and between $\mu - x$ and $\mu + x$ increases as σ^2 decreases.

Of course, it can be shown that the area under the curve, above the x-axis, and between $-\infty$ and $+\infty$, is 1. Unfortunately, we cannot give an explicit formula for F_X. Nevertheless, we can, with the aid of tables, find the probability that X is in any given interval.

2.1 Using the Normal Table

Table A on page 307 contains all the information we need in order to obtain probabilities for any normal random variable. This is how we use it. Denote by Z and Φ a normal random variable with mean 0 and variance 1 and its cumulative distribution function, respectively; Z is called the *standard normal random variable* and Φ is called the *standard normal distribution*. In order to find $P(Z \leq z)$, where $z \geq 0$ is some given number, find the row and column that corresponds to z. For example, if $z = 1.47$, then z corresponds to the fifteenth row and eighth column in the table. The entry in the table at this position is .9292, which is the probability (accurate to four decimal places) that $Z \leq z$—that is, $\Phi(z)$. Since f_Z is symmetric about zero and $\Phi(0) = .5$, Table A can be used to find the probability that Z is in *any* given interval. Moreover, with the aid of Theorem 1 below, we can use Table A to find probabilities for any normal random variable and not just for Z. (If this were not true, we would need a separate table for each normal random variable.)

THEOREM 1

If X is $N(\mu, \sigma^2)$, then $Z \equiv (X - \mu)/\sigma$ is $N(0, 1)$.

We omit the proof that Z is normal, as it involves the use of calculus,[5] while we can employ Corollary 1 and Theorem 4 of Chapter 2 to show that $E(Z) = 0$ and $\mathrm{Var}(Z) = 1$.

EXAMPLE 1

Find $P(Z \geq 1)$.

SOLUTION: In finding probabilities for a normal random variable, it is useful to draw the associated picture [see Figure 3.5(a)]. The entry in Table A corresponding to $z = 1$ is .8413, so

$$P(Z \geq 1) = 1 - P(Z \leq 1) = 1 - .8413 = .1587. \bullet$$

[5] For an elementary proof see Hogg and Craig, *Introduction to Mathematical Statistics*, 2d ed. (New York: The Macmillan Company, 1965), pages 98–99.

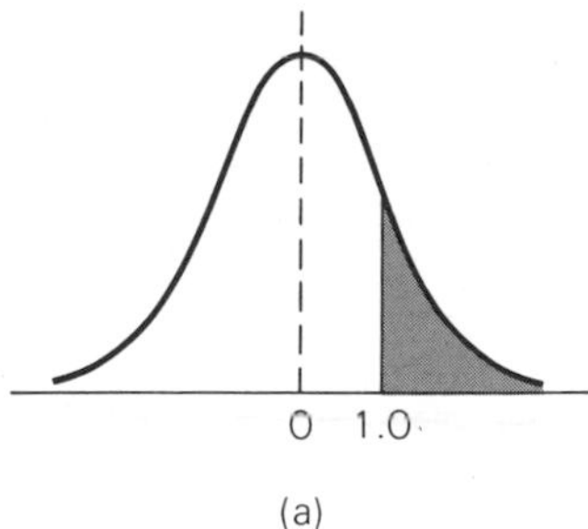

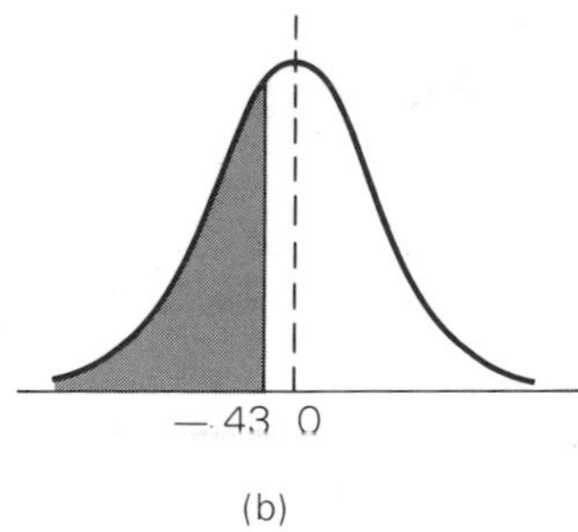

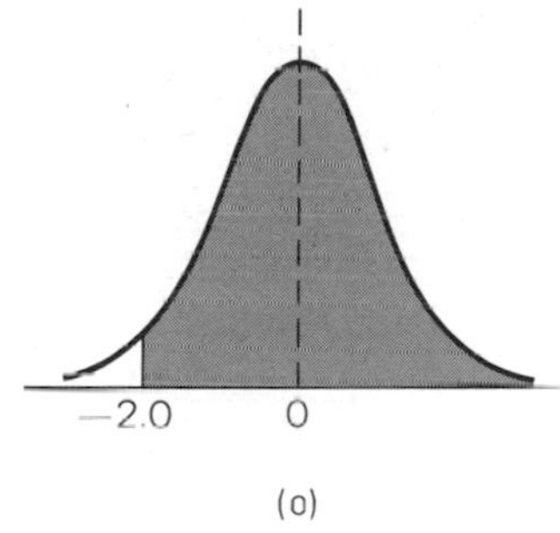

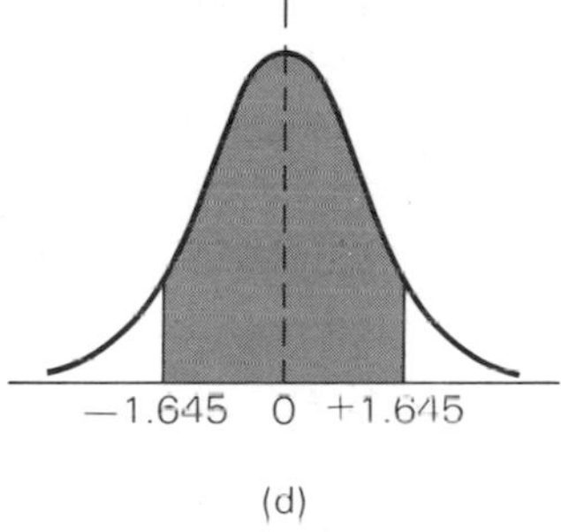

Figure 3.5

EXAMPLE 2

Find $P(Z \leq -.43)$.

SOLUTION: Here we make use of the symmetry of f_Z about $\mu = 0$ so that [see Figure 5(b)]

$$P(Z \leq -.43) = P(Z \geq .43) = 1 - \Phi(.43) = 1 - .6664 = .3336. \bullet$$

EXAMPLE 3

Find z such that $P(Z \geq z) = .9772$.

SOLUTION: Clearly z must be a negative number [since if it were positive, $P(Z \geq z)$ would be less than .5], so

$$.9772 = P(Z \geq z) = P(Z \leq -z) = P(Z \leq 2.00)$$

as .9772 is the entry in Table A corresponding to 2.00. Consequently $z = -2.00$. [See Figure 5(c).] •

EXAMPLE 4

Find $P(-1.645 \leq Z \leq 1.645)$.

SOLUTION: The table contains no row and column corresponding to 1.645. Consequently we average the entries of the two values closest to 1.645, namely 1.64 and 1.65. Thus, the average is

$$\tfrac{1}{2}(.9495) + \tfrac{1}{2}(.9505) = .9500,$$

so $P(Z \leq 1.645) = .9500$ and by symmetry, $P(Z \leq -1.645) = .05$, so $P(-1.645 \leq Z \leq 1.645) = .9000$.

Similarly, if $z = 1.642$, then we get

$$P(Z \leq 1.642) = \tfrac{8}{10}(.9495) + \tfrac{2}{10}(.9505) = .9497$$

so

$$P(-1.642 \leq Z \leq 1.642) = .8994.$$

We use the factors $\frac{8}{10}$ and $\frac{2}{10}$, since 1.642 is "4 times as close to 1.64 as to 1.65." This method of averaging is called *interpolation*. •

EXAMPLE 5

Suppose X is $N(4, 25)$. Find $P(-5 \leq X \leq 7)$.

SOLUTION: We use Theorem 1 and Table A to obtain

$$\begin{aligned} P(-5 \leq X \leq 7) &= P(-9 \leq X - 4 \leq 3) = P\left(-\frac{9}{5} \leq \frac{X-4}{5} \leq \frac{3}{5}\right) \\ &= P(-1.8 \leq Z \leq .6) = .7257 - (1 - .9641) \\ &= .6898. \bullet \end{aligned}$$

The next theorem will be very useful to us in our study of statistics. It states that the sum of independent (but not necessarily identical) normal random variables is itself a normal random variable.[6]

THEOREM 2

If $X_1, X_2, \ldots, X_n$ are independent random variables and X_i is $N(\mu_i, \sigma_i^2)$, $1 \leq i \leq n$, then

$$\sum_{i=1}^{n} X_i \quad \text{is} \quad N\left(\sum_{i=1}^{n} \mu_i, \sum_{i=1}^{n} \sigma_i^2\right).$$

2.2 The Central Limit Theorem

The central limit theorem is, as its name implies, one of the most important and remarkable theorems in probability theory. Not only does it provide a simple method for computing (approximate) probabilities for

[6] The proof of this fact can be found on pages 138–139 in Hogg and Craig.

sums of independent random variables; it also explains the astonishing, but too seldom justified, fact that the empirical distribution of so many natural "populations" exhibits a normal distribution.

THEOREM 3. (The Central Limit Theorem)

Let $X_1, X_2, \ldots$ be independent, identically distributed random variables each with mean μ and variance σ^2. Then the distribution of $Z_n \equiv \left(\sum_{i=1}^{n} X_i - n\mu\right)\Big/\sigma\sqrt{n}$ tends to the standard normal as n increases. That is, for each real number x,

$$P\left(\frac{X_1 + X_2 + \cdots + X_n - n\mu}{\sigma\sqrt{n}} \leq x\right) \to \Phi(x) \qquad \text{as } n \to \infty.$$

Note that, just like the Chebychev inequality, this theorem holds for *any* distribution; herein lies its power.

EXAMPLE 6

Recall Example 11 of Chapter 2, where Y_i was the outcome corresponding to the ith roll of a fair die, $1 \leq i \leq 420$, $X = \sum_{i=1}^{420} Y_i$ and $W = X/420$. We sought $P(3 \leq W \leq 4)$. Using the facts that $E(Y_i) = 3\frac{1}{2}$ and $\operatorname{Var}(Y_i) = \frac{35}{12}$ for each i, it follows that $E(W) = 3\frac{1}{2}$ and

$$\operatorname{Var}(W) = \frac{420 \cdot \frac{35}{12}}{420^2} = \frac{1}{144},$$

so, using the Chebychev inequality, we obtain

$$\begin{aligned} P(3 \leq W \leq 4) &= P(|W - E(W)| \leq \tfrac{1}{2}) = 1 - P(|W - E(W)| > \tfrac{1}{2}) \\ &\geq 1 - P(|W - E(W)| \geq \tfrac{1}{2}) \geq 1 - \frac{\operatorname{Var}(W)}{\frac{1}{4}} \\ &= \tfrac{35}{36} = .9723. \end{aligned}$$

On the other hand, by invoking the central limit theorem we obtain

$$\begin{aligned} P(3 \leq W \leq 4) &= P\left(\frac{3 - 3\frac{1}{2}}{\frac{1}{12}} \leq \frac{W - E(W)}{\sqrt{\operatorname{Var}(W)}} \leq \frac{4 - 3\frac{1}{2}}{\frac{1}{12}}\right) \approx P(-6 \leq Z \leq 6) \\ &= 1 - 2P(Z \geq 6). \end{aligned}$$

An approximation known as Mill's ratio[7] shows that

$$P(Z > 6) \approx \frac{1}{\sqrt{2\pi}\,6}\, e^{-6^2/2} < (\tfrac{1}{10})^9,$$

[7] See Feller, p. 175.

so

$$P(3 \leq W \leq 4) \approx 1 - 2(\tfrac{1}{10})^9 = .999999998.$$

In a similar manner, the Chebychev inequality can be used to obtain the bounds

$$P(X \geq 1470) \leq 1, \qquad P(X \geq 2520) \leq .583,$$

and

$$P(1400 \leq X \leq 1540) \geq .75.$$

Likewise, use of the central limit theorem yields the approximations

$$P(X \geq 1470) \approx \tfrac{1}{2}, \qquad P(X \geq 2520) \approx (\tfrac{1}{10})^{195},$$

and

$$P(1400 \leq X \leq 1540) \approx .9544. \bullet$$

It can be shown that the approximations obtained by use of the central limit theorem is within .005 of the true probabilities except for $P(X \geq 1470)$—see discussion of the correction factor below. Moreover, even with the aid of a high-speed electronic computer, the computations needed to find the exact (to 32 decimal places) probabilities would cost thousands of dollars. Evidently the normal approximation is quite valuable!

When using the normal distribution to approximate the distribution of the sum X of independent, identical, integer-valued random variables X_i, we must take into account the fact that X assumes integer values only. We do this by introducing a *correction factor*. To illustrate this, suppose that we wanted to find $P(X = 8)$; then we might write

$$\begin{aligned} P(X = 8) &= P(X \leq 8) - P(X \leq 7) \text{ ``}\approx\text{'' } P(Y \leq 8) - P(Y \leq 7) \\ &= P(7 \leq Y \leq 8), \end{aligned}$$

where Y is $N(E(X), \text{Var}(X))$. This can result in a poor approximation. By examining the histogram of the random variable X (see Example 7 below) upon which is superimposed the p.d.f. of Y, where Y is $N(E(X), \text{Var}(X))$ in Figure 3.6, you will notice that $P(7.5 \leq Y \leq 8.5)$ is closer to $P(X = 8)$ than is $P(7 \leq Y \leq 8)$. This illustrates the fact that (in general) the best possible approximation (with a normal) of the probability that an integer-valued random variable X equals some integer k is obtained by using $P(k - .5 \leq Y \leq k + .5)$, where Y is $N(E(X), \text{Var}(X))$. The $+.5$ and $-.5$ is the correction factor mentioned earlier.

EXAMPLE 7

Let X_i be the outcome corresponding to the ith roll of a fair die, $1 \leq i \leq 3$, and let $X = X_1 + X_2 + X_3$. Then we can show [see Exercise

21, Chapter 2] that $f_X(k) = n(k)/216$, where $n(k) = 1, 3, 6, 10, 15, 21, 25, 27, 27, 25, 21, 15, 10, 6, 3$, and 1 for $k = 3, 4, 5, \ldots, 18$, respectively (see Figure 3.6). We shall use the normal approximation and the correction factor to find $P(X = 8)$. First note that $E(X) = 10.5$ and $\text{Var}(X) = 3(\frac{35}{12})$, so

$$P(X = 8) \approx P\left(\frac{7.5 - 10.5}{\sqrt{3 \cdot 35/12}} \leq Z \leq \frac{8.5 - 10.5}{\sqrt{3 \cdot 35/12}}\right)$$
$$= P(-1.014 \leq Z \leq -.676) = .0944.$$

Of course, the true value is $\frac{21}{216} = .0972$. Thus, by using the normal approximation and the correction factor we understate the true probability by 3.91 percent of the true value, whereas we understate the true probability by 16.97 percent if we do not use the correction factor (see Exercise 9). For most applications, 3 or 4 percent represents a very minor discrepancy indeed. •

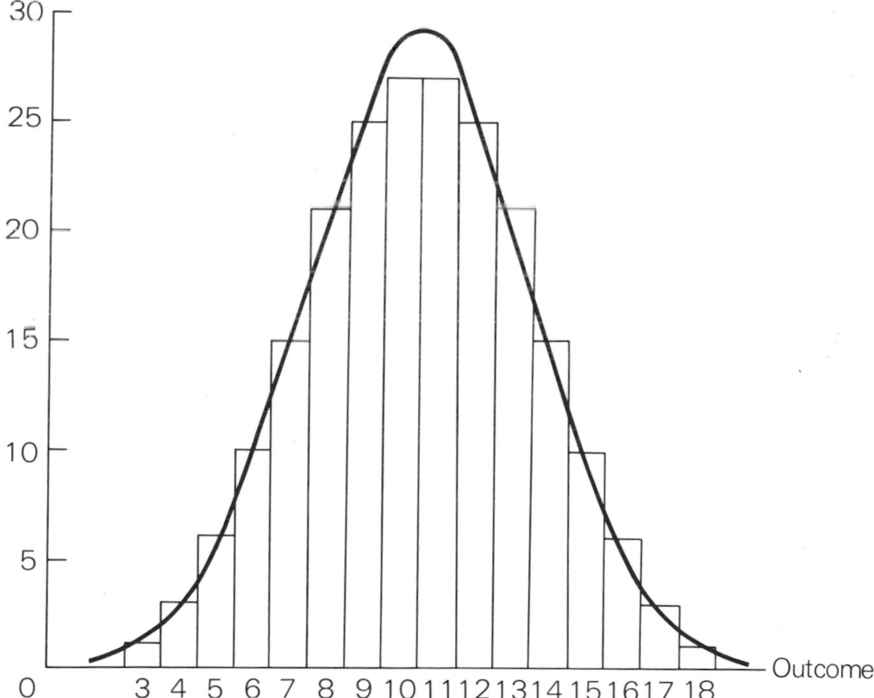

Figure 3.6 The Normal Approximation to X

EXAMPLE 8

Let X_i be independent, uniform random variables (see Exercises 1 and 4), $1 \leq i \leq 3$, and let

$$S_3 \equiv \frac{X_1 + X_2 + X_3 - 3(\frac{1}{2})}{\sqrt{3(\frac{1}{12})}}.$$

Then $E(S_3) = 0$, $\text{Var}(S_3) = 1$, and the range of S_3 is $\{x: -3 < x \leq 3\}$. Of course, no correction factor would be used here, since S_3 is a continuous random variable. The table and accompanying graph[8] (Figure 3.7) show how close the normal distribution is to the distribution of S_3. •

These examples show that the normal approximation may be good even for very small values of n.

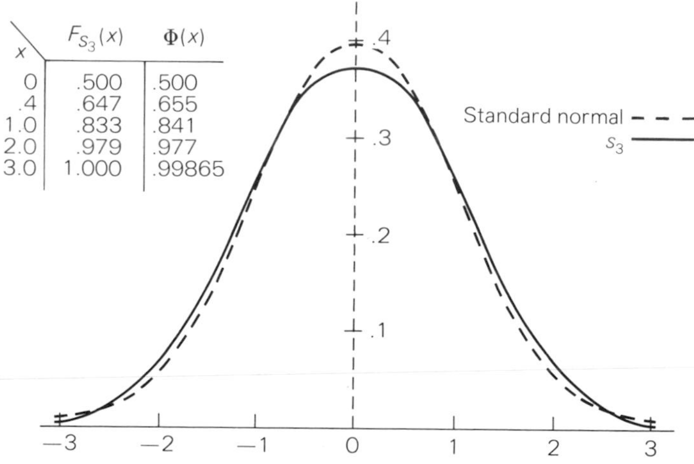

x	$F_{S_3}(x)$	$\Phi(x)$
0	.500	.500
.4	.647	.655
1.0	.833	.841
2.0	.979	.977
3.0	1.000	.99865

Figure 3.7 The Normal Approximation to S_3

If X is $B(n, p)$, then as shown in Section 5 of Chapter 2 we can write X as the sum of n independent random variables, each of which is $B(1, p)$. Consequently, it follows from the central limit theorem that a normal distribution provides a good approximation to X if n is large. More formally we have the following

[8] John P. Hoyt, "A Simple Approximation to the Standard Normal Probability Density Function," *American Statistician*, Vol. 22, No. 2 (1968), pp. 25–26.

COROLLARY 1

If F_n is the cumulative distribution function of

$$\frac{X_n - np}{\sqrt{np(1-p)}}, \text{ where } X_n \text{ is } B(n, p),$$

then for each x, $F_n(x)$ approaches $\Phi(x)$ as n approaches infinity.

Actually, the goodness of the approximation depends upon both n and p. If p is close to either 0 or 1, then f_X is very asymmetrical (skewed). As a rough guideline, the normal approximation to the binomial (with a correction factor) will be quite good if $np(1-p) \geq 10$.

EXAMPLE 9

Let X be the number of times out of $n = 40$, 20, and 10 flips that a fair coin lands heads. Find the probability that $X = n/2$.

SOLUTION: Since X is $B(n, \frac{1}{2})$, we can, after much calculation, show that these probabilities are .1268, .1762, and .2461. If instead we approximate these probabilities by using a correction factor and the normal distribution, we get .1256, .1772, and .2480. For example, if $n = 40$, then $E(X) = 20$ and $\text{Var}(X) = 10$, so

$$\begin{aligned} P(X = 20) &= P\left(\frac{19.5 - 20}{\sqrt{10}} \leq \frac{X - 20}{\sqrt{10}} \leq \frac{20.5 - 20}{\sqrt{10}}\right) \\ &\approx P(-.158 \leq Z \leq .158) = .1256. \bullet \end{aligned}$$

2.3 The Role of the Normal Distribution in Statistics

We now ask why the normal distribution plays such an important role in probability and statistics—and we proffer three reasons.

First, by virtue of Theorem 2, the central limit theorem, and the fact that we can tabulate $\Phi(\cdot)$, the normal distribution provides us with a very easy method of computing probabilities for sums of independent, identically distributed random variables. Moreover, our work in statistics is highly dependent upon such sums.

Second, as an illustration consider the heights of all adult persons in the United States, and let $f(x)$ be the proportion of these people having height x where height is measured to the nearest $\frac{1}{100}$ inch. We refer to this set of heights as our population and $f(\cdot)$ as the empirical distribution associated with the population. Observe that an individual's height is determined by a great number of factors such as heredity (the millions of genes), diet, exercise, climate, and so on. Evidently each individual factor has little effect, and while these variables are admittedly not unrelated (independent), we choose to treat them as such. Thus, in view of the central limit theorem, it is quite reasonable to argue that f, the empirical distribution of the population, can be well approximated by a normal density. Of course, we must have $f(x) = 0$ for $x \leq 0$, as height cannot be negative. Nevertheless, the fact of the matter is that height does exhibit approximately a normal distribution. We don't mean to imply that the empirical distributions of most populations encountered in the real world are normal, but nearly normal populations are encountered frequently.

Third, if we assume that the distribution of a population in which we are interested is normal, then the purely mathematical problem of obtaining the distributions of certain functions of "samples from the population" is rendered tractable.

Hopefully these arguments not only lend credence to the important role played by the normal distribution but also justify our somewhat lengthy study of it.

EXERCISES:

8. If Z is $N(0, 1)$, find
   (a) $P(Z \leq 1)$.
   (b) $P(Z \leq -2)$.
   (c) $P(Z \geq -1.43)$.
   (d) $P(-3 \leq Z \leq 3)$.
   (e) $P(-1.28 \leq Z \leq 1.28)$.
   (f) $P(-.84 \leq Z \leq .84)$.
   (g) $P(Z \geq 1.39 \text{ or } Z \leq -2.12)$.
   (h) $P(Z \geq 3 \text{ or } Z \leq -3)$.

9. Find $P(7 \leq Y \leq 8)$, where Y is $N(10.5, 3 \cdot \frac{35}{12})$ (see Example 7).

10. Find the value of z such that the area under the standard normal curve
    (a) to the left of z is .9699.
    (b) to the right of z is .9699.
    (c) to the left of z is .5000.
    (d) between $-z$ and z is .90.
    (e) between $-z$ and z is .95.
    (f) between $-z$ and z is .99.
    (g) between $-z$ and $3z$ is .7530.

11. Let X be $N(\mu, \sigma^2)$ and find
    (a) $P(X \leq 0)$ if $\mu = -2, \sigma^2 = 1$.
    (b) $P(X \geq -11)$ if $\mu = 5, \sigma^2 = 64$.
    (c) $P(-7 \leq X \leq 4)$ if $\mu = 5, \sigma^2 = 4$.
    (d) $P(-5 \leq X \leq 5)$ if $\mu = 1, \sigma^2 = 9$.

12. Let X be the total number of points that appear when 6 fair dice are rolled. Find the probability that (a) $X = 6$, (b) $X = 21$, (c) $20 \leq X \leq 25$.

13. A multiple-choice examination has 100 questions. If .6 is the probability of a correct answer on any given question and the questions are each quite

different, find (in decimal form) the probability that there are 59, 60, or 61 correct answers. What is the probability that there are between 50 and 70, between 55 and 75, and between 52 and 72 correct answers?

14. Find the probability that a fair coin lands heads half of the time if there are n flips with $n = 4$, 40, and 400.

15. It is known that a person's weight (in pounds) is roughly proportional to the cube of his height (in inches). That is, there is a number a, $0 < a < 1$, such that weight $\approx a(\text{height}^3)$. On the basis of this information and the fact that height (in the United States) is nearly normally distributed, speculate as to whether or not weight is normally distributed. Justify your conclusion.

16. In Example 20 of Chapter 1 we found that the probability of winning in craps is .493. What is the probability that you are no more than \$5 (\$10) behind if you wagered \$1 for 20 games?

17. You play 10 games of roulette and on each game you win either \$35 or $-\$1$ with probability $\frac{1}{38}$ and $\frac{37}{38}$, respectively. What is the probability that you end up ahead? What if you play 50 games?

18. Find the probability that X is between 70 and 80 if X is χ^2_{70} (see definition of chi-square random variable in Chapter 5). Note that Table B (p. 308) does not give such probabilities if $n > 30$.

19. Indicate when and why the normal distribution provides a good approximation to the (a) hypergeometric, (b) negative binomial (see Exercise 69 of Chapter 2).

3. RANDOM NUMBERS AND SIMULATION

Often we are interested in actually performing a random experiment. Recall Example 14 of Chapter 1, where the lender of a book picked one of the numbers 1, 2, 3, whence Alan, Barry, and Carl made guesses at the number chosen. In that example we saw that each of the boys had the same probability of winning, namely $\frac{1}{3}$. Of course, if the lender had a favorite number and it was known, then Alan would have had a distinct advantage. On the assumption that the lender did, in fact, have a favorite number and that $\frac{1}{3}$ was the desired probability of each boy's winning, the lender would have to find some other method of choosing the numbers 1, 2, and 3 with equal probabilities. The simplest thing for him to do is to perform a random experiment with three equally likely events E_1, E_2, and E_3 that partition the sample space Ω. One way of achieving this is to roll a fair die and let $E_1 = \{1, 2\}$, $E_2 = \{3, 4\}$, and $E_3 = \{5, 6\}$. Of course, the number X chosen by the lender is a random variable with $X(\omega) = i$ if $\omega \in E_i$, $i = 1, 2, 3$. (We choose to ignore the fact that our die will not be perfectly fair. Ignoring this fact is a natural part of the process of constructing this mathematical model.)

What we accomplished here was to draw a "random number" from the p.m.f. f_X, where $f_X(i) = \frac{1}{3}$, $i = 1, 2, 3$. A somewhat more general situation is that of performing a random experiment wherein there are n equally likely outcomes. This is easily done with the use of a table of random digits (see Table 3.1).

Table 3.1 Table of Random Digits

66175	60394	48929	20234	63370	03299	53662	81355	99416	87432
09628	21873	65620	96793	20817	21407	49168	93986	89766	44553
85759	41790	76036	21818	83907	08739	25722	32030	62684	73058
15985	51980	21451	93993	64949	60306	81049	19152	42043	04645
59845	29644	25273	39164	46882	28869	79154	94512	04146	89390
00972	54034	14312	19671	54824	58770	06543	12184	46704	39767
47506	42899	08337	87921	47012	58499	78888	11950	36037	42011
60205	05177	40800	25526	41677	39989	48079	70804	25866	11572
69260	09475	01298	41067	92685	76557	27781	80982	52903	19132
36553	88938	20180	26903	14825	04664	00005	53209	66101	49650
52194	07808	42926	94516	37928	13559	86004	73655	53116	73181
33324	07693	46622	04057	78984	82116	35103	41138	32406	07842
99980	97764	43095	09089	58607	30994	41226	58081	60698	05113
69712	26991	44346	72937	73872	48937	91780	06844	96990	09122
93016	35903	12963	13483	70789	97870	18877	68090	10140	43176
68732	99857	71317	88810	87139	64913	73348	85836	27833	45918
24879	98919	34405	54186	50960	64158	98175	34069	87190	61799
21276	47719	11147	98783	09950	38558	11865	29720	50678	92762
16586	37121	32712	37734	63663	10696	56259	44784	26349	44119
57464	96881	83341	40602	61312	20724	28579	50613	56548	10146

Table 3.1 consists of the outcomes of 1000 independent, identical random experiments where each of the 10 outcomes, 0, 1, 2, . . . , 9, has probability $\frac{1}{10}$. Specifically, ten numbered pennies were placed in a jar and mixed, and one of them was withdrawn. The number on the penny withdrawn was the outcome of that experiment. (After having beein withdrawn, the penny was replaced.) The outcomes of the experiments were written down in order;[9] thus the 9 in the 23rd column of the 4th row means that the outcome of the 173rd experiment was 9.

[9] The author is indebted to Kris Brown, who graciously performed this burdensome experiment.

Now suppose we want to perform an experiment wherein

$$P(\text{outcome is } i) = \tfrac{1}{153}, \qquad i = 1, 2, \ldots, 153.$$

Choosing a starting place in the table, say the 8th column of the 19th row, we look at the number formed by the first three digits, which is 121. In this case, the outcome of the experiment was 121. Had we started at the 9th instead of the 8th column, however, the number formed by the first three digits would have been 213, a number larger than 153. In this case, we simply ignore these three digits and move on to the next three. But the next number is 271, so we ignore it too. We continue in this manner, ignoring the numbers 237, 734, 636, and 631, until we finally get 69 as the outcome of our experiment. Thus, to perform a random experiment with n equally likely outcomes we choose a number m such that $10^m \geq n$ ($m = 3$ for $n = 153$) and a starting place in the table. We then consider the number formed by the first m digits in the table and call this number the outcome of the experiment if it is between 1 and n.[10] If this number is larger than n, then we discard it and consider the next m digits in the table as prescribed above.

If we wanted an ordered random sample of size s with replacement from a population of size n, we would simply go through the procedure (experiment) outlined above s times. Of course, after we found the first element in our random sample, we would not start at the same place in the table to find the second element. (Why not?) We would simply start at the place in the table where we left off.

If we wanted an ordered random sample of size s without replacement, we would use the procedure for finding ordered random samples of size s with replacement with one exception: if the procedure generated a number that had already been included as an element in our sample, we simply would not include it; that is, we would treat it as if it were larger than n.

The preceding discussion leads us to ask how to generate random numbers according to a distribution, such as the normal, that does not arise from an experiment with equally likely outcomes. For the most part, we cannot generate such random numbers. However, with the aid of modern electronic computers truly remarkable approximations can be obtained. As we shall see shortly, it would suffice to be able to perform an experiment with outcome U such that $P(a \leq U \leq b) = b - a$, $0 \leq a \leq b \leq 1$. Of course, U is the uniform random variable (see Exercise 4).

Fortunately, there is a method (known as the multiplicative-congruential method) that generates on the computer a sequence of (pseudo-) random numbers (almost) having the uniform distribution. Using these

[10] If $n = 10^m$, we say that the outcome is n if the number formed by the first m digits is 0.

(pseudo-) random numbers and Theorem 4 below, we can, in effect, perform any random experiment. That is, given a random variable X with cumulative distribution function F_X, we can perform a random experiment so that $P(\text{outcome} \leq x) = F_X(x)$.

▶▶
THEOREM 4

Let X be a random variable with cumulative distribution function F, define F^{-1} to be that function such that[11] $F^{-1} \circ F(x) = F \circ F^{-1}(x) = x$ for all x, and define the random variable T by $T(x) = F^{-1} \circ U(x)$ for $0 \leq x \leq 1$, where U is the uniform random variable. Then $F_T(x) = F(x)$ for all x.

PROOF. Using the definition of F^{-1} and U, we have for each x

$$\begin{aligned} F_T(x) &= P(T \leq x) = P(F^{-1} \circ U \leq x) = P(F \circ F^{-1} \circ U \leq F(x)) \\ &= P(U \leq F(x)) = F(x). \;\blacklozenge \end{aligned}$$

▶

We illustrate the use of Theorem 4 in the next two examples.

EXAMPLE 10

Suppose we want to draw random numbers according to the distribution of X, where its distribution corresponds to the placing of a magnet under the dial (see Figure 3.3) so $F_X(x) = x^2$ for $0 \leq x \leq 1$. By Theorem 4 it suffices to draw random numbers from the uniform distribution. If the outcome of the experiment is ξ, then $F_X^{-1}(\xi)$ is the random number drawn according to the distribution of X. We shall approximate the uniform by U' where $P(U' = i/10^6) = 1/10^6$ for $i = 0, 1, \ldots, 10^6 - 1$. Having graphed F_X (see Figure 3.8) and found ξ, we find $F_X^{-1}(\xi)$ as follows: first, draw a horizontal line from ξ on the vertical axis to the point where it intersects the curve $F_X(x)$; next, draw a vertical line from this point to the x-axis. Then the point on the x-axis where this vertical line cuts the x-axis is the desired random number, namely, $F_X^{-1}(\xi)$. (It so happens that in this particular case F_X^{-1} is easily found: $F_X^{-1}(x) = \sqrt{x}$, $0 \leq x \leq 1$.)

Let us start from the tenth row and first column of Table 3.1. The first six digits encountered give rise to the number $\xi = .365538$, so $\sqrt{.365538}$ is the desired random number, as shown in Figure 3.8. •

[11] If no such function exists—that is, if F is not strictly increasing and continuous—then define F^{-1} by $F^{-1}(x) = \max\{t\colon F(t) \leq x\}$. In this case, a more sophisticated argument is needed to prove the theorem.

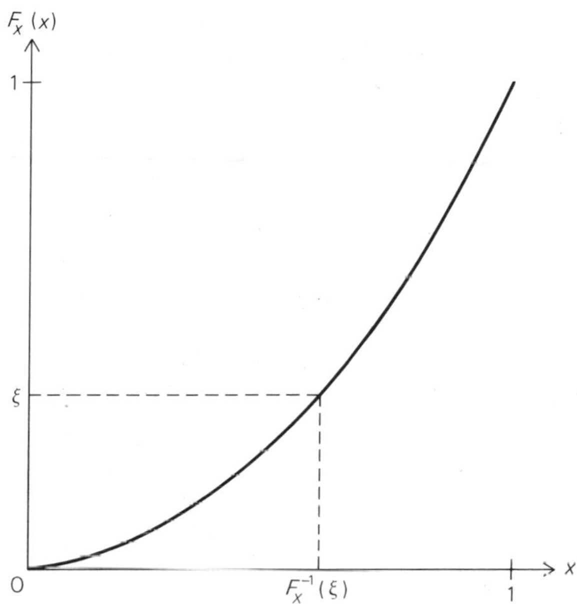

Figure 3.8

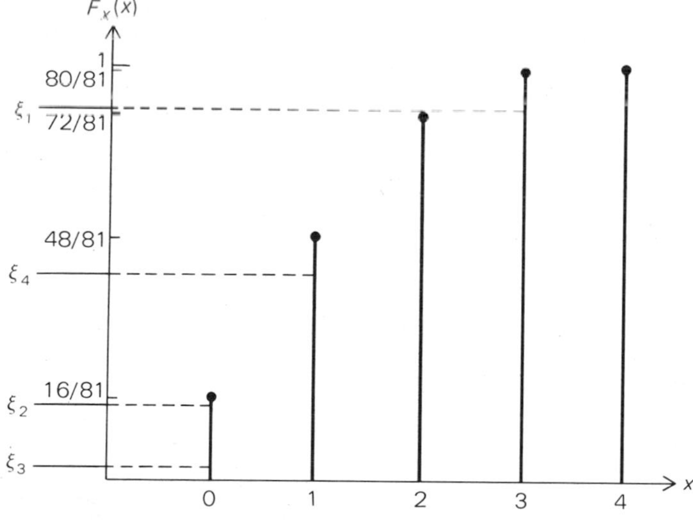

Figure 3.9

EXAMPLE 11

Next suppose X is $B(4, \frac{1}{3})$ (recall Example 3 of Chapter 2). Then, using the approach outlined in Example 10, and continuing from the 10th row and 7th column, the 4 uniform random numbers $\xi_1 = .893820$, $\xi_2 = .180269$, $\xi_3 = .031482$, and $\xi_4 = .504664$ result in $F_X^{-1}(\xi_1) = 3$, $F_X^{-1}(\xi_2) = 0$, $F_X^{-1}(\xi_3) = 0$, and $F_X^{-1}(\xi_4) = 1$, respectively. •

3.1 Simulation

Real-world problems are often so complicated and complex that they cannot be solved mathematically. Luckily, many of these problems, although not amenable to mathematical analysis, can be "solved" experimentally by *simulation* using a large table of random digits or more generally an electronic computer. Thus, we solve the problem by artificially performing the experiment embodied in the mathematical model and drawing our conclusions from the empirical evidence.

To be specific, suppose we wanted to find out whether or not a particular rat had any memory (in terms of running a maze). To do so we have decided to have him run the maze in Figure 3.10. As you can see, every time he makes an incorrect choice, he returns to the start. Of course, he gets to the cheese by making two *consecutive* correct decisions (turns). If the rat has no memory, then upon assuming that he favors neither right nor left (the maze can be so arranged to achieve this) we see he always chooses the decision "make a left" with probability $\frac{1}{2}$. Stated mathematically, the total number X of decisions made by our rat with no memory and no preference for right or left is the number of independent, identical, Bernoulli trials with probability $\frac{1}{2}$ of success needed until 2 consecutive successes are obtained.

In order to test the rat's memory, we do three things: first, we find the distribution of X; second, we run the rat through the maze once; and third, we infer whether or not the rat has any memory by comparing the theoretical and empirical results. Our first problem is to actually find the distribution of X. (We leave the rat-running to the psychologist and defer the question of inference until Chapter 4.) Although this problem can be solved mathematically,[12] simulation provides an easier alternative.

[12] The author has shown that

$$P(X = 1 + i) = \frac{k(i)}{2^{i+1}}, \qquad i = 1,2,\ldots; \quad \text{and} \quad E(X) = 6,$$

where $k(1) = k(2) = 1$, and $k(i) = k(i-1) + k(i-2)$ for $i \geq 3$. The numbers $k(i)$ are the so-called Fibonacci numbers.

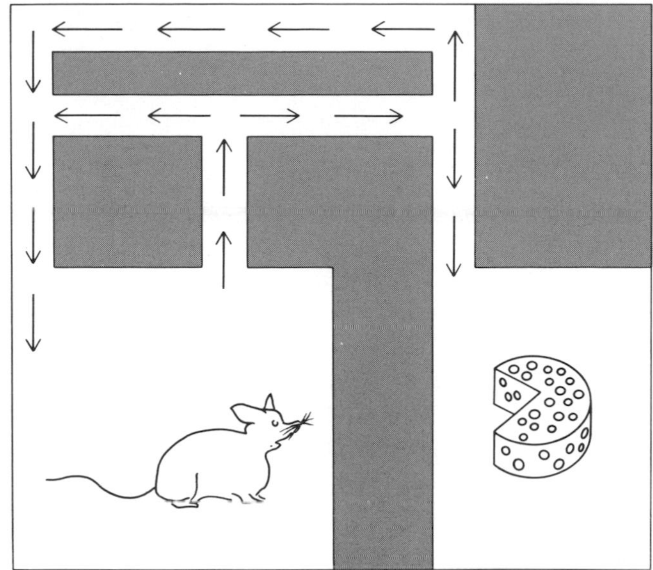

Figure 3.10 Rat in a Maze

Table 3.2 Simulation Results

Outcomes of each experiment

2	6	2	2	2	2	9	6	4	2
4	4	12	7	6	1	13	1	3	3
2	4	9	3	7	18	7	9	7	5
28	4	2	11	4	4	2	5	4	2
5	6	2	7	3	6	15	7	10	4
3	2	2	6	4	2	11	6	2	2
4	2	6	11	2	14	4	2	16	6
6	2	13	10	4	6	3	4	7	2
4	2	13	10	2	2	2	8	8	7
4	6	3	12	7	10	2	4	10	9

Total number of i's

$i =$	2	3	4	5	6	7	8	9	10	>10
	26	7	19	3	12	9	2	4	5	13

6.00 = expected number of trials per experiment
5.87 = average number of trials per experiment

To begin the simulation experiment, identify the digits 0–4 and 5–9 with success and failure, respectively. Next pick a place to start in Table 3.1, say the sixth row and first column, and count the number of trials (digits) needed until 2 consecutive successes are obtained. This then is the result of the first experiment. We repeated this experiment 100 times. The simulation experiment consists of these 100 experiments. The results are shown in Table 3.2. Owing to the relatively small number of experiments (in this case 100), the relative frequencies do not correspond too closely to the true probabilities given in footnote 12. However, the average number of trials per experiment was quite close to the expected number of trials.

EXAMPLE 11. (Draft Lottery)

On December 1, 1969, the Selective Service people placed 366 allegedly identical blue capsules into an urn. Each capsule contained a slip of paper with one of the days of the year written on it. Then the capsules were withdrawn one by one from the urn. The order of the days drawn determined the vulnerability to the draft of men who were between 19 and 26 years old as of January 1, 1970. Moreover, it was announced that the men whose birthdays were among the first 120 days drawn were almost certain to be drafted.

If the drawing were random—that is, if each of the ordered samples of size 366 without replacement were equally likely to be observed—then we should expect to find roughly 10 of the days from each month to be among the first 120 days drawn. Consequently, we could conclude that the drawing was not random if there was too much disparity between 10 and the actual number of days drawn from each of the months.

We will use the random variable D, defined by

$$D = \sum_{i=1}^{12} \frac{(N_i - e_i)^2}{e_i},$$

to measure the disparity between the actual observed results and the theoretical results which are based upon the assumption of a random drawing. Here, the random variable N_i represents the number of days from the ith month among the first 120 days drawn, and e_i is the expected value of N_i under the assumption of a random drawing. Notice that if there are r_i days in the ith month (for example, $r_1 = 31$) and if the drawing is random, then N_i is a hypergeometric random variable with parameters 120, r_i, and 366, so that $e_i = 120r_i/366$. For example, $e_1 = 120 \cdot 31/366 = 10.17$, which is roughly 10.

In Chapter 6, we motivate the use of the random variable D to measure the disparity between the theoretical and the actual results, but let it suffice for now to note that D takes into account the disparity for each of the individual months, since the value of D increases as the disparity for any given month increases.

We now have a measure of disparity, and all we have to know is the distribution of D in order to judge whether or not the drawing was, in fact, random. It turns out that it is next to impossible to discover the distribution of D. Therefore, we must simulate it.

A simulation experiment was performed on a high-speed electronic computer as follows. A random sample of 120 days without replacement from the population of 366 days was generated. Next, the number of days in the sample from each of the twelve months was determined. Then these observed values of the N_i's were substituted in the expression for D, and the resulting value of D was calculated. This simulation experiment was performed 399 times; the experimental results together with the value of D obtained from the results of the drawing done by the Selective Service people are presented in Figure 3.11.

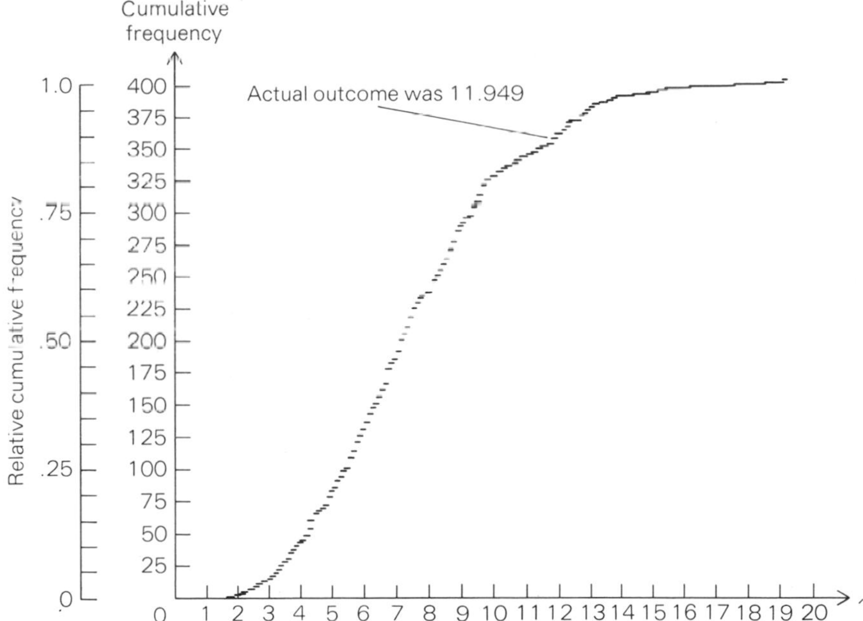

Figure 3.11

From these figures we see that if the drawing is, in fact, a random drawing, we should expect to observe a disparity even larger than the one obtained by the Selective Service people 10.75 percent of the time. Thus, it would appear that the drawing done on December 1, 1969, was random. •

EXERCISES:

20. Starting from the northwest corner of Table 3.1, generate
    (a) a random number between 1 and 50,
    (b) a random number between 7 and 11,
    (c) an ordered random sample of size 5 with replacement from a population of size 14,
    (d) an ordered random sample of size 5 without replacement from a population of size 14.
21. Suppose that instead of listing the outcomes of the 1000 independent, identical experiments in Table 3.1 we (a) listed all of the 0's first, the 1's next, . . . , and the 9's last, (b) listed them in reverse order. What effect would these changes have on our use of the table?
22. Why must we *not* start at the same place in the table to generate the second element of our ordered random sample of size s with replacement?
23. Explain in detail why the $\binom{n}{s} s!$ ordered samples of sizes s without replacement are equally likely if we use the method described in the text?
24. How could you use the central limit theorem and Table 3.1 to generate normal random numbers? [*Hint:* Consider Example 8.]
25. Use the idea of conditional expectation (see page 54) to show that $E(X) = 6$, where X is the number of decisions made by our rat with no memory.

part II

Statistical Inference

INTRODUCTION

The primary goal of many readers has been to learn about statistical inference rather than about probability theory. These readers are now about to reap their reward. We shall now see that statistical inference is simply a straightforward application of probability theory. The foundations for the study of statistical inference have been firmly laid, and the major difficulties are behind us.

Up to now we have concerned ourselves only with probability theory, making but a few scant references to statistical inference. A few preliminary remarks about statistical inference thus are in order at this point.

In our study of probability, we began by postulating three axioms. Using these, we are able to solve many problems and prove many theorems. For example, we found that the probability of getting a blackjack is $128/(52 \cdot 51)$, and we proved that if (Ω, P) is a probability model with $A \subset \Omega$ and $B \subset \Omega$, then $P(A \cup B) = P(A) + P(B) - P(A \cap B)$. These are just two of the many examples we encountered where there was no uncertainty as to the validity of our conclusions.[1] This is the very nature of *deductive inference*. However, many of the truly interesting problems one encounters in life involve drawing general conclusions from specific empirical evidence. This process is called *inductive inference*. In contrast to deductive inference, inductive inference necessarily results in uncertainty as to whether or not the conclusions drawn are valid.

We have already encountered such problems: using the empirical evidence that the motorcycle "quit suddenly" in Example 22 of Chapter 1, Mac the mechanic had to decide whether it was the distributor, fuel pump, or carburetor that caused the motorcycle to malfunction, while in Example 26 of Chapter 1 we wanted to know whether or not the youngest of three sisters was clumsy, using the empirical evidence that she was responsible for breaking 3 out of 4 dishes.

These two examples are typical of the problems involving inductive inference with which we shall be dealing, and they are called problems of *statistical inference*. Implicit in these two examples and in all problems of statistical inference is the existence of some probabilistic mechanism (probability model) that governs the situation. Our goal, of course, is to discover the true nature of this mechanism and to draw conclusions accordingly. For example, we would like to know the true proportion (probability) of the dishes broken by the youngest sister, and we would agree that she is, in fact, clumsy if this proportion is larger than $\frac{1}{3}$.

Of fundamental importance in statistical inference is the information or empirical evidence itself. If we are to draw "good" conclusions from the data, then the data must be relevant to and representative of the problem at hand. For example, suppose we wanted to know the average height of males in the United States. The only way we could be absolutely sure of obtaining the correct answer would be to find the height of each of the approximately 100 million males in the United States. Obviously, this would be much too costly, hence not feasible. However, we could observe the heights of a few males and on the basis of this sample of heights make an estimate of the average height of males in the United States. Naturally we would neither go to another country and note the heights of a few men there nor go to the men's gym and note the heights of the various mem-

[1] Of course, the conclusions were usually probabilistic statements.

bers of the basketball team, because their heights are obviously neither relevant to nor representative of the population of heights of males in the United States.

This example leads us to ask the important question: How do we obtain a representative sample? Unfortunately, we shall have to completely ignore this question.[2] We will always assume that our samples are, in fact, representative. (In the section on random numbers in Chapter 3, we indicated how to obtain random samples with or without replacement.)

In summary, the general problem to which we shall henceforth address ourselves is that of drawing "good" conclusions about a characteristic of a certain population, such as average height of males in the United States. Since it is generally not feasible to examine the entire population, we shall have to content ourselves with drawing our conclusions by examining a representative sample of the population. The remainder of the book is devoted to answering the question of how to draw "good" conclusions when we have a representative sample.

[2] Good books dealing with this question are available; see W. G. Cochran, *Sampling Techniques*, 2d ed. (New York: John Wiley & Sons, Inc., 1953) and M. H. Hansen, W. N. Hurwitz, and W. G. Madow, *Sample Survey Methods and Theory*, Vols. I and II (New York: John Wiley & Sons, Inc., 1953).

4

HYPOTHESIS TESTING

1. BASIC NOTIONS

Once again let us consider the plight of the youngest of three sisters who has been called clumsy. The evidence before us is that she broke the first, second, and fourth dishes. On the basis of this evidence, we must draw our conclusion, namely, she is clumsy or she is not clumsy.

While this example is artificial, the student should not be misled as to its value. We use it not because there is a paucity of real-world problems, but rather because its simplicity renders it an ideal vehicle from which to present new ideas and concepts.

As is true of all problems of statistical inference, an underlying probabilistic mechanism governs the situation—that is, determining which dishes the youngest sister will break. In this particular case, we start to analyze the problem by making the initial basic assumption that there is some probability p, $0 \leq p \leq 1$, that she is responsible for breaking the ith dish, $i = 1, 2, 3, 4, 5, \ldots$.[1] Moreover, we assume that p is the probability that she breaks the ith dish regardless of which of the first $i - 1$ dishes she broke. Restated in probabilistic terminology, we assume that $X_1, X_2, \ldots$ is a sequence of independent random variables such that $P(X_i = 1) = p$ and $P(X_i = 0) = 1 - p$ for each i, $i = 1, 2, \ldots$. Of course, $X_i = 1$ means that she broke the ith dish, and $X_i = 0$ means that she did not break the ith dish. We have now fully specified the underlying probabilistic mechanism except for giving the numerical value of p.[2]

[1] Theoretically, the girls could break any number of dishes. As it happened, we took note only of the first four dishes they broke.

[2] We have specified the *structure* of the mechanism, but we have not specified the value of its *parameter* p.

Many people would quickly criticize the assumptions underlying our probabilistic mechanism. For instance, after breaking the first two dishes the youngest sister may be more nervous and hence more prone to break the third dish; or, after breaking the first two dishes, she may be very careful and thus less likely to break the third dish. We choose to simplify the problem and ignore such possibilities as the random variables' not being independent or identical. Hopefully, we still have a model that closely approximates the situation we are trying to analyze.

In analyzing our problem, we consider two sets of values for the parameter p: $p = \frac{1}{3}$ and $\frac{1}{3} < p \leq 1$. Quite naturally we associate the conclusion that she is not clumsy with $p = \frac{1}{3}$ and the conclusion that she is clumsy with $\frac{1}{3} < p \leq 1$. Thus, we have reduced, or rather restated, our problem to that of concluding that p lies in the set $\{\frac{1}{3}\}$ or in the set $\{x: \frac{1}{3} < x \leq 1\}$.

As noted before, no matter what methods we employ to draw our conclusions, we may very well make an error. Using our restatement of the problem, we see that there are two types of errors we can make: we can say that $\frac{1}{3} < p \leq 1$ when, in fact, $p = \frac{1}{3}$—that is, say she is clumsy when she is not; or we can say that $p = \frac{1}{3}$ when, in fact, $\frac{1}{3} < p \leq 1$—that is, say she is not clumsy when she is. These possibilities are represented pictorially in Figure 4.1.

Conclusion \ Truth	$p = 1/3$	$p > 1/3$
say $p = 1/3$	No error	Second type of error
say $p > 1/3$	First type of error	No error

Figure 4.1

Clearly, there is a loss associated with making an error. Suppose we make an error of the first type; that is, we say that $p > \frac{1}{3}$ when, in fact, $p = \frac{1}{3}$. Then the youngest sister will be incorrectly labeled as clumsy and therefore will be unjustly ridiculed. (Moreover, mother may then decide to save her dishes by not having her wash them for awhile—this would be unfair to the two older daughters, as they would then have to do more than their share of the dishwashing.) Next, suppose we make an error of the second kind; that is, we say that $p = \frac{1}{3}$ when, in fact, $p > \frac{1}{3}$. In this case, mother will continue to have the youngest daughter wash her share of the dishes, which will almost surely result in mother's ending up with more broken dishes than if she had decided not to have her wash dishes for awhile.

This brings us to the problem of deciding which is the more important error to avoid, for no matter which choice we make, we run the risk of making the wrong choice. While this choice is more a matter of opinion than of fact, two reasons present themselves for considering the first to be the more important error to avoid. First, and more importantly, it appears that falsely labeling her as clumsy is a more egregious or serious error than labeling her as not clumsy; for in the former case, the "cost" accruing to this mistake includes her hurt feelings, whereas in the latter case the "cost" is merely a few extra broken dishes. Second, in testing statistical hypotheses, we must take the approach of the conservative empiricist, which is to prefer the status quo. That is, we will not change our opinion in favor of a new theory unless the preponderance of evidence suggests that the new theory provides a better model of reality than does the old theory. Taken together, these two ideas represent the framework within which our criminal courts work—and quite appropriately so. The court operates from the viewpoint that it is more important to avoid convicting an innocent man than it is to avoid acquitting a guilty man. Thus, the status quo can be roughly stated as "a man is presumed innocent until proven guilty beyond a reasonable doubt."[3] Evidently, the current accepted theory in this case is $p = \frac{1}{3}$, not $p > \frac{1}{3}$.

In our study of hypothesis testing, there will always be two distinguished conclusions or hypotheses, which we label the *null hypothesis* and the *alternative hypothesis*, denoted H_0 and H_A, respectively. We refer to the error of rejecting (or equivalently not concluding in favor of) the null hypothesis when it is, in fact, true as an error of the first kind or a *type I error*, and we refer to the error of accepting the null hypothesis when it is, in fact, false as an error of the second type or a *type II error*. After having decided which of the two kinds of errors is the more important to avoid, we choose the null hypothesis, so that making this (the more important) error corresponds to making a type I error. In our problem we have already decided that saying $p > \frac{1}{3}$ when, in fact, $p = \frac{1}{3}$ is the more important error to avoid, so that if $p = \frac{1}{3}$ is the null hypothesis, this error is the type I error. In symbols, we have

$$H_0\colon p = \tfrac{1}{3} \qquad \text{and} \qquad H_A\colon p > \tfrac{1}{3}.$$

So far we have reduced our original problem to that of deciding whether or not to accept the null hypothesis $p = \frac{1}{3}$ in favor of the alternative hypothesis $p > \frac{1}{3}$. Now we must see how to take the empirical evidence

[3] We will have more to say shortly as to what constitutes "reasonable doubt" within the framework of hypothesis testing.

into account so as to draw a "good" conclusion. We defer our discussion of what "good" means until the next section, and satisfy ourselves for now with a description of how to take the empirical evidence into account in arriving at a conclusion. Before describing this process, we need to introduce some new terminology and notation.

For any fixed positive integer n, an ordered set of n objects is said to be an *n-tuple*. Thus, an ordered pair is a 2-tuple. If $X_1, X_2, \ldots, X_n$ are independent random variables each having the same distribution as a random variable X, then the n-tuple $\mathbf{X}$ defined by $\mathbf{X} \equiv (X_1, X_2, \ldots, X_n)$ is said to be a *random sample of size n* from the distribution of X.[4] To avoid cumbersome notation, henceforth, when we write x_i we shall mean that the particular value of X_i which was observed was the number x_i. (Don't confuse this with the old notation where x_i was one of the members of the range of X!) Moreover, we shall denote by $\mathbf{x}$ the n-tuple $(x_1, x_2, \ldots, x_n)$, and we shall refer to $\mathbf{x}$ as the *empirical evidence* or as the *outcome of "the experiment."* Finally, we define $\mathcal{S}$ to be the set of all possible outcomes of the experiment. Thus, $s \in \mathcal{S}$ if and only if $s = (s_1, s_2, \ldots, s_n)$ with s_i being in the range of X_i, $1 \leq i \leq n$.

No matter what the outcome of the experiment is, our decision process must lead us to either accept H_0 or reject H_0. Naturally, the decision of whether or not to accept H_0 will depend upon the actual outcome $\mathbf{x}$ of the experiment; that is, the decision will be a function of the outcome of the experiment. Thus, with each $\mathbf{x} \in \mathcal{S}$ we must associate the decision accept H_0 or reject H_0. Rather than specify the function that accomplishes this, we specify a region or subset R of $\mathcal{S}$ such that we reject H_0 if $\mathbf{x} \in R$ and we accept H_0 if $\mathbf{x} \notin R$. Such a region is called a *rejection region* or critical region, and sometimes we refer to this region as a *test* of H_0 against H_A.

Our task will be complete upon specifying a "good" rejection region. In our particular example, the outcome of the experiment is a 4-tuple, and

$$\mathcal{S} = \{(x_1, \ldots, x_4) \colon x_i \in \{0, 1\}, 1 \leq i \leq 4\}.$$

Without further ado, let us choose as an example of a rejection region the region R_2, where

$$R_2 = \left\{\mathbf{x} \colon \sum_{i=1}^{4} x_i \geq 3\right\}.$$

Since the empirical evidence is $\mathbf{x} = (1, 1, 0, 1)$, we see that $\mathbf{x} \in R_2$, so using this test we reject H_0; that is, we conclude that the youngest sister is indeed clumsy.

[4] This is a simple extension of the idea of a random sample with replacement introduced in Section 3 of Chapter 1.

In the next section we shall take up the problem of choosing a "good" test, and we shall see just what a "good" region is for testing H_0 against H_A. We now give a brief summary and formally define the concepts that have been presented to this point.

After having motivated and explained new concepts, we often present them again in the format of formal definitions. This is done for two pedagogically sound reasons: they serve to remove possible ambiguities, and they are useful in referencing. If, however, the reader finds that some of the definitions confuse already clear concepts or obscure the intuitive content of the concept being defined, then he should simply pay no further attention to these definitions.

DEFINITION

A *random sample of size n* is an n-tuple $\mathbf{X}$ of n independent and identically distributed random variables. We denote by $\mathbf{x}$ and $\mathcal{S}$ the n-tuple of observed values of $\mathbf{X}$ and the set of possible values of $\mathbf{x}$, respectively. Also, $\mathbf{x}$ is often called the *empirical evidence* or the outcome of the experiment.

In the case of the three sisters, X_i is $B(1, p)$, $1 \leq i \leq 4$,

$$\mathcal{S} = \{\mathbf{x}: x_i \in \{0, 1\}, 1 \leq i \leq 4\} \text{ and } \mathbf{x} = (1, 1, 0, 1).$$

DEFINITION

A *probabilistic mechanism* simply (1) specifies the (functional) form of the distribution of each of the random variables in which we are interested; (2) specifies the interdependence relationships between these random variables—that is, specifies their joint distribution; and (3) specifies the set of possible values that the unknown parameter(s) can assume.

DEFINITION

The set of possible values that the unknown parameter can assume is called the *parameter set*, and we denote it by Θ.[5]

[5] The symbols "θ" and "Θ" are the lower- and upper-case Greek letter theta. A more precise and mathematical definition of a probabilistic mechanism is: A *probabilistic mechanism* is a family $\mathfrak{F}$ of (joint) distribution functions indexed by a set Θ called the *parameter set*. For each fixed n we have

$$\mathfrak{F} = \{f_{\mathbf{X}}^{\theta}: \theta \in \Theta\},$$

where $\mathbf{X} = (X_1, X_2, \ldots, X_n)$ and for each fixed value of θ in Θ, $f_{\mathbf{X}}^{\theta}$ is the joint distribution of $\mathbf{X}$.

In the case of the three sisters, the random variables of interest are X_1, X_2, X_3, and X_4, where $X_i = 1$ if she breaks the ith dish and $X_i = 0$ if she does not break the ith dish, $i = 1, 2, 3, 4$. We assume that (1) each X_i is a binomial random variable with parameters 1 and p (so they are identically distributed), (2) the random variables X_1, X_2, X_3, X_4 are mutually independent, and (3) the unknown parameter p is a number between 0 and 1—that is, $\Theta = \{p: 0 \leq p \leq 1\}$.

In order to indicate that the probability mass or density function of X_i depends upon the unknown parameter θ, we will sometimes write $f_{X_i}^{\theta}$ instead of f_{X_i} and $f_{\mathbf{X}}^{\theta}$ instead of $f_{\mathbf{X}}$, where $\mathbf{X} = (X_1, X_2, \ldots, X_n)$.

Sometimes we refer to the probabilistic mechanism as "the experiment," and it is then natural to refer to $\mathcal{S}$ as the sample space of the experiment.

At this point in our development of the theory of hypothesis testing, the reader would be perfectly justified in asking why we have bothered to introduce the definition of a probabilistic mechanism. The answer is twofold. First, by going to all the trouble of writing down the probabilistic mechanism we are forced to spell out and make explicit all assumptions. This enables us to make a better model of the real-world problem under study by exposing unjustified implicit assumptions. Second, it forces us to bear in mind the fact that our inquiry centers around the question: "In which of the two subsets of Θ does the true value of the parameter lie?" while we (almost) never test the validity of the probabilistic mechanism itself.

Finally, we remark that in order to maintain notational consistency, we will *always* label the parameter set by Θ—even though we may label the unknown parameter by p, N, μ, or σ^2 when appropriate.

DEFINITION

Given a probabilistic mechanism, we distinguish two disjoint subsets of Θ. We refer to these subsets as the *null hypothesis* and the *alternative hypothesis*, labeled H_0 and H_A, respectively.

In the case of the three sisters, $H_0 = \{\frac{1}{3}\}$ and $H_A = \{p: \frac{1}{3} < p \leq 1\}$. Notice that we need not have $H_0 \cup H_A = \Theta$.

DEFINITION

A *type I error* consists in rejecting H_0 when it is true. A *type II error* consists in accepting H_0 when it is false.

DEFINITION

A *rejection region* or a *test* is a subset R of $\mathcal{S}$ such that we reject H_0 if $\mathbf{x} \in R$, and we accept H_0 if $\mathbf{x} \notin R$.

SUMMARY

We start by describing the structure of a probabilistic mechanism that we believe fairly represents the real situation we want to analyze. We then distinguish two disjoint subsets of the parameter set; we label one of them H_0 and the other H_A, and we consider the errors concomitant with deciding that the true value of the parameter(s) lies in one or the other of these two sets. Two rough rules of thumb for determining which set of parameter values is to be labeled the null hypothesis are as follows: (1) choose the null hypothesis so as to make the more egregious error correspond to rejecting H_0 when, in fact, H_0 is true [that is, when the true value of the parameter(s) lies in the set H_0], and (2) make accepting H_0 correspond to preserving the status quo. We refer to rejecting H_0 when it is true and accepting H_0 when it is false as making a type I error and a type II error, respectively. Finally, we choose a subset R of the sample space $\mathcal{S}$ of the experiment. If the outcome $\mathbf{x}$ of the experiment is in R, then we reject H_0; otherwise, we accept H_0. We refer to R as the rejection region.

CAVEAT: Throughout our treatment of hypothesis testing the reader should bear in mind the fact that our inquiry centers around the question: "In which of the two subsets of Θ does the true value of the parameter lie?" We (almost) never consider the validity of the probabilistic mechanism itself—although an effort along these lines is made in Chapters 6 and 9.

EXAMPLE 1. (Spinach)[6]

School officials have noticed that when spinach is served as part of the school lunch, only half the students eat the spinach, although they all at least taste it; moreover, contrary to what one might expect, the children's decisions of whether or not to eat the spinach are (almost) independent. A salesman of frozen spinach claims that 80 percent of the children will eat his brand of spinach, which, however, is slightly more expensive than the brand usually served. In order to test his contention, the school board

[6] This example is due to Jerzey Neyman and appears as an exercise in his book *First Course in Probability and Statistics* (New York: Holt, Rinehart and Winston, Inc., 1950), p. 315.

randomly places samples of the new spinach on five of the luncheon trays, and notices that only the first, fourth, and fifth samples are eaten. What conclusion should they draw?

SOLUTION: We employ the method prescribed in the summary to begin analyzing this problem.[7] Let $X_i = 1$ if the ith child eats the new frozen spinach and 0 otherwise. Since the children (or equivalently the trays) are chosen at random and they do not affect each other (very much), it is reasonable to assume that $X_1, X_2, \ldots$ are independent random variables and that each is $B(1, p)$. Clearly, the parameter set is $\Theta = \{p: 0 \leq p \leq 1\}$. We have now determined the probabilistic mechanism. Also, the two distinguished subsets of Θ are $\{\frac{1}{2}\}$ and $\{\frac{4}{5}\}$. If the school board says $p = \frac{1}{2}$ when $p = \frac{4}{5}$, then they will continue to serve the old brand of spinach, thereby denying the children the nutritional benefits of this leafy green vegetable. On the other hand, if they say $p = \frac{4}{5}$ when $p = \frac{1}{2}$, then they will use the new frozen brand and unnecessarily incur extra expenses—not to mention the fact that they will have been duped. Thus, we take the more egregious error to be saying $p = \frac{4}{5}$ when $p = \frac{1}{2}$. Also, they should have a tendency to accept the current theory that $p = \frac{1}{2}$ and be at least a little skeptical of the salesman's claim (new theory) that $p = \frac{4}{5}$. Therefore, the school officials should choose $p = \frac{1}{2}$ as the null hypothesis—that is, H_0: $p = \frac{1}{2}$ and H_A: $p = \frac{4}{5}$. The sample space of this experiment is the set

$$\mathcal{S} = \{\mathbf{x}: x_i \in \{0, 1\}, 1 \leq i \leq 5\}. \bullet$$

EXAMPLE 2. (Fish)

Recently, the population of lake fish has been depleted by fishermen, and in order to maintain the ecological balance in the lake, the game warden has decided not to allow any fishing this summer. He made his decision after performing the following experiment: 100 fish were caught, tagged, and released back into the lake. A few days later, 120 fish were caught, and 5 of these were tagged (that is, had been caught previously). Moreover, just prior to performing this experiment he said that if he thought there were as many as 2000 fish in the lake, then he would have allowed some fishing this summer. Did he make the correct decision?

SOLUTION: Let X be the number of tagged fish caught, and let N be the number of fish in the lake. If we assume that (i) the number of fish in the lake had not changed between catches, (ii) none of the tagged fish

[7] At this point we only "set up" the problem. This will help us to solve the problem and actually answer the question "What conclusion should they draw?" in later sections. This comment also holds for Examples 2 and 3.

died before the second catch, and (iii) all the fish, including the tagged ones, were equally likely to be caught the second time, then X, the random variable of interest, is $H(120, 100, N)$. Also, the set Θ of possible values of N is given by $\Theta = \{120, 121, 122, \ldots\}$. Thus, we have now determined our probabilistic mechanism.[8] The two distinguished sets (hypotheses) are $N = 2000$ and $N > 2000$. If the game warden says $N > 2000$ when in fact $N = 2000$, then he will allow fishing, and so the ecological balance will be threatened. On the other hand, if he says $N = 2000$ when in fact $N > 2000$, then the fishermen will be denied the joy of one summer's fishing. Once again, the choice as to which is the more egregious error is more a matter of opinion than of fact. In view of his position as protector of the ecology, the game warden decided that the worst error is to say $N > 2000$ when in fact $N = 2000$. Thus, $H_0: N = 2000$ and $H_A: N > 2000$. The sample space of this experiment is the set

$$\mathcal{S} = \{x: x = 0, 1, 2, \ldots, 100\}. \bullet$$

EXAMPLE 3. (Tanks)

During World War II the allied command gathered information about the monthly production of German tanks by noting the coded number on each captured tank. Although each tank produced had a special coded number, the code was cracked; furthermore, it was discovered that the numbers on the tanks (that is, the cracked codes), were consecutive, so that if the last tank produced in February was numbered 1005, the first and second tanks produced in March were numbered 1006 and 1007, respectively. From other intelligence sources it was known that one particular factory of interest produced 200 tanks per month when undamaged and fewer when damaged. Of course, the allied command would bomb an undamaged factory and, owing to limited resources, would not bomb a damaged factory. Suppose that this was the first month of production, the factory had been bombed, and tanks numbered 2, 13, and 99 were captured. Should the allied command bomb this factory again?

[8] More formally (see footnote 6), we have

$$\mathfrak{F} = \{f_X^N: N \in \Theta = \{120, 121, 122, \ldots\}\},$$

where

$$f_X^N(i) = \frac{\binom{100}{i}\binom{N-100}{120-i}}{\binom{N}{120}}, \qquad i = 0, 1, 2, \ldots, 100.$$

SOLUTION: Let X_i be the number on the ith tank captured, $i = 1, 2, 3$, and let N be the number of tanks produced the first month. Note that X_1, X_2, and X_3 are not independent, since, for example,

$$P(X_3 = 1 \mid X_1 = 1, X_2 = 2) = 0 \neq \frac{1}{N} = P(X_3 = 1).$$

However, $\mathbf{X} = (X_1, X_2, X_3)$ constitutes an ordered random sample of size 3 without replacement from a population of size N, so that

$$f_{\mathbf{X}}^{N}(\mathbf{x}) = \frac{(N-3)!}{N!}, \qquad 1 \leq x_1, x_2, x_3 \leq N, \quad x_1 \neq x_2 \neq x_3 \neq x_1.$$

Here $\Theta = \{1, 2, \ldots, 200\}$. Although it is open to dispute whether or not the more egregious error is to say $N < 200$ when $N = 200$ or to say $N = 200$ when $N < 200$, the current state of knowledge (excluding the empirical evidence of course) is $N = 200$. Consequently, the allies would choose $H_0: N = 200$ and $H_A: N \in \{0, 1, 2, \ldots, 199\}$. The sample space of the experiment is the set of 3-tuples whose objects are distinct positive integers—that is,

$$\mathcal{S} = \{(x_1, x_2, x_3): x_i \in \{1, 2, \ldots, 200\}, 1 \leq i \leq 3, x_1 \neq x_2 \neq x_3 \neq x_1\}. \bullet$$

EXERCISES:

1. Describe a situation wherein the two rules of thumb for determining which hypothesis is to be chosen as the null hypothesis are in conflict. How would you make your choice in this case?

2. How many possible rejection regions are there if $\mathcal{S}$ has 3 points? What if $\mathcal{S}$ has n points?

3. Realizing that H_0 and H_A are subsets of the parameter set Θ, is it possible to have $H_A \cap H_0 \neq \varnothing$? Can we ever have $H_A \cup H_0 \neq \Theta$?

In the following exercises, (1) describe the probabilistic mechanism and the sample space, (2) list the null and alternative hypotheses, (3) tell how you decided which was to be chosen as the null, and (4) discuss the real-life assumptions implicit in the probabilistic mechanism. Do not actually attempt to give an answer.

4. Grandpa claims that more often than not he can tell which side is facing up when a penny is placed in his flat open hand. In order to test Grandpa's claim we have decided to flip a penny and then place it on Grandpa's hand. We then repeat this 6 more times. What would we conclude if Grandpa was correct only the first and third time?

5. What would you say about our coin's being fair if we flipped it 6 times and heads appeared on the first, second, third, and fifth flip?

6. John claims no particular preference for blondes, yet 5 out of the last 8 girls he dated have been with blondes whereas only 20 percent of all girls are blondes. Do you think John prefers blondes?

7. Ordinarily, John hits the first shot he takes each day 7 out of 10 times. Would you say John hasn't been shooting as well as usual if he only made his first shot twice in the last 5 days?

8. Consider Exercise 62 of Chapter 2 and assume the person being tested is not unwilling to write down the same suit more than once. What would you say about his clairvoyance if he named 3 of the 5 suits correctly?

9. On a certain multiple-choice examination there are 5 questions. If a student knows the subject material on the test, he will be able to give the correct answer to any question on the test with probability $\frac{3}{4}$. If he does not know the material he will have a probability of $\frac{1}{5}$ of answering any question. In the past, Jack has been a very poor student. Would you conclude he was just a lucky guesser or that he knew the material on the test if he answered 3 out of the 5 questions correctly (numbers 1, 2, and 4)?

10. John was told he would be given use of the family car if he improved his spelling. John used to misspell every other word, but today his father had to ask him to spell 6 words before he misspelled one. Was this evidence of improved spelling ability?

11. On the average, Fred catches one big fish every third day he goes deep sea fishing. Fred told his wife that he would return home the very next day after he caught his first big fish. He didn't return home until the 8th day; that is, he caught his first big fish on the 7th day. When asked if he hadn't really caught a fish before the 7th day, Fred pleaded innocence. Should Fred's wife have believed him? What if he had returned home on the 20th day?

12. In the exercise above, what would you have answered if he had agreed to return home the day after he caught his 3rd fish and he returned home on the 16th day? [*Hint:* See Exercise 69 of Chapter 2.]

13. A large jar contains 20 marbles. Jill said that 7 of them were the beautiful red kind; Jack said there were not that many red ones in the jar. With whom would you agree if 4 marbles were withdrawn and only 1 of them was red?

14. (Continuation) With whom would you have agreed if the marbles had been drawn with replacement instead of without replacement?

15. (Continuation) After some dispute, Jack and Jill agreed that there were 7 of the beautiful red marbles in the jar, but Jack said that there were

more than 20 marbles in all, whereas Jill still thought there were 20. With whom would you agree if 3 marbles were withdrawn and none of them were red?

16. (Continuation) What if the marbles were drawn with replacement?

17. A stack of cards was numbered serially starting was number 1. For convenience, they were placed in a bag where they were (accidentally) randomly mixed. The seller of the cards claimed that there were 100 cards in all, but the buyer took 2 cards out of the bag and called the seller a liar when he noted the numbers 2 and 19 on these cards. The seller said the buyer was mistaken. Who is to be believed?

18. (Continuation) What would you have concluded if the 2 cards had been drawn with replacement?

19. One jar contains 7 peanut butter and 3 raisin cookies while a second jar contains 5 peanut butter and 5 raisin cookies. Mother saw Laura spill 4 cookies from a jar, and mother had a hunch that it was the first jar. However, 2 of each kind of cookie spilled from the jar. Do you think mother was correct?

20. (Continuation) What would you conclude if there were a third jar which contained 4 peanut butter and 3 raisin cookies?

21. The plant foreman came in Monday morning looking for trouble. He went directly to the end of the assembly line and waited until he found a defective part. He didn't have to wait long; in fact, he found the 5th part to be defective. He then heaped abuse upon the workmen and told them they weren't doing their jobs properly, as there are usually 10 percent defective parts but today (and evidently in the future) there are 20 percent defectives. Do you think the foreman's criticism was justified?

2. DEFINING A GOOD TEST

We continue to illustrate new ideas with the example of the three sisters washing dishes. Our problem is to find a "good" test of $H_0\colon p = \frac{1}{3}$ against $H_A\colon p > \frac{1}{3}$. For purely pedagogical reasons let us temporarily modify the problem so that the alternative hypothesis is $p = \frac{4}{5}$ instead of $p > \frac{1}{3}$. (Any number between $\frac{1}{3}$ and 1 would suffice.)

Choosing a test is equivalent to choosing a rejection region R so everyone would agree that a perfect or fabulous test is a region, call it R_{fab}, such that

$$P_{1/3}(R_{\text{fab}}) = 0 \quad \text{and} \quad P_{4/5}(\tilde{R}_{\text{fab}}) = 0,$$

where $P_{p_0}(E)$ denotes the probability of the event $E \subset \mathcal{S}$ when the true value of p is, in fact, p_0. Thus, $P_{1/3}(R_{\text{fab}})$ is the probability of committing a type I error and $P_{4/5}(\tilde{R}_{\text{fab}})$ is the probability of committing a type II error.

Four of the possible candidates for R_{fab} are

$$R_0 = \varnothing, \quad R_1 = \left\{\mathbf{x}: \sum_{i=1}^{4} x_i = 4\right\}, \quad R_2 = \left\{\mathbf{x}: \sum_{i=1}^{4} x_i \geq 3\right\}, \quad R_3 = \mathcal{S}.$$

Since $\sum_{i=1}^{4} X_i$ is $B(4, p)$ with either $p = \frac{1}{3}$ or $p = \frac{4}{5}$, it is easy to see that

$$P_{1/3}(R_0) = 0, \quad P_{1/3}(R_1) = \tfrac{1}{81}, \quad P_{1/3}(R_2) = \tfrac{9}{81}, \quad P_{1/3}(R_3) = 1,$$

and

$$P_{4/5}(\tilde{R}_0) = 1, \quad P_{4/5}(\tilde{R}_1) = \tfrac{369}{625}, \quad P_{4/5}(\tilde{R}_2) = \tfrac{113}{625}, \quad P_{4/5}(\tilde{R}_3) = 0.$$

(Verify that these values are correct just to double-check your grasp of the situation!) Notice that $R_0 \subset R_1 \subset R_2 \subset R_3$ and that

$$0 = P_{1/3}(R_0) \leq P_{1/3}(R_1) \leq P_{1/3}(R_2) \leq P_{1/3}(R_3) = 1$$

and

$$1 = P_{4/5}(\tilde{R}_0) \geq P_{4/5}(\tilde{R}_1) \geq P_{4/5}(\tilde{R}_2) \geq P_{4/5}(\tilde{R}_3) = 0.$$

This leads us to the correct conclusion (see Exercise 23) that as we enlarge the rejection region, the probability of committing a type I error increases while that of committing a type II error decreases. Consequently, there is no perfect region R_{fab}.

It is now evident that whatever region we pick (with the exception of $\varnothing$ and $\mathcal{S}$), the probability of making either a type I or a type II error will be greater than zero. Also, any decrease in the probability of making a type II error—which must be due to the enlargement of the rejection region—is necessarily accompanied by an increase in the probability of making a type I error. In view of this tradeoff between the probabilities of committing type I and type II errors, and considering that it is more important to avoid making a type I error, we narrow the field of regions we might choose as the "best test" by picking a level of probability of type I error that we are willing to tolerate. Each region in this narrowed field of candidates is called a *feasible region*. Having chosen this probability, which is called the *level of significance* of our test and is denoted by α, we see that each feasible region R satisfies

$$P_{1/3}(R) \leq \alpha.$$

That is, R is a feasible region if and only if the probability of committing a type I error when R is the rejection region is no larger than α.

In view of the consequences concomitant with committing a type I error, somehow mother was able to decide on a level of significance of $\alpha = .12$. There are $2^4 = 16$ points in $\mathcal{S}$, and 1, 4, 6, 4, and 1 of them lie in the sets

$$N_i \equiv \left\{\mathbf{x}: \sum_{j=1}^{4} x_j = i\right\}, \qquad i = 0, 1, 2, 3, \text{ and } 4, \text{ respectively.}$$

Furthermore, each point in N_i has probability (when $p = \frac{1}{3}$) $\frac{16}{81}$, $\frac{8}{81}$, $\frac{4}{81}$, $\frac{2}{81}$, and $\frac{1}{81}$ for $i = 0, 1, 2, 3$ and 4, respectively. Hence, we can show—although it is far from obvious—that there are 202 feasible regions among which to choose. (Originally there were $2^{16} = 65{,}536$ different regions among which to choose.)

It now remains only to choose among these 202 feasible regions. Since all of these regions satisfy our criterion of not committing a type I error too often, this leaves us with the natural criterion of choosing the feasible region that has the smallest probability of type II error. Thus, we seek a region, called the *most powerful test* or best test, that

$$\text{minimizes} \quad P_{4/5}(\tilde{R})$$

subject to the constraint that

$$P_{1/3}(R) \leq .12.$$

Luckily, it turns out that finding the most powerful test is quite easy. In fact, $R_2 = \left\{\mathbf{x}: \sum_{i=1}^{4} x_i \geq 3\right\}$ is the best region. In the next section we verify this, and we show how to construct the most powerful test in similar settings. We now give a brief summary and formally define the concepts presented in this section.

DEFINITION

The *level of significance*, denoted by α, is the largest permissible probability of committing a type I error.

For $R \subset \mathcal{S}$ and $\theta \in \Theta$, we define $P_\theta(R)$ to be the probability of the event R when the true value of the unknown parameter is θ. [Thus $P_\theta(R) = \sum_{\mathbf{x} \in R} f_{\mathbf{X}}^{\theta}(\mathbf{x})$.[9]] Sometimes we shall write $P_H(R)$ where $H \subset \Theta$. In

[9] If $f_{\mathbf{X}}^{\theta}$ is a p.d.f. rather than a p.m.f., then replace the sum by an integral.

this case, we will say that $P_H(R) \leq \alpha$ if and only if $P_\theta(R) \leq \alpha$ for each $\theta \in H$.

For example, in the case of the three sisters consider the region $R = \left\{\mathbf{x} \in \mathcal{S}: \sum_{i=1}^{4} x_i \leq 1\right\}$ and the hypothesis $H = \{p: \frac{1}{3} < p \leq 1\}$. Then for each $p \in H$ we have

$$P_p(R) = \binom{4}{1} p(1-p)^3 + (1-p)^4.$$

Notice that it is not true that $P_p(R) \leq .12$ for each $p \in H$, but it is true that $P_p(R) \leq \frac{8}{9}$ for each p in H so that $P_H(R) \leq \frac{8}{9}$.

DEFINITION

Given a level of significance α, a *feasible region* is a region R such that

$$P_{H_0}(R) \leq \alpha;$$

that is, $P_\theta(R) \leq \alpha$ for each $\theta \in H_0$.

In the three-sisters example with $\alpha = .12$, examples of feasible regions include (see Exercise 22)

$$\left\{\mathbf{x} \in \mathcal{S}: \sum_{i=1}^{4} x_i \geq 3\right\}, \qquad \{(0, 0, 0, 0); (1, 1, 1, 1)\},$$

and

$$\{(0, 0, 0, 0); (1, 0, 0, 0); (1, 1, 0, 0)\}.$$

DEFINITION

If H_0 or H_A is a set with one member, then H_0 or H_A is said to be a *simple hypothesis*. A hypothesis that is not simple is said to be *composite*.

For example, the hypothesis $p = \frac{1}{3}$ is simple, whereas $p > \frac{1}{3}$ is composite.

DEFINITION

Given a level of significance α, a feasible region R_α is said to be the *most powerful test* of the simple hypothesis H_0 against the simple alternative H_A if

$$P_{H_A}(\tilde{R}_\alpha) = \text{minimum } P_{H_A}(\tilde{R}) \tag{4.1}$$

subject to

$$P_{H_0}(R) \leq \alpha. \tag{4.2}$$

Furthermore, if H_A is not simple and (4.1) and (4.2) hold, then R_α is said to be the *uniformly most powerful test* of H_0 against H_A.

In the case of the three sisters, a level of significance of $\alpha = .12$ was chosen, and

$$P_{H_0}(R_2) = \tfrac{9}{81} \leq .12,$$

so that R_2 is seen to be a feasible region. The hypothesis H_0: $p = \frac{1}{3}$ is simple, whereas H_A: $\frac{1}{3} < p \leq 1$ is not. In the next section it is shown that R_2 is the uniformly most powerful test of H_0 against H_A—that is, for each fixed p in H_A, the probability of committing a type II error when using the feasible region R_2 is smaller than the probability of committing a type II error when using any other feasible region.

SUMMARY

The smaller we make the probability of committing a type I error, the larger the probability of committing a type II error will be. Therefore, we choose a number α, called the level of significance, and we limit our consideration to a set of regions, called the feasible regions, such that the probability of committing a type I error when using one of these regions is no larger than α. A "good" test, or rather the best or most powerful test, is then defined to be a region R_α such that the probability of committing a type II error when using R_α is smaller (no larger) than when using any other feasible region.

EXAMPLE 4. (Vaccine)

A local pharmaceutical company produces a polio vaccine that contains live virus. After the vaccine has been made, it is heated to a certain temperature for a particular length of time. Hopefully, the heat has attenuated the live virus so that they will prevent rather than induce polio in humans. This is checked out by inoculating laboratory rats with the serum. If too many rats contract polio, then the serum is called dangerous; otherwise it is called safe. Clearly the null hypothesis is "the vaccine is dangerous" and the alternative is "the vaccine is not dangerous." If a type I error is made, then the dangerous vaccine will be sold and some people might contract polio. Humanitarian considerations aside, this error will lead to court claims and cost the firm enormous sums of money. To be concrete, let us say that the cost of making a type I error is $\$10^6$. On the other hand, if a type II error is made, the firm will discard the perfectly good vaccine. The cost of this error is the cost of producing a batch of vaccine. Let us take this cost to be \$200. In view of the relative sizes of these costs, it is clear that α should be chosen to be quite small, say $\alpha = .0001$. •

Thus, the more important it is to avoid making a type I error compared to a type II error, the smaller we ought to choose α. We illustrate this point further in the next two examples.

EXAMPLE 5. (Spinach)

In the spinach example, α need not be chosen particularly small, as the cost of a type I error relative to the cost of a type II error is not large. Thus, $\alpha = .20$ appears to be a reasonable choice for the level of significance. We will show that $R_\alpha = \{\mathbf{x}\colon \sum_{i=1}^{5} x_i \geq 4\}$ is the most powerful test of H_0 against H_A. •

EXAMPLE 6. (Fish)

In the fish example, it is particularly hard to choose a value of α as both errors are very costly. The warden chose $\alpha = .015$ which appears to be reasonable. It turns out that the uniformly most powerful test of H_0 against H_A with $\alpha = .015$ is $R_{.015} = \{x\colon x = 0 \text{ or } x = 1\}$. •

EXAMPLE 7. (Tanks)

In the tank example, the cost of making a type I error appears to be substantially larger than the cost of making a type II error, so α should be choosen small, say $\alpha = .03$.

With $\alpha = .03$, the uniformly most powerful test of H_0 against H_A is quite difficult to describe and, incidentally, is not unique. Consequently, we shall choose

$$\alpha = \frac{62(61)(60)}{200(199)(198)} = .0288.$$

With this choice of α, the uniformly most powerful test turns out to be $R_\alpha = \{\mathbf{x} \in \mathcal{S}\colon \text{maximum}(x_1, x_2, x_3) \leq 62\}$. •

EXERCISES:

22. Verify that the values of $P_{1/3}(\cdot)$ and $P_{4/5}(\cdot)$ given on page 124 are correct.

23. Show that if $R \subset R' \subset \mathcal{S}$, then $P_{H_0}(R) \leq P_{H_0}(R')$ and $P_{H_A}(\tilde{R}) \geq P_{H_A}(\tilde{R}')$.

24. Show that there are 202 feasible regions for $\alpha = .12$ in the example of the three sisters.

25. Consider the example of the three sisters and suppose that only 3 (instead of 4) dishes were broken. Using $\alpha = .12$,
    (a) List all of the 8 feasible regions.
    (b) Find the probability of committing a type II error for each of the feasible regions (use $H_A\colon p = \frac{4}{5}$).
    (c) Find the most powerful test of H_0 against H_A. Is it unique?

26. In Exercises 4 through 21, tell whether H_0 and/or H_A are simple or composite.

27. Discuss the choice of α in Exercises 4 through 21, and suggest a reasonable value of α to use.

28. Consider the problem described in Example 4, and suppose that the probability that a given rat will contract polio even though the vaccine is, in fact, safe is $\frac{1}{10}$; this probability is $\frac{1}{2}$ otherwise. If the laboratory uses 10 rats per batch, and if they use the serum only if no rats contract polio, then what value of α will they have chosen (implicitly)? What is the associated probability of type II error?

3. CONSTRUCTING A MOST POWERFUL TEST

Now that we have succeeded in defining a "good" or most powerful test, we must describe a procedure that will enable us to easily find one.

In order to construct a most powerful test, it would suffice to have at hand a function (whose domain is $\mathcal{S}$) that properly measures the credibility of the null hypothesis in light of the evidence, namely $\mathbf{x}$. Let us label this function L. We would then choose the most powerful test R_α so that only the least credible (as measured by L) points of $\mathcal{S}$ are in R_α.

Thus, as a first thought, our criterion leads us to suggest as our measure of credibility the function L defined by $L(\mathbf{x}) \equiv P_{H_0}(\mathbf{X} = \mathbf{x})$ for each $\mathbf{x} \in \mathcal{S}$. Using this measure of credibility, R_α is composed of those outcomes that are least likely (that is, have the smallest probability) under the null hypothesis. In the case of the three sisters, this leads us to choose $R_{.12} = \left\{\mathbf{x} \in \mathcal{S}\colon \sum_{i=1}^{4} x_i \geq 3\right\}$, which is, indeed, the most powerful test. However, this selection method would not distinguish among the feasible regions in either the spinach or the tank example, since each of the $2^5 = 32$ points of $\mathcal{S}$ has probability $\frac{1}{32}$ when H_0 is true in the spinach example and each of the 200(199)(198) points of $\mathcal{S}$ has equal probability when H_0 is true in the tank example. Moreover, if we were to test the hypothesis $H_0\colon p = \frac{2}{3}$ (she is moderately clumsy) against the hypothesis $H_A\colon p = \frac{4}{5}$ (she is extremely clumsy) in the case of the three sisters, then we would obtain $R_{.12} = \left\{\mathbf{x}: \sum_{i=1}^{4} x_i \leq 1\right\}$ using this method. But these points are easily seen to be the most credible, not the least credible; that is, if $\sum_{i=1}^{4} x_i$ is small, then this is not evidence of being extremely clumsy but rather evidence of being only a little clumsy. Consequently, $L(\mathbf{x}) \equiv P_{H_0}(\mathbf{X} = \mathbf{x})$ is not the measure of credibility we are seeking.

The trouble with the above measure is that in no way does it take into account the alternative hypothesis. One way of taking the alternative hypothesis into account is to define $L(\mathbf{x})$ as the ratio of $P_{H_0}(\mathbf{X} = \mathbf{x})$ to $P_{H_A}(\mathbf{X} = \mathbf{x})$. In this case, $\mathbf{x} \in R_\alpha$ only if $L(\mathbf{x})$ is small relative to other points in $\mathcal{S}$. Roughly speaking, $L(\mathbf{x})$ small means that the null hypothesis does not "explain" the evidence $\mathbf{x}$ as well as the alternative hypothesis does.

DEFINITION

If H_0 is a simple hypothesis and $\theta_1 \in H_A$, then we define the *likelihood ratio* L_{θ_1} at the point $\mathbf{x} \in \mathcal{S}$ by[10]

$$L_{\theta_1}(\mathbf{x}) = \frac{P_{H_0}(\mathbf{X} = \mathbf{x})}{P_{\theta_1}(\mathbf{X} = \mathbf{x})}.$$

In the case of the three sisters where $\alpha = .12$, $H_0\colon p = \frac{1}{3}$, and $H_A\colon p = .8$, we have (verify that these values are correct!)

$$L_{.8}(\mathbf{x}) = \tfrac{625}{81} \cdot \tfrac{1}{256} \qquad \text{if } \sum_{i=1}^{4} x_i = 4;$$

$$L_{.8}(\mathbf{x}) = \tfrac{625}{81} \cdot \tfrac{1}{32} \qquad \text{if } \sum_{i=1}^{4} x_i = 3;$$

$$L_{.8}(\mathbf{x}) = \tfrac{625}{81} \cdot \tfrac{1}{4} \qquad \text{if } \sum_{i=1}^{4} x_i = 2; \tag{4.3}$$

$$L_{.8}(\mathbf{x}) = \tfrac{625}{81} \cdot 2 \qquad \text{if } \sum_{i=1}^{4} x_i = 1;$$

and

$$L_{.8}(\mathbf{x}) = \tfrac{625}{81} \cdot 16 \qquad \text{if } \sum_{i=1}^{4} x_i = 0.$$

Thus, $L_{.8}(\mathbf{x})$ gets smaller as $\sum_{i=1}^{4} x_i$ gets larger, which agrees with our

The smaller P(event) becomes, the smaller L(x) becomes

[10] If the random variables under consideration are continuous, then we replace $P_\theta(\mathbf{X} = \mathbf{x})$ by $f_{\mathbf{X}}^{\theta}(\mathbf{x})$.

intuition that the more dishes she breaks $\left(\text{that is, the larger is } \sum_{i=1}^{4} x_i\right)$, the more certain we are that she is clumsy. Consequently, if we construct $R_{.12}$ sequentially by using the credibility or likelihood ratio $L_{.8}$ defined above, we will obtain $R_{.12} = \left\{\mathbf{x} \in \mathcal{S}: \sum_{i=1}^{4} x_i \geq 3\right\}$ as desired. This is seen as follows: First we make a list of the 16 sample points of $\mathcal{S}$, listing the points with the smallest values of L_{θ_1} first:

$$\begin{aligned} \text{List} \equiv\ & (1, 1, 1, 1); (1, 1, 1, 0), (1, 1, 0, 1), (1, 0, 1, 1), (0, 1, 1, 1); \\ & (1, 1, 0, 0), (1, 0, 1, 0), (1, 0, 0, 1), (0, 1, 1, 0), \\ & \qquad (0, 1, 0, 1), (0, 0, 1, 1); \\ & (1, 0, 0, 0), (0, 1, 0, 0), (0, 0, 1, 0), (0, 0, 0, 1); (0, 0, 0, 0). \end{aligned}$$

Second, we proceed through the list until we reach the farthest place in the list where the region formed by the points at or before this place is feasible. Now the region formed by the first 5 points in this list is feasible, as its probability is $\frac{9}{81} < .12$ when H_0 is true. However, the region formed by the first 6 points in this list is not feasible, as its probability is $\frac{13}{81} > .12$ when H_0 is true. Thus, the region constructed (sequentially) using the likelihood function is the first 5 points in the list—that is,

$$\left\{\mathbf{x} \in \mathcal{S}: \sum_{i=1}^{4} x_i \geq 3\right\}.$$

The next theorem—which is known both as *the fundamental theorem of hypothesis testing* and as the *Neyman-Pearson lemma*, after its codiscoverers—asserts that the most powerful test is obtained by constructing a region using the likelihood ratio L_θ as described above. Most readers will want to skip its proof.

THEOREM 1. (Neyman-Pearson Lemma)

Suppose H_0 is a simple hypothesis, $\theta_1 \in H_A$, and k_α is a positive number, and let R_α be a subset of the sample space $\mathcal{S}$ such that

(i) $L_{\theta_1}(\mathbf{x}) \leq k_\alpha$ for each point $\mathbf{x} \in R_\alpha$;
(ii) $L_{\theta_1}(\mathbf{x}) \geq k_\alpha$ for each point $\mathbf{x} \notin R_\alpha$; and
(iii) $P_{H_0}(R_\alpha) = \alpha$.

Then R_α is the most powerful test of H_0 against $\theta_1 \in H_A$ at level of significance α.

Conditions (i) and (ii) simply state that only the most credible—as measured by L_θ—points of $\mathcal{S}$ are in R_α, whereas condition (iii) states that the probability of committing a type I error is precisely α if R_α is the rejection region.

▶▶

PROOF. Let R_α be a subset of $\mathcal{S}$ satisfying (i), (ii), and (iii), and let R be any other feasible region. We wish to show that

$$P_{\theta_1}(R_\alpha) \geq P_{\theta_1}(R). \tag{4.4}$$

First note that

$$R_\alpha = (R_\alpha \cap R) \cup (R_\alpha \cap \tilde{R})$$

and

$$R = (R \cap R_\alpha) \cup (R \cap \tilde{R}_\alpha),$$

so that

$$\begin{aligned} P_{\theta_1}(R_\alpha) - P_{\theta_1}(R) &= P_{\theta_1}(R_\alpha \cap R) + P_{\theta_1}(R_\alpha \cap \tilde{R}) - P_{\theta_1}(R \cap R_\alpha) \\ &\qquad - P_{\theta_1}(R \cap \tilde{R}_\alpha) \\ &= P_{\theta_1}(R_\alpha \cap \tilde{R}) - P_{\theta_1}(R \cap \tilde{R}_\alpha). \end{aligned} \tag{4.5}$$

By hypothesis (i) we have $L_{\theta_1}(\mathbf{x}) \leq k_\alpha$ or equivalently $P_{\theta_1}(\{\mathbf{x}\}) \geq (1/k_\alpha)P_{H_0}(\{\mathbf{x}\})$ at each point of R_α, and hence at each point of $R_\alpha \cap \tilde{R}$; thus,

$$P_{\theta_1}(R_\alpha \cap \tilde{R}) \geq \frac{1}{k_\alpha} P_{H_0}(R_\alpha \cap \tilde{R}).$$

Similarly, $L_{\theta_1}(\mathbf{x}) \geq k_\alpha$ at each point of $\tilde{R}_\alpha$, so

$$P_{\theta_1}(R \cap \tilde{R}_\alpha) \leq \frac{1}{k_\alpha} P_{H_0}(R \cap \tilde{R}_\alpha).$$

Combining these two inequalities and recalling that R is feasible and $P_{H_0}(R_\alpha) = \alpha$ yields

$$\begin{aligned} P_{\theta_1}(R_\alpha \cap \tilde{R}) - P_{\theta_1}(R \cap \tilde{R}_\alpha) &\geq \frac{1}{k_\alpha}[P_{H_0}(R_\alpha \cap \tilde{R}) - P_{H_0}(R \cap \tilde{R}_\alpha)] \\ &= \frac{1}{k_\alpha}[P_{H_0}(R_\alpha \cap \tilde{R}) + P_{H_0}(R_\alpha \cap R) - P_{H_0}(R_\alpha \cap R) - P_{H_0}(R \cap \tilde{R}_\alpha)] \\ &= \frac{1}{k_\alpha}[P_{H_0}(R_\alpha) - P_{H_0}(R)] \\ &\geq \frac{1}{k_\alpha}[\alpha - \alpha] = 0. \end{aligned}$$

Substituting this result in (4.5), we obtain (4.4) as desired. ◆

▶

Thus, if R_α is one of the regions[11] chosen in accord with Theorem 1, then

$$L_{\theta_1}(\mathbf{x}) \leq L_{\theta_1}(\mathbf{x}^0) \qquad \text{for any pair } (\mathbf{x}, \mathbf{x}^0) \text{ with } \mathbf{x} \in R_\alpha \text{ and } \mathbf{x}^0 \notin R_\alpha \tag{4.6}$$

[11] There may be more than one; see Example 3.

and

$$P_{H_0}(R_\alpha \cup E) > \alpha$$
$$\text{for any event } E \subset \mathcal{S} \text{ such that } E \cap R_\alpha = \varnothing \text{ and } P_{H_0}(E) > 0. \quad (4.7)$$

Consequently, a most powerful test is obtained by performing two simple steps:

I: Make a list of all of the sample points of $\mathcal{S}$, and order this list by placing those sample points with the smallest values of L_{θ_1} first—that is, at the "top" of the list. For example, if $\mathbf{x}$ and $\mathbf{x}^0$ are two points of $\mathcal{S}$, then $\mathbf{x}$ precedes $\mathbf{x}^0$ on the list if $L_{\theta_1}(\mathbf{x}) < L_{\theta_1}(\mathbf{x}^0)$. If $L_{\theta_1}(\mathbf{x}) = L_{\theta_1}(\mathbf{x}^0)$, then either one may precede the other.

II: Start from the beginning or top of the list and proceed down the list until a place on the list is reached such that if we were to define a region to be those sample points in the list that appear at or above this place, then this region would be feasible; however, if we were to add additional points to this region (that is, proceed further down the list), then this region would not be feasible.

Then a most powerful test is that region defined in Step II.

EXAMPLE 8. (Spinach)

Recall in Example 1 that the X_i's are independent random variables, that each is $B(1, p)$, and that we want to test H_0: $p = .5$ against H_A: $p = .8$ at level of significance $\alpha = .20$. In order to construct the best critical region, we start by ordering the points of $\mathcal{S}$ according to L_{θ_1} with $\theta_1 = .8$. We have

$$L_{.8}(\mathbf{x}) = \frac{\prod_{i=1}^{5} (.5)^{x_i}(1 - .5)^{1-x_i}}{\prod_{i=1}^{5} (.8)^{x_i}(1 - .8)^{1-x_i}}$$

$$= \prod_{i=1}^{5} (\tfrac{5}{8})^{x_i}(\tfrac{5}{2})^{1-x_i}$$

$$= (\tfrac{5}{2})^5(\tfrac{5}{8})^{\sum_{i=1}^{5} x_i}(\tfrac{5}{2})^{-\sum_{i=1}^{5} x_i}$$

$$= (\tfrac{5}{2})^5(\tfrac{5}{8}/\tfrac{5}{2})^{\sum_{i=1}^{5} x_i}$$

$$= (\tfrac{5}{2})^5(\tfrac{1}{4})^{\sum_{i=1}^{5} x_i}.$$

Consequently, $L_{.8}(\mathbf{x})$ gets smaller as $\sum_{i=1}^{5} x_i$ gets larger. Thus, the following list is as described in Step I above:

List ≡ (1, 1, 1, 1, 1); (1, 1, 1, 1, 0), (1, 1, 1, 0, 1), (1, 1, 0, 1, 1), (1, 0, 1, 1, 1), (0, 1, 1, 1, 1);

(1, 1, 1, 0, 0), (1, 1, 0, 1, 0), (1, 0, 1, 1, 0), (0, 1, 1, 1, 0), (1, 1, 0, 0, 1), (1, 0, 1, 0, 1), (0, 1, 1, 0, 1), (1, 0, 0, 1, 1), (0, 1, 0, 1, 1), (0, 0, 1, 1, 1);

(1, 1, 0, 0, 0), (1, 0, 1, 0, 0), (1, 0, 0, 1, 0), (1, 0, 0, 0, 1), (0, 1, 1, 0, 0), (0, 1, 0, 1, 0), (0, 1, 0, 0, 1), (0, 0, 1, 1, 0), (0, 0, 1, 0, 1), (0, 0, 0, 1, 1);

(1, 0, 0, 0, 0), (0, 1, 0, 0, 0), (0, 0, 1, 0, 0), (0, 0, 0, 1, 0), (0, 0, 0, 0, 1); (0, 0, 0, 0, 0).

Since each of the 32 points in the sample space has probability $\frac{1}{32}$ under H_0, a feasible region can contain at most 6 points as $\frac{6}{32} < \alpha < \frac{7}{32}$. Consequently, the most powerful test of H_0 against H_A with $\alpha = .20$ is the region formed by the first 6 points in the list—that is,

$$R_{.20} = \left\{\mathbf{x} \in \mathcal{S}\colon \sum_{i=1}^{5} x_i \geq 4\right\}.$$

If we had chosen $\alpha = .50$ instead, then we would have had

$$R_{.50} = \left\{\mathbf{x} \in \mathcal{S}\colon \sum_{i=1}^{5} x_i \geq 3\right\}.$$

Note that the probability of committing a type II error is .26272 and .05792 when $\alpha = .20$ and .50, respectively, as

$$\begin{aligned} P_{.8}(\tilde{R}_{.20}) &= 1 - P_{.8}(R_{.20}) = 1 - \left[\binom{5}{5}(.8)^5(.2)^{5-5} + \binom{5}{4}(.8)^4(.2)^{5-4}\right] \\ &= 1 - [(.8)^5 + 5(.8)^4(.2)^1] = .26272 \end{aligned}$$

and

$$\begin{aligned} P_{.8}(\tilde{R}_{.50}) &= 1 - \left[\binom{5}{5}(.8)^5(.2)^{5-5} + \binom{5}{4}(.8)^4(.2)^{5-4} + \binom{5}{3}(.8)^3(.2)^{5-3}\right] \\ &= .05792. \end{aligned}$$

Finally, since the actual outcome of the experiment was $\mathbf{x} = (1, 0, 0, 1, 1)$, we accept H_0 when $\alpha = .20$ and reject H_0 when $\alpha = .50$, since $\mathbf{x} \notin R_{.20}$ and $\mathbf{x} \in R_{.50}$. •

EXAMPLE 9. (Fish)

In Example 2, the alternative hypothesis was composite. Therefore, let us choose a value of $N \subset H_A$, say $N = 5000$. We now seek the most powerful test of H_0: $N = 2000$ against H_A: $N = 5000$ with $\alpha = .015$. We have

$$L_{5000}(x) = \left[\frac{\binom{100}{x}\binom{1900}{120-x}}{\binom{2000}{120}}\right] \Big/ \left[\frac{\binom{100}{x}\binom{4900}{120-x}}{\binom{5000}{120}}\right] = \frac{\binom{5000}{120}\cdot 1900!}{\binom{2000}{120}\cdot 4900!}\cdot\frac{(4780+x)!}{(1780+x)!},$$

which is easily seen to get smaller as x gets smaller. Thus, the following list is as described in Step I above: List $\equiv 0, 1, 2, 3, \ldots, 99, 100$. Using a computer to evaluate $P_{H_0}(X = i)$,[12] we see that the most powerful test of H_0 against H_A with $\alpha = .015$ is $R_\alpha = \{x: x = 0 \text{ or } x = 1\}$. •

▶▶
EXAMPLE 10. (Tanks)

In Example 3, the alternative hypothesis was composite, so let us choose a value of $N \in H_A$, say $N = 150$. We now seek the most powerful test of H_0: $N = 200$ against H_A: $N = 150$, where

$$\mathcal{S} = \{(x_1, x_2, x_3): x_1 \neq x_2 \neq x_3 \neq x_1,\ x_i \in \{1, 2, \ldots, 200\},\ 1 \leq i \leq 3\},$$

$$P_{H_0}(\{\mathbf{x}\}) = \frac{1}{200(199)(198)} \qquad \text{if } \mathbf{x} \in \mathcal{S},$$

$$P_{150}(\{\mathbf{x}\}) = \frac{1}{150(149)(148)} \qquad \text{if } \mathbf{x} \in \mathcal{S}_{150} \equiv \mathcal{S} \cap \{\mathbf{x}: x_i \leq 150,\ 1 \leq i \leq 3\},$$

and

$$P_{150}(\{\mathbf{x}\}) = 0 \qquad \text{if } \mathbf{x} \notin \mathcal{S}_{150}.$$

Thus,

$$L_{150}(\{\mathbf{x}\}) = \begin{cases} \dfrac{150(149)(148)}{200(199)(198)}, & \text{if } \mathbf{x} \in \mathcal{S}_{150}, \\ \infty, & \text{if } \mathbf{x} \notin \mathcal{S}_{150}, \end{cases}$$

[12] $P_{H_0}(X = 0 \text{ or } X = 1) = .0136$ and $P_{H_0}(X = 2) = .0388$.

so L_{150} is smallest when all of the components of $\mathbf{x}$ are small. Consequently, [if $\alpha \leq 1 - [150(149)(148)]/[200(199)(198)] \approx 1 - (\frac{3}{4})^3$] a best test of $N = 200$ against $N = 150$ is a subset R_α of $\mathcal{S}$ such that each point $\mathbf{x}$ in R_α is a member of $\mathcal{S}_{150}$ and R_α is as large as possible.

For reasons that will become apparent in Section 5, we make our list so that $\mathbf{x}$ precedes $\mathbf{x}^0$ if maximum$\{x_1, x_2, x_3\}$ < maximum$\{x_1^0, x_2^0, x_3^0\}$. Also, with $\alpha = (62)(61)(60)/(200)(199)(198)$, we can choose (62)(61)(60) points of $\mathcal{S}$ to be in R_α, since each of the 200(199)(198) points of $\mathcal{S}$ is equally likely when $N = 200$. In view of this and the fact that there are (62)(61)(60) points of $\mathcal{S}$ with maximum $\{x_1, x_2, x_3\} \leq 62$, we see that R_α as described in Example 7 is a best critical region. •

▶

EXERCISES:

29. Show that the credibility function $L(\mathbf{x}) = P_{H_0}(\mathbf{X} = \mathbf{x})$ yields $\left\{\mathbf{x}\colon \sum_{i=1}^{4} x_i \leq 1\right\}$ as the rejection region for testing $H_0\colon p = \frac{2}{3}$ against $H_A\colon p = \frac{4}{5}$ in the case of the three sisters.

30. Suppose $\mathcal{S}$ is given. In what way does the empirical evidence $\mathbf{x}$ affect the choice of a best critical region?

31. Verify that the values of $L_{.8}$ given on page 130 are correct.

32. Verify the fact that $R_{.50} = \left\{\mathbf{x} \in \mathcal{S}\colon \sum_{i=1}^{5} x_i \geq 3\right\}$ in the spinach example.

33. In Exercises 4 through 21 use your good sense to guess the "form" of the best critical region—that is to guess the "list" produced by L_θ. Then use the Neyman-Pearson lemma to find the best critical regions and decide whether or not to accept H_0. Also, when possible, find the probability of committing a type II error for any particular value of the alternative hypothesis. Use the following values of α for the various exercises:

(a) Exercise 4, $\alpha = .23$.
(b) Exercise 5, $\alpha = .12$.
(c) Exercise 6, $\alpha = .25$.
(d) Exercise 7, $\alpha = .17$.
(e) Exercise 8, $\alpha = .10$.
(f) Exercise 9, $\alpha = .05$.
(g) Exercise 10, $\alpha = .04$.
(h) Exercise 11, $\alpha = .20$.
(i) Exercise 12, $\alpha = .30$.
(j) Exercise 13, $\alpha = .15$.
(k) Exercise 14, $\alpha = .15$.
(l) Exercise 15, $\alpha = .26$.
(m) Exercise 16, $\alpha = .30$.
(n) Exercise 17, $\alpha = .10$.
(o) Exercise 18, $\alpha = .10$.
(p) Exercise 19, $\alpha = .25$.
(q) Exercise 20, $\alpha = .25$.
(r) Exercise 21, $\alpha = .20$.

34. In a law suit, the department of recreation claimed damages against the CRP chemical company for dumping pollutants into the lake and thereby drastically reducing its population of giant turtles. It is known that previously there were many turtles in the lake, and legal precedent is such that only if the population fell to as little as 24 has there been a "drastic reduction." Moreover, the court operates on the premise that a person or firm is "innocent until proven guilty" and that "guilt beyond a reasonable doubt" means that the probability of convicting an innocent party is less than or equal to .01. The department of recreation's claim rested upon the following facts: 6 turtles were caught, tagged, and released back into the lake. Later, 5 turtles were caught, and 3 of these were tagged. Did the department of recreation win the suit? (Assume all samples are equally likely.)

35. One student hypothesized that the test scores would be uniformly distributed on the integers 1, 2, . . . , 100, while a second student hypothesized that the test scores would be uniformly distributed on the integers 1, 2, . . . , 50. Design a good test for testing the first hypothesis against the second for samples of sizes 4. Use $\alpha = .15$. For which of the following samples of size 4 would you reject H_0: 11, 25, 24, 45; 50, 72, 17, 56; 5, 61, 25, 57; or 35, 39, 89, 46? Find the probability of type II error.

36. (Vaccine) A random variable X is said to have the *Poisson distribution* with parameter $p \geq 0$, written "X is Po(p)," if

$$P(X = n) = \frac{e^{-p}p^n}{n!}, \qquad n = 0, 1, 2, \ldots; \quad [E(X) = p].$$

The pharmaceutical company has determined that for a batch of vaccine to be considered safe, it must contain no more than 10,000 live viruses, or equivalently 3 live viruses per cubic centimeter. The danger level is 4 live viruses per cc. What should they conclude if a 1-cc sample contains 2 live viruses? What if a 3-cc sample contains 8 live viruses? Use $\alpha = .10$.

37. (Randomized Tests) By defining a test—that is, a critical region—in a slightly more general way, we sometimes can obtain a lower probability of type II error with a given level of significance. Given a level of significance α, we define a *randomized test* to be a function r whose domain is the sample space $\mathcal{S}$ of the experiment and whose range is contained in $[0, 1] \equiv \{x: 0 \leq x \leq 1\}$, and r is said to be feasible if

$$P_{H_0}(r) \equiv \sum_{\mathbf{x} \in \mathcal{S}} r(\mathbf{x}) P_{H_0}(\{\mathbf{x}\}) \leq \alpha.$$

(If $\mathcal{S}$ is not discrete, replace Σ by $\int$.) Thus, if the outcome of the experiment is $\mathbf{x}$, then $r(\mathbf{x})$ is the probability of rejecting H_0. As described in the text, r would be a feasible test if r satisfied the inequality above and if $r(\mathbf{x}) \in \{0, 1\}$ for each $\mathbf{x} \in \mathcal{S}$; that is, we reject H_0 if $r(\mathbf{x}) = 1$ and accept H_0 if $r(\mathbf{x}) = 0$.

(a) Show that if R_α is the most powerful test of H_0 against H_A and $P_{H_0}(R_\alpha) < \alpha$, then there is a feasible randomized test r_α that satisfies (i) $P_{H_A}(\tilde{r}_\alpha) < P_{H_A}(\tilde{R}_\alpha)$, (ii) $r_\alpha(\mathbf{x}) = 1$ if $\mathbf{x} \in R_\alpha$, and (iii) $P_{H_A}(\tilde{r}_\alpha) \leq P_{H_A}(\tilde{r})$ for any other feasible randomized test r.

(b) Suppose there are numbers c_{I} and c_{II} such that c_i is the cost of making a type i error for $i = \mathrm{I}$ and II, respectively. If we wanted to choose a test to minimize the expected cost, would there be any advantage in considering randomized tests; that is, is there a randomized test r such that

$$c_{\mathrm{I}}P_{H_0}(r)s + c_{\mathrm{II}}P_{H_A}(\tilde{r})(1 - s) < c_{\mathrm{I}}\alpha^*s + c_{\mathrm{II}}P_{H_A}(\tilde{R}_{\alpha^*})(1 - s),$$

where α^* satisfies

$$c_{\mathrm{I}}\alpha^*s + c_{\mathrm{II}}P_{H_A}(\tilde{R}_{\alpha^*})(1 - s) \leq c_{\mathrm{I}}\alpha s + c_{\mathrm{II}}p_{H_A}(\tilde{R}_\alpha)(1 - s)$$

for all α, $0 \leq \alpha \leq 1$, and s is the probability that H_0 is true?

*4. POWER CURVES AND SAMPLE SIZE

Up to this point, we have concentrated on finding the feasible region R_α whose associated probability of committing a type II error is a minimum among those regions whose associated probability of committing a type I error is no larger than a given number α. We have, however, neglected to find the probability of committing a type II error associated with R_α.

4.1 Power Curves

Finding the probability of committing a type II error is quite simple. A type II error is made only if we accept H_0 when H_A is true. Since accepting H_0 means that the outcome of the experiment is not in R_α, we see that the probability of committing a type II error is given by

$$P_{H_A}(\tilde{R}_\alpha).$$

We hasten to note that if H_A is composite rather than simple, then this probability may differ for the various values of the parameter in the set H_A.

To be more explicit, consider the case of the three sisters. There we found that the uniformly most powerful test of $p = \frac{1}{3}$ against $p > \frac{1}{3}$ with $\alpha = .12$ is

$$R_{.12} = \left\{\mathbf{x} \in \mathcal{S}: \sum_{i=1}^{4} x_i \geq 3\right\}.$$

Thus, if $p = \frac{4}{5}$, then

$$\beta_{4/5} \equiv P_{4/5}(\tilde{R}_{.12}) = 1 - P_{4/5}(R_{.12})$$

$$= 1 - \binom{4}{4}\left(\frac{4}{5}\right)^4 - \binom{4}{3}\left(\frac{4}{5}\right)^3\left(\frac{1}{5}\right) = \frac{113}{625}$$

is the probability of committing a type II error—that is, of saying she is not clumsy when, in fact, she is clumsy. Since there is more than one value of p in H_A, we must be careful to distinguish between the probabilities of committing a type II error for the various values of p in H_A. For example, if $p = \frac{3}{4}$, then the probability of committing a type II error is

$$\beta_{3/4} \equiv P_{3/4}(\tilde{R}_{.12}) = 1 - \left(\frac{3}{4}\right)^4 - \binom{4}{3}\left(\frac{3}{4}\right)^3\frac{1}{4} = \frac{67}{256},$$

which is not the same as when $p = \frac{4}{5}$. More generally, given any value of p, say p_0, in H_0 or H_A, we have

$$\beta_{p_0} \equiv P_{p_0}(\tilde{R}_\alpha) = 1 - p_0^4 - 4p_0^3(1 - p_0),$$

so β_{p_0}, the probability of accepting H_0 when the true value of p is p_0, differs for each value of p_0 in H_0 or H_A.

DEFINITION

Given a level of significance α, the function π defined by

$$\pi_\theta = 1 - \beta_\theta = P_\theta(R_\alpha)$$

for each θ in either H_0 or H_A is called the *power curve* or the *operating characteristic* (OC-curve) of the test. For any value $\theta \in H_0 \cup H_A$, π_θ gives the probability of rejecting H_0. Of course, $\pi_\theta \leq \alpha$ for each $\theta \in H_0$.

The power curves for the three sisters, the spinach, the tank, and the fish examples are shown in Figures 4.2, 4.3, 4.4, and 4.5, respectively. Notice that π_θ increases as the difference between θ and θ_0 increases, where $\theta_0 \in H_0$. This is typical of the problems we treat. Unfortunately, there are probabilistic mechanisms for which we cannot find π_θ for $\theta \notin H_0$ (for example, the Student's t and χ^2 distributions).

In Subsection 4.2 we will use the function β_θ to determine how large a sample (how many observations) to take.

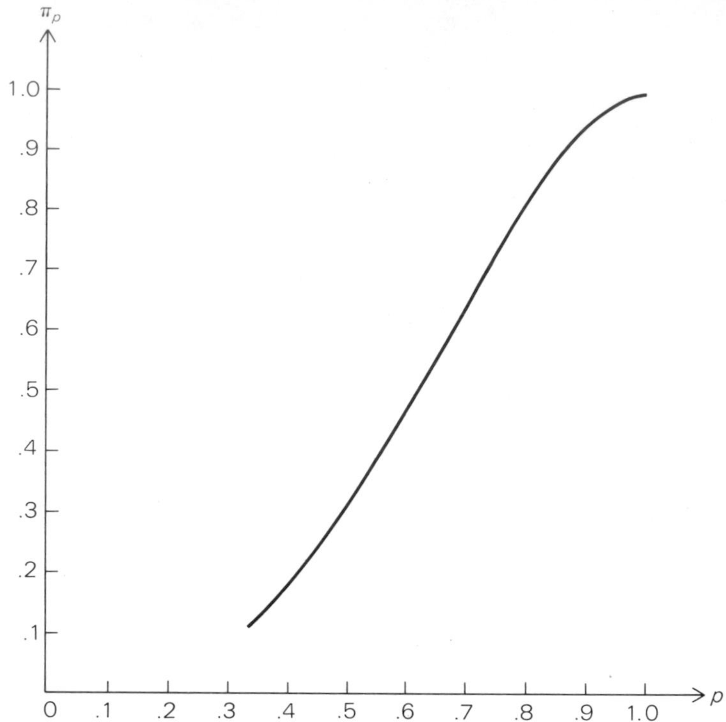

Figure 4.2 π_p **for the Three-Sisters Example**

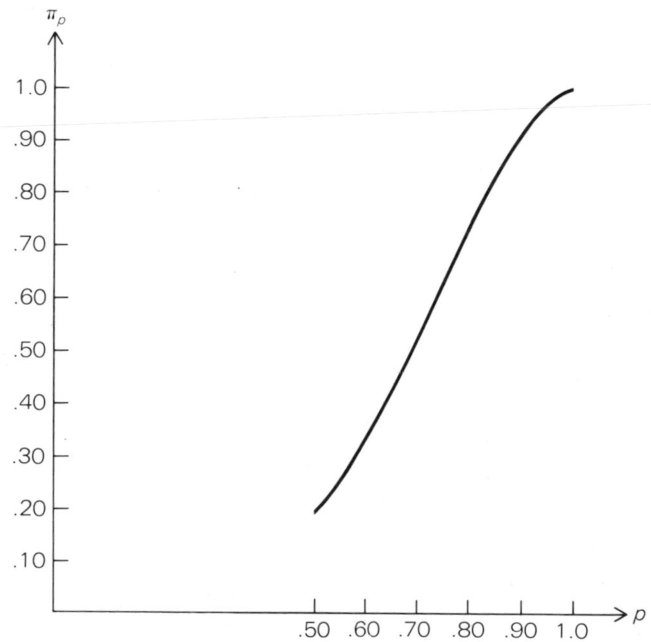

Figure 4.3 π_p **for the Spinach Example**

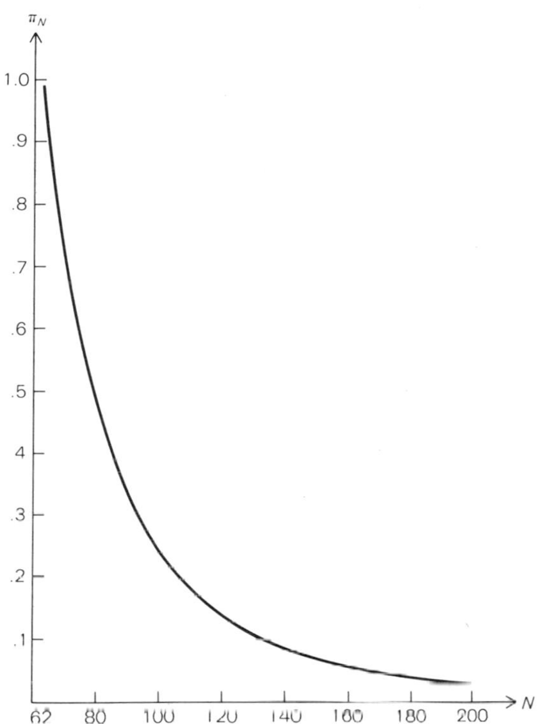

Figure 4.4 π_N for the Tank Example

EXERCISES:

38. In the case of the three sisters with $\alpha = .12$ find π_p for $p = \frac{1}{3}, \frac{2}{3}, \frac{3}{4}, \frac{4}{5}, \frac{5}{6}$, and 1. Do the same with $\alpha = \frac{33}{81}$. Sketch these two curves. What conclusions does this lead you to?

39. In Exercise 5, find π_p for $p = .55, .60, .70$, and 1.0.

40. In Exercise 10, find π_p for $p = .55, .6, .7, .8$, and 1.0.

41. Given Θ, what relationship can you establish between f_0 and f_1, where $f_0 = \beta_\theta$ when $\alpha = \alpha_0$, $f_1 = \beta_\theta$ when $\alpha = \alpha_1$, and $\alpha_0 < \alpha_1$?

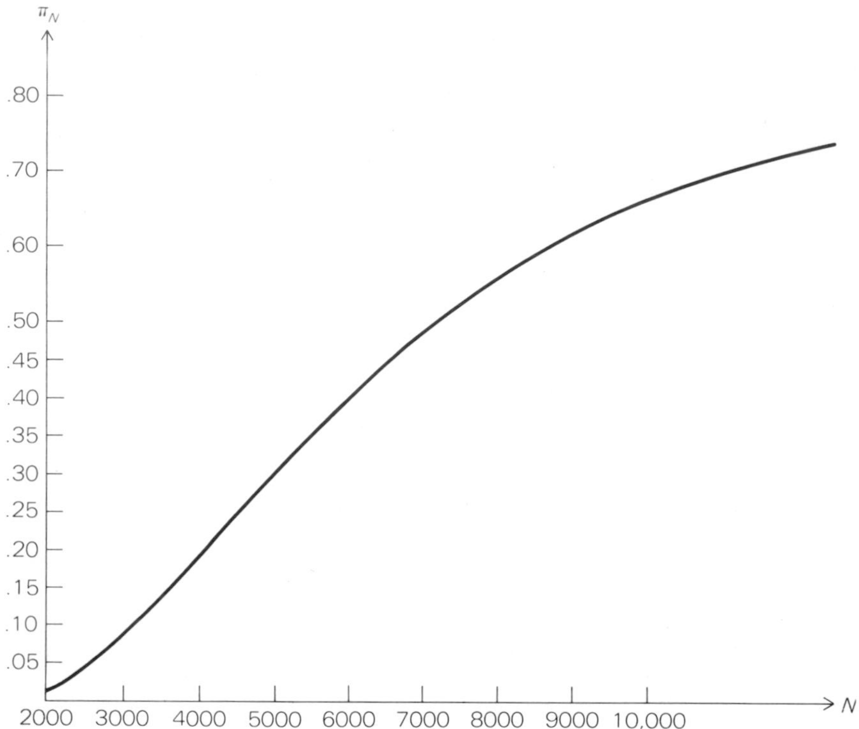

Figure 4.5 π_N **for the Fish Example**

4.2 Sample Size

In many instances, it is important to control not only α, the probability of committing a type I error, but also β_θ, the probability of committing a type II error. That is, we may want both α and β_θ to be small.

In Section 2 we showed that "any decrease in the probability of making a type II error is necessarily accompanied by an increase in the probability of making a type I error." Thus, it would appear that our goal of having both probabilities small is unattainable. Luckily, this is not the case. For implicit in the remark from Section 2 is the idea that the sample size is fixed. Thus, this statement does not contradict the following rather obvious fact. Given a level of significance α, the probability of committing a type II error decreases (to zero) as the sample size increases (to infinity).

Hence, our only problem is how to determine the sample size n. That is, having specified a level of type I and a level of type II error that we are willing to tolerate, how large must we choose n? We answer this question for two special, but important, cases.

First suppose that we wish to test $H_0: \mu = \mu_0$ against $H_A: \mu = \mu_1$, where X is $N(\mu, \sigma^2)$, σ^2 is known, and $\mu_1 > \mu_0$. As will be shown in Chapter 5, we have $R_\alpha = \{\mathbf{x}: \bar{x} \geq \mu_0 + z_{1-\alpha}(\sigma/\sqrt{n})\}$. Thus, we must choose n sufficiently large so that β, the probability of committing a type II error, satisfies

$$\begin{aligned}\beta = P_{H_A}(\tilde{R}_\alpha) &= P_{H_A}\left(\bar{X} \leq \mu_0 + z_{1-\alpha}\frac{\sigma}{\sqrt{n}}\right) \\ &= P_{H_A}\left(\frac{\bar{X} - \mu_1}{\sigma/\sqrt{n}} \leq \frac{\mu_0 - \mu_1}{\sigma/\sqrt{n}} + z_{1-\alpha}\right) \\ &= \Phi\left(\frac{\mu_0 - \mu_1}{\sigma/\sqrt{n}} + z_{1-\alpha}\right).\end{aligned}$$

Hence,

$$z_\beta = \frac{\mu_0 - \mu_1}{\sigma/\sqrt{n}} + z_{1-\alpha}.$$

Equivalently,

$$n = \left[\frac{z_\beta - z_{1-\alpha}}{\mu_0 - \mu_1}\sigma\right]^2. \tag{4.8}$$

[When the right-hand side of equation (4.8) is not an integer, choose n to be the next larger integer.] Thus, if $\mu_0 = 1$, $\mu_1 = 3$, $\sigma = 4$, $\beta = .15$, and $\alpha = .05$, then the right-hand side of equation (4.8) is $(5.37)^2 = 28.6369$, so we would choose $n = 29$.

Next, suppose that we wish to test $H_0: p = p_0$ against $H_A: p = p_1$, where X is $B(1, p)$ and $p_0 > p_1$. As shown in Example 12, there is a number c such that $R_\alpha = \{\mathbf{x}: \bar{x} \leq c\}$. Using the normal approximation to the binomial, we have $\bar{X}$ is $N(p, p(1 - p)/n)$, where

$$\bar{X} = \frac{1}{n}(X_1 + \cdots + X_n).$$

Thus

$$\begin{aligned}\alpha = P_{H_0}(\bar{X} \leq c) &= P_{H_0}\left[\frac{\bar{X} - p_0}{\sqrt{p_0(1 - p_0)/n}} \leq \frac{c - p_0}{\sqrt{p_0(1 - p_0)/n}}\right] \\ &= \Phi\left[\frac{c - p_0}{\sqrt{p_0(1 - p_0)/n}}\right],\end{aligned}$$

so $z_\alpha = (c - p_0)/\sqrt{p_0(1 - p_0)/n}$ or equivalently

$$c = p_0 + z_\alpha\sqrt{p_0(1 - p_0)/n}. \tag{4.9}$$

Also,

$$\beta = P_{H_A}(\bar{X} \geq c) = P_{H_A}\left[\frac{\bar{X} - p_1}{\sqrt{p_1(1 - p_1)/n}} \geq \frac{c - p_1}{\sqrt{p_1(1 - p_1)/n}}\right]$$
$$= 1 - \Phi\left[\frac{c - p_1}{\sqrt{p_1(1 - p_1)/n}}\right],$$

so $z_{1-\beta} = (c - p_1)/\sqrt{p_1(1 - p_1)/n}$ or equivalently

$$c = p_1 + z_{1-\beta}\sqrt{p_1(1 - p_1)/n}. \tag{4.10}$$

Combining equations (4.9) and (4.10) yields

$$p_0 + z_\alpha \sqrt{p_0(1 - p_0)/n} = p_1 + z_{1-\beta}\sqrt{p_1(1 - p_1)/n}.$$

Hence,

$$n = \left[\frac{z_\alpha \sqrt{p_0(1 - p_0)} - z_{1-\beta}\sqrt{p_1(1 - p_1)}}{p_1 - p_0}\right]^2. \tag{4.11}$$

For instance, if $p_0 = .5$, $p_1 = .2$, $\alpha = .10$, and $\beta = .05$, then the right-hand side of (4.9) is $(4.451)^2 = 19.8$, so we would choose $n = 20$.

Note that in both of these special cases, the increase in n is proportional to the square of the decrease in either α or β.

EXERCISES:

42. Suppose in testing the mean μ that H_0: $\mu = \mu_0$, $\sigma^2 = \sigma_0^2$ and H_A: $\mu = \mu_1$, $\sigma^2 = \sigma_1{}^2$, where X is $N(\mu, \sigma^2)$, and $\mu_1 > \mu_0$. How large a sample must we take to ensure that the probabilities of committing type I and type II errors do not exceed α and β, respectively? [*Hint:*

$$R_\alpha = \left\{\mathbf{x}: \bar{x} \geq \mu_0 + z_{1-\alpha}\frac{\sigma_0}{\sqrt{n}}\right\}.\Big]$$

What value of n does this yield if $\sigma_0^2 = 9$, $\sigma_1^2 = 16$, $\alpha = .025$, $\beta = .01$, $\mu_0 = 1$, and $\mu_1 = 5$?

43. Suppose that we wish to test H_0: $p = p_0$ against H_A: $p = p_1$, where X is $B(1, p)$ and $p_0 < p_1$. Determine the sample size needed to ensure that the type I and type II errors do not exceed α and β, respectively. What value of n does this yield if $\alpha = .10$, $\beta = .05$, $p_0 = \frac{1}{2}$, and $p_1 = \frac{3}{4}$?

44. Suppose that we wish to test H_0: $p = p_0$ against H_A: $p \geq p_1$, where X is $B(1, p)$ and $p_1 > p_0$. Determine the sample size needed to ensure that the type I and type II errors do not exceed α and β, respectively. Note that H_A

is composite. [*Hint:* Note that if $p \geq p' \geq \frac{1}{2}$, then $p(1-p) \leq p'(1-p')$, and consider whether $p_1 \geq \frac{1}{2}$.]

45. A manufacturer asserts that no more than 5 percent of the electrical items he produces are defective. A wholesaler sells these items in lots of $N = 500$, and he is anxious that not more than 15 percent of the items be defective. The two parties agree to inspect n of the 500 items. They agree that the wholesaler need not accept the lot if the number X of defectives found is c or larger. We say a lot of 500 items is good if no more than 5 percent are defective and bad if 15 percent or more are defective. What values must c and n be so that the manufacturer's risk of having a good lot rejected is $\alpha = .10$ and the wholesaler's risk of accepting a bad lot is only $\beta = .05$? [*Hint:* H_0: X is $H(n, 25{,}500)$, H_A: X is $H(n, 75{,}500)$. Use the normal approximation.] [*Answer:* $n = 65.4$ and $c = 5.88$ or $n = 66$ and $c = 6$.]

*5. UNIFORMLY MOST POWERFUL TESTS

We have now seen how to construct a most powerful test of a simple hypothesis against a simple alternative. Sometimes, however, we need to construct a uniformly most powerful test of a simple hypothesis against a composite alternative. For example, in the case of the three sisters, the most powerful test of $H_0\colon p = \frac{1}{3}$ against $H_A\colon p = \frac{4}{5}$ at level of significance $\alpha = .12$ was seen to be $R_{.12} = \left\{\mathbf{x}\colon \sum_{i=1}^{4} x_i \geq 3\right\}$. This does not solve the original problem, since the alternative hypothesis was not really H_A: $p = \frac{4}{5}$ but rather $H_A\colon \frac{1}{3} < p \leq 1$. In order to solve this problem, we need to find the uniformly most powerful test of $H_0\colon p = \frac{1}{3}$ against $H_A\colon \frac{1}{3} < p \leq 1$ with $\alpha = .12$. If there is such a test (in general there is not), then surely it must be $R_{.12}$, since $R_{.12}$ is the only region that is the most powerful test of $p = \frac{1}{3}$ against $p = \frac{4}{5}$ with $\alpha = .12$. We now show that $R_{.12}$ is the uniformly most powerful test of $H_0\colon p = \frac{1}{3}$ against $H_A\colon \frac{1}{3} < p \leq 1$ with $\alpha = .12$.

Pick a value of $p \in H_A$; then

$$L_p(\mathbf{x}) = \prod_{i=1}^{4} \frac{(\frac{1}{3})^{x_i}(1-\frac{1}{3})^{1-x_i}}{p^{x_i}(1-p)^{1-x_i}}$$

$$= \frac{(\frac{1}{3})^{\sum_{i=1}^{4} x_i}(1-\frac{1}{3})^{4-\sum_{i=1}^{4} x_i}}{p^{\sum_{i=1}^{4} x_i}(1-p)^{4-\sum_{i=1}^{4} x_i}} = \left(\frac{1-\frac{1}{3}}{1-p}\right)^4 \left[\frac{\frac{1}{3}\cdot(1-p)}{\frac{2}{3}\cdot p}\right]^{\sum_{i=1}^{4} x_i}.$$

Observe that the term in brackets is less than 1, since $p > \frac{1}{3}$. Hence, $L_p(\mathbf{x})$ decreases as $\sum_{i=1}^{4} x_i$ increases. Consequently, no matter what value of $p \in H_A$ we choose, the list we obtain in conjunction with L_p (see Step I, page 133) is the same. Therefore, $R_{.12}$ is the uniformly most powerful test of H_0 against H_A with $\alpha = .12$.

More generally, if there is to be a uniformly most powerful test of a simple hypothesis against a composite alternative at a given level of significance α, then for each $\theta_1 \in H_A$ we must be able to produce the same list via L_{θ_1}.

EXAMPLE 11

It is not always possible to obtain a uniformly most powerful test. For example, consider the spinach example and suppose we wanted to test $H_0: p = \frac{1}{2}$ against $H_A: p \neq \frac{1}{2}$ (that is, $H_A: p \in \{x: 0 \leq x \leq 1, x \neq \frac{1}{2}\}$) with $\alpha = .20$. As shown in Example 8, the most powerful test of $p = \frac{1}{2}$ against $p = \frac{4}{5}$ with $\alpha = .20$ is $\left\{\mathbf{x} \in \mathcal{S}: \sum_{i=1}^{5} x_i \geq 4\right\}$. However, it is easy to show that $\left\{\mathbf{x} \in \mathcal{S}: \sum_{i=1}^{5} x_i \leq 1\right\}$ is the most powerful test of $p = \frac{1}{2}$ against, say, $p = \frac{1}{10}$ with $\alpha = .20$. Consequently, in this example there is no uniformly most powerful test. •

EXAMPLE 12

Suppose that our random variables X_i are independent, each is $B(1, p)$, $H_0: p = p_0$, and $H_A: p_0 < p \leq 1$. Then for any $p \in H_A$, $L_p(\mathbf{x})$ gets smaller as $\sum_{i=1}^{n} x_i$ gets larger. This is seen as follows: choose $p \in H_A$; then

$$L_p(\mathbf{x}) = \frac{\prod_{i=1}^{n} p_0^{x_i}(1 - p_0)^{1-x_i}}{\prod_{i=1}^{n} p^{x_i}(1 - p)^{1-x_i}}$$

$$= \left(\frac{1 - p_0}{1 - p}\right)^n \left[\frac{p_0(1 - p)}{p(1 - p_0)}\right]^{\sum_{i=1}^{n} x_i};$$

this verifies our assertion, as the term in brackets is less than 1. Thus, given α, a uniformly most powerful test is obtained by choosing any value of $p \in H_A$ and using Steps I and II. •

EXAMPLE 13. (Fish)

In Example 9 we found that

$$L_{5000}(x) = K\frac{(4780 + x)!}{(1780 + x)!}, \qquad \text{for } x = 0, 1, 2, \ldots, 100,$$

where K is some constant. Similarly, given any value of $N > 2000$, there is a constant K_N such that

$$L_N(x) = K_N\frac{(N - 120 + x)!}{(1780 + x)!}, \qquad \text{for } x = 0, 1, 2, \ldots, 100.$$

Thus, no matter what value of $N \in H_A$ we choose, $L_N(x)$ gets smaller as x does. Consequently, the list remains unchanged for all values of $N \in H_A$, so $\{x: x = 0 \text{ or } x = 1\}$ is the uniformly most powerful test of H_0 against H_A with $\alpha = .015$. •

EXAMPLE 14. (Tanks)

In Example 10 we showed that

$$L_N(\mathbf{x}) = \begin{cases} \dfrac{N(N-1)(N-2)}{200(199)(198)}, & \text{if maximum}(x_1, x_2, x_3) \leq N, \\ \infty, & \text{otherwise.} \end{cases}$$

Define $E_N = \{x \subset \mathcal{S}: \text{maximum}(x_1, x_2, x_3) = N\}$, $3 \leq N \leq 200$. Then it is clear upon reflection that if we make our list so that every point of E_N precedes every point of E_{N+1}, $N = 3, 4, \ldots, 199$, then no matter what value of $N \in H_A$ we choose, our list is the one produced by L_N. Hence, the region described in Example 7 is the uniformly most powerful test of H_0 against H_A. •

EXERCISES:

46. Show that $[\frac{1}{3}(1 - p)/(\frac{2}{3}p)] < 1$ for $p > \frac{1}{3}$. More generally, show that

$$\left[\frac{p_0(1-p)}{p(1-p_0)}\right] < 1 \qquad \text{if } 1 \geq p > p_0 \geq 0.$$

What if $0 \leq p < p_0 \leq 1$?

47. Show that $(N - 120 + x)!/(1780 + n)!$ increases as x increases if $N > 2000$. What if $N < 2000$?

48. Show that $[(1 - p_0)/(1 - p)]^x$ increases as x increases if $1 \geq p > p_0 \geq 0$. What if $0 \leq p < p_0 \leq 1$?

49. Discuss uniformly most powerful tests (UMPT) when the random variables $X_1, X_2, \ldots, X_n$ are independent, H_0 is simple, and
    (a) X_i is $B(1, p)$, $1 \leq i \leq n$ (see Example 12 and Exercise 46).
    (b) X_1 is $H(m, r, N)$, r is unknown, and $n = 1$ (see Example 13 and Exercise 46).
    (c) X_1 is $H(m, r, N)$, N is unknown, and $n = 1$.
    (d) X_1 is $G(p)$ and $n = 1$ (see Exercise 48).
    (e) X_i is $G(p)$, $1 \leq i \leq n$.
    (f) X_i is $\text{Po}(p)$, $1 \leq i \leq n$ (see Exercise 36).

50. Discuss uniformly most powerful tests when $\mathbf{X} \equiv (X_1, X_2, \ldots, X_n)$ constitutes a random sample of size n without replacement from the population $\{1, 2, \ldots, N\}$ where N is the unknown parameter and H_0 is simple (see Example 14). What if $\mathbf{X}$ is a random sample with replacement?

51. Extend the definition of UMPT's to the case when H_0 is not simple. Solve Exercises 49 and 50 when H_0 is not simple—for example, H_0: $0 \leq p \leq \frac{1}{3}$ and H_A: $\frac{1}{3} < p \leq 1$.

5

NORMAL TESTS

In light of the fact that the normal distribution arises frequently in real problems, it is not surprising to find that statisticians have paid a great deal of attention to the case where the random variables in the probabilistic mechanism are normal. We now present some methods for treating this case. Unless stated otherwise, we shall assume throughout this chapter that the random variables $X_1, X_2, \ldots, X_n$ are independent and that they are each normally distributed with mean μ and variance σ^2. Here,

$$\Theta = \{(\mu, \sigma^2): -\infty < \mu < \infty, 0 < \sigma^2 < \infty\}$$

and

$$\mathcal{S} = \{\mathbf{x}: -\infty < x_i < \infty, 1 \leq i \leq n\}.$$

In performing the various tests of hypotheses concerning the mean and variance of normal random variables, we need to know the distribution of certain other, but related, random variables. The reason is that the best critical regions are defined in terms of these other random variables. Consequently, we now turn briefly to a discussion of chi-square and Student's-t random variables.

1. CHI-SQUARE AND STUDENT'S-t RANDOM VARIABLES

Two continuous random variables very closely related to the normal are the chi-square and Student's-t. These two random variables are of considerable practical importance in many areas of statistics other than hypothesis testing.

DEFINITION

If $X_1, X_2, \ldots, X_n$ are independent normal random variables each having mean μ and variance σ^2, then

$$X \equiv \sum_{i=1}^{n} \left(\frac{X_i - \mu}{\sigma} \right)^2$$

is said to be a *chi-square random variable with n degrees of freedom*, written "X is χ^2_n."[1] Its p.d.f. is given by

$$f(x) = \frac{1}{[(n/2) - 1]!2^{n/2}} x^{(n/2)-1} e^{-x/2}, \qquad x > 0,$$

which is depicted in Figure 5.1.

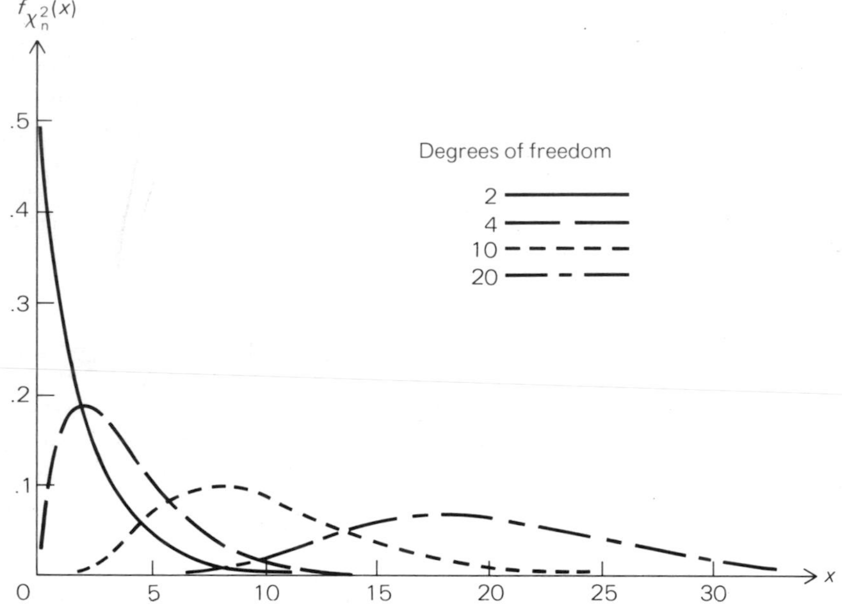

Figure 5.1 Chi-Square Distribution

Probabilities related to chi-square random variables can be found with the aid of Table B (p. 308). Also, if X is χ^2_n, then $E(X) = n$ and $\mathrm{Var}(X) = 2n$.

[1] The symbol "χ" is the lower-case Greek letter chi.

DEFINITION

If X is $N(\mu, \sigma^2)$, Y is χ^2_n, and X and Y are independent, then

$$T_n \equiv \frac{[(X - \mu)/\sigma]\sqrt{n}}{\sqrt{Y}}$$

is said to be a *Student's-t random variable*[2] *with n degrees of freedom*, written "T_n is τ_n."[3] Its p.d.f. is given by

$$f(x) = \frac{[(n-1)/2]!}{\sqrt{n\pi}\,[(n-2)/2]![1 + x^2/n]^{(n+1)/2}}, \qquad \text{for all } x,$$

which is shown in Figure 5.2.

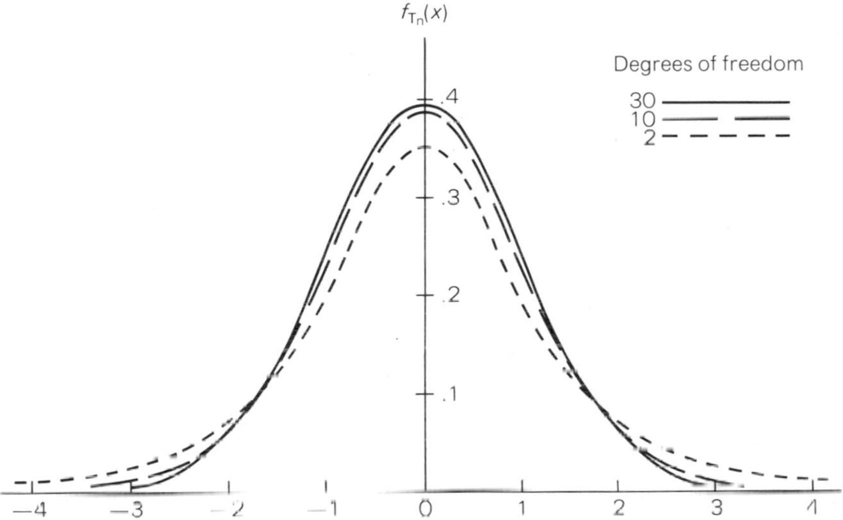

Figure 5.2 Student's-t Distribution

Probabilities related to Student's-t random variables are obtained by using Table C (p. 309); moreover, when n is 30 or larger, the normal distribution provides an excellent approximation. If X is τ_n, then $E(X) = 0$ and $\text{Var}(X) = n/(n-2)$ (if $n > 2$).

[2] We note with amusement the manner in which this important random variable got its name. In 1908 William Gosset published his famous paper showing that the p.d.f. of T_n is as given above. At that time, he was working for an Irish brewery that didn't permit its employees to publish their research findings. Consequently, Gosset published under the pen name of Student.

[3] The symbol "τ" is the lower-case Greek letter tau.

Before examining the various situations that commonly arise when testing hypotheses about normal random variables, we present three facts about chi-square and Student's-t random variables that we will find quite useful in these situations. Henceforth, we reserve the letter T_n for a random variable that has the Student's-t distribution with n degrees of freedom.

Also, for notational simplicity, we define the random variables $\bar{X}$ and S^2 by

$$\bar{X} = \frac{1}{n}\sum_{i=1}^{n} X_i \quad \text{and} \quad S^2 = \frac{1}{n-1}\sum_{i=1}^{n}(X_i - \bar{X})^2,$$

and we denote their actual observed values by $\bar{x}$ and s^2, respectively. As we will see later (in Chapter 7), $\bar{x}$ and s^2 are good "guesses" of the mean μ and the variance σ^2.

THEOREM 1

If $X_1, X_2, \ldots, X_n$ are independent normal random variables each with mean μ_X and variance σ^2, if $Y_1, Y_2, \ldots, Y_m$ are independent normal random variables each with mean μ_Y and variance σ^2, and if the X and the Y random variables are independent, then

(a) $$\sum_{i=1}^{n}\left(\frac{X_i - \bar{X}}{\sigma}\right)^2 \text{ is } \chi^2_{n-1} \quad \text{and} \quad \sum_{i=1}^{m}\left(\frac{Y_i - \bar{Y}}{\sigma}\right)^2 \text{ is } \chi^2_{m-1},$$

(b) $$\frac{\bar{X} - \mu_X}{\sqrt{S_X^2/n}} \text{ is } \tau_{n-1} \quad \text{and} \quad \frac{\bar{Y} - \mu_Y}{\sqrt{S_Y^2/m}} \text{ is } \tau_{m-1},$$

and

(c) $$\frac{(\bar{X} - \bar{Y}) - (\mu_X - \mu_Y)}{\sqrt{\dfrac{(n-1)S_X^2 + (m-1)S_Y^2}{n+m-2}\left(\dfrac{1}{n} + \dfrac{1}{m}\right)}} \text{ is } \tau_{n+m-2}.$$

▶▶

PROOF. Before giving the proof, we state two other facts about chi-square random variables. First, if U is χ^2_s, if V is χ^2_t, and if U and V are independent, then $U + V$ is χ^2_{s+t}.[4] Second, if $U_1, U_2, \ldots, U_s$ are independent normal random variables, then $\bar{U}$ and S_U^2 are independent.[5]

[4] This is shown on page 139 of Hogg and Craig (Reference 17).

[5] This is shown on page 233 of Hogg and Craig and on page 238 of Mood and Graybill (Reference 18).

Note that

$$\Sigma_{i=1}^{n}(X_i - \mu_X)^2 = \Sigma_{i=1}^{n}(X_i - \bar{X} + \bar{X} - \mu_X)^2$$
$$= \Sigma_{i=1}^{n}(X_i - \bar{X})^2 + n(\bar{X} - \mu_X)^2$$

since $2\Sigma_{i=1}^{n}(X_i - \bar{X})(\bar{X} - \mu_X) = 0$, so

$$\sum_{i=1}^{n}\left(\frac{X_i - \mu_X}{\sigma}\right)^2 = \sum_{i=1}^{n}\left(\frac{X_i - \bar{X}}{\sigma}\right)^2 + \left(\frac{\bar{X} - \mu_X}{\sigma/n}\right)^2.$$

We know from the definition of a chi-square random variable that

$$\sum_{i=1}^{n}\left(\frac{X_i - \mu_X}{\sigma}\right)^2 \text{ is } \chi_n^2 \quad \text{and} \quad \left(\frac{\bar{X} - \mu_X}{\sigma/n}\right)^2 \text{ is } \chi_1^2,$$

so part (a) follows from the above two facts about chi-square random variables.

Using the definition of a Student's-t random variable, part (a), and the independence of $\bar{X}$ and S_X^2, we have

$$\frac{\bar{X} - \mu_X}{\sqrt{S_X^2/n}} = \frac{\dfrac{\bar{X} - \mu_X}{\sqrt{\sigma^2/n}}}{\sqrt{\displaystyle\sum_{i=1}^{n}\left(\frac{X_i - \bar{X}}{\sigma}\right)^2}}\sqrt{n - 1} \text{ is } \tau_{n-1},$$

which establishes part (b).

Next, note that

$$U = \frac{(\bar{X} - \bar{Y}) - (\mu_X - \mu_Y)}{\sigma\sqrt{\dfrac{1}{n} + \dfrac{1}{m}}} \text{ is } N(0, 1)$$

and that

$$V = \frac{(n - 1)S_X^2 + (m - 1)S_Y^2}{\sigma^2} \text{ is } \chi_{n+m-2}^2$$

by the first fact about chi-square random variables. Moreover, U and V are independent by the second fact about chi-square random variables. Hence,

$$\frac{U}{V}\sqrt{n + m - 2} \text{ is } \tau_{n+m-2}$$

which establishes part (c). ♦

▶

2. NORMAL TESTS

We now consider seven distinct situations that arise when testing the mean and variance of a normal random variable.

Case 1: Testing the Mean with Known Variance

Sometimes we know the true value of σ^2, say $\sigma^2 = \sigma_0^2$, and we want to decide, for example, if $\mu = \mu_0$ or if $\mu > \mu_0$. Then our problem is to test

$$H_0\colon \mu = \mu_0, \sigma^2 = \sigma_0^2 \quad \text{against} \quad H_A\colon \mu > \mu_0, \sigma^2 = \sigma_0^2,$$

where μ_0 and σ_0^2 are given numbers. Since H_0 is simple, we can attempt to employ the Neyman-Pearson lemma to find a uniformly most powerful (UMP) test of H_0 against H_A. As $e^a \cdot e^b = e^{a+b}$ and $e^a/e^b = e^{a-b}$, we have

$$\begin{aligned}
L_\mu(\mathbf{x}) &= \frac{\prod_{i=1}^{n} \frac{1}{\sqrt{2\pi}\,\sigma_0} e^{-(x_i-\mu_0)^2/(2\sigma_0^2)}}{\prod_{i=1}^{n} \frac{1}{\sqrt{2\pi}\,\sigma_0} e^{-(x_i-\mu)^2/(2\sigma_0^2)}} \\
&= \exp\left\{\left(-\frac{1}{2\sigma_0^2}\right)\left[\sum_{i=1}^{n} (x_i - \mu_0)^2 - \sum_{i=1}^{n} (x_i - \mu)^2\right]\right\} \\
&= \exp\left\{\left(-\frac{1}{2\sigma_0^2}\right)\left[\sum_{i=1}^{n} x_i^2 - 2\mu_0 \sum_{i=1}^{n} x_i + n\mu_0^2 - \sum_{i=1}^{n} x_i^2\right.\right. \\
&\qquad\qquad \left.\left. + 2\mu \sum_{i=1}^{n} x_i - n\mu^2\right]\right\} \\
&= \exp\left(-\frac{n(\mu_0^2 - \mu^2)}{2\sigma_0^2}\right) \exp \frac{1}{\sigma_0^2}\left[(\mu_0 - \mu) \sum_{i=1}^{n} x_i\right].
\end{aligned}$$

Since $\mu_0 - \mu < 0$ for $\mu \in H_A$ and e^t decreases as t decreases, we see that for each $\mu \in H_A$, $L_\mu(\mathbf{x})$ decreases as $\Sigma_{i=1}^n x_i$ increases. Consequently, the list induced by L_μ starts with $\Sigma_{i=1}^n x_i$ [or equivalently $(1/n)\Sigma_{i=1}^n x_i$] large—that is, $R_\alpha = \{\mathbf{x}\colon \bar{x} \geq c\}$. To perform Step II in constructing the UMP test, we find that number c such that $\alpha = P_{H_0}(\bar{X} \geq c)$. Since $X_1, X_2, \ldots, X_n$ are independent and each has mean μ_0 and variance σ_0^2 when H_0 is true, it follows from Theorem 2 of Chapter 3 that $\bar{X}$ is $N(\mu_0, \sigma_0^2/n)$. Consequently, c satisfies

$$\frac{c - \mu_0}{\sigma_0/\sqrt{n}} = z_{1-\alpha} \quad \text{or} \quad c = \mu_0 + \frac{\sigma_0}{\sqrt{n}} \cdot z_{1-\alpha},$$

where $z_{1-\alpha}$ is defined by $1 - \alpha \equiv \Phi(z_{1-\alpha})$. This follows since

$$\alpha = P_{\mu_0}(\bar{X} \geq c) = P_{\mu_0}\left(\frac{\bar{X} - \mu_0}{\sigma_0/\sqrt{n}} \geq \frac{c - \mu_0}{\sigma_0/\sqrt{n}}\right) = P\left(Z \geq \frac{c - \mu_0}{\sigma_0/\sqrt{n}}\right)$$
$$= 1 - \Phi\left(\frac{c - \mu_0}{\sigma_0/\sqrt{n}}\right),$$

so

$$\Phi\left(\frac{c - \mu_0}{\sigma_0/\sqrt{n}}\right) = 1 - \alpha = \Phi(z_{1-\alpha}).$$

For example, suppose $\mu_0 = 4$, $\sigma_0^2 = 9$, $n = 25$, and $\alpha = .10$. Then $.9 = 1 - \alpha$, so $z_{1-\alpha} = 1.282$ from Table A on p. 307. Hence,

$$c = 4 + \tfrac{3}{5}(1.282) = 4.77,$$

so the UMP test of H_0 against H_A with $\alpha = .10$ is

$$R_\alpha = \left\{\mathbf{x}: \frac{1}{25}\sum_{i=1}^{25} x_i \geq 4.77\right\}.$$

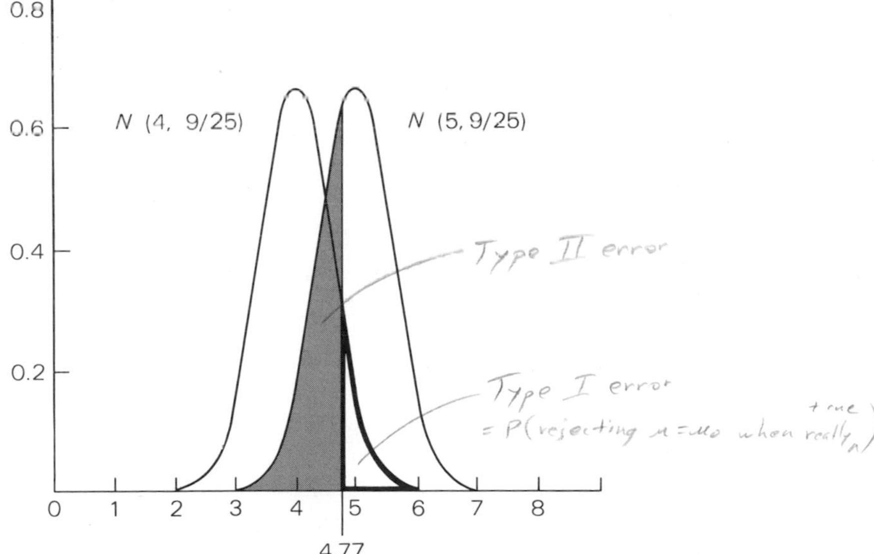

Figure 5.3 Pictorial Representation of Type I and Type II Errors

Given a value $\mu \in H_A$, the probability of committing a type II error is

$$P_\mu(\bar{X} \leq c) = P_\mu\left(\frac{\bar{X} - \mu}{\sigma_0/\sqrt{n}} \leq \frac{c - \mu}{\sigma_0/\sqrt{n}}\right)$$
$$= P\left(Z \leq \frac{c - \mu}{\sigma_0/\sqrt{n}}\right).$$

For example, if $\mu = 5$ in the example above, then the probability of committing a type II error is .3509, since $(c - \mu)/(\sigma_0/\sqrt{n}) = -.383$.

We depict this in Figure 5.3, where the probabilities of committing a type I and type II error are represented by the area enclosed by the heavy lines and the shaded area, respectively.

EXAMPLE 1. (Height)

Some doctors believe that, on the average, men from California are taller than men from other parts of the country. (They attribute this to the fine climate there!) You know that 5 ft 10 in. is the average height of men in the country and 3 in. is the standard deviation; moreover, your 4 California friends have heights of 5 ft 7 in., 6 ft 2 in., 5 ft $9\frac{1}{2}$ in., and 5 ft 11 in. In light of this, would you concur with the doctors' opinion? Use $\alpha = .15$.

SOLUTION: We want to test H_0: $\mu = 70$, $\sigma^2 = 9$ against H_A: $\mu > 70$, $\sigma^2 = 9$, where μ is the average height in inches of men from California. As discussed in Chapter 3, there is good reason to believe that height is normally distributed. Using Table A, we find that $z_{.85} = 1.04$, so

$$c = 70 + \sqrt{\tfrac{9}{4}}\,(1.04) = 71.56.$$

Thus, we will reject H_0 if $\bar{x} \geq 71.56$. Since, in fact, $\bar{x} = 70\frac{3}{8} < 71.56$, we accept H_0. The probability of committing a type II error is .6455 if $\mu = 71$, since $(c - \mu)/(\sigma_0/\sqrt{n}) = .373$. •

Next, suppose we wanted to test

$$H_0\colon \mu = \mu_0, \sigma^2 = \sigma_0^2 \quad \text{against} \quad H_A\colon \mu < \mu_0, \sigma^2 = \sigma_0^2,$$

where μ_0 and σ_0^2 are given numbers. Then, employing the Neyman-Pearson lemma as before, we find that $L_\mu(\mathbf{x})$ decreases as $\bar{x}$ decreases, so that

$$R_\alpha = \{\mathbf{x}\colon \bar{x} \leq c\}$$

for some number c. This result should have been anticipated, for, intuitively, a small value of $\bar{x}$ is evidence in favor of H_A. This time c satisfies (verify this!)

$$c = \mu_0 + \frac{\sigma_0}{\sqrt{n}} \cdot z_\alpha = \mu_0 - \frac{\sigma_0}{\sqrt{n}} \cdot z_{1-\alpha}.$$

Finally, we consider a third possibility. Suppose we wanted to test

$$H_0\colon \mu = \mu_0, \sigma^2 = \sigma_0^2 \quad \text{against} \quad H_A\colon \mu \neq \mu_0, \sigma^2 = \sigma_0^2,$$

where μ_0 and σ_0^2 are given numbers. Then employing the Neyman-Pearson lemma as before, we find that $L_\mu(\mathbf{x})$ decreases as $\bar{x}$ increases if $\mu > \mu_0$, whereas it decreases as $\bar{x}$ decreases if $\mu < \mu_0$. Thus, there is no uniformly most powerful test. It does, however, seem reasonable to reject H_0 if $\bar{x}$ is either very large or very small. And this is just what we shall do. Moreover, there is no reason to give preference to either $\{\mu\colon \mu < \mu_0\}$ or to $\{\mu\colon \mu > \mu_0\}$, so we choose as our rejection region the region R_α defined by $R_\alpha = \{\mathbf{x}\colon \bar{x} \geq u \text{ or } \bar{x} \leq v\}$, where $P_{\mu_0}(\bar{X} \geq u) = P_{\mu_0}(\bar{X} \leq v) = \alpha/2$. Calculations similar to those above yield

$$R_\alpha = \left\{\mathbf{x}\colon \bar{x} \geq \mu_0 + \frac{\sigma_0}{\sqrt{n}} z_{1-(\alpha/2)} \quad \text{or} \quad \bar{x} \leq \mu_0 - \frac{\sigma_0}{\sqrt{n}} z_{1-(\alpha/2)}\right\}.$$

Such a critical region is often called a *two-tailed test.* The reason for this terminology is made evident in Figure 5.4, where R_α is shown when $\mu_0 = 4$, $\sigma_0^2 = 9$, $n = 25$, and $\alpha = .10$. (Compare Figures 5.3 and 5.4.)

Case 2: Testing the Mean with Unknown Variance

Suppose we wish to test

$$H_0\colon \mu = \mu_0, 0 < \sigma^2 < \infty \quad \text{against} \quad H_A\colon \mu > \mu_0, 0 < \sigma^2 < \infty,$$

where μ_0 is a given number. Notice that H_0 is composite as σ^2 is unknown, so in this case we cannot apply the Neyman-Pearson lemma. Using a slightly more sophisticated technique called the generalized likelihood ratio[6], however, we can show that a reasonably good critical region[7] is the one obtained by rejecting H_0 if $W \equiv (\bar{X} - \mu_0)/(S/\sqrt{n})$ is large—that is, reject H_0 if $\mathbf{x}$ is a member of

$$R_\alpha \equiv \left\{\mathbf{x}\colon \frac{\bar{x} - \mu_0}{s/\sqrt{n}} \geq c\right\}.$$

[6] See Mood and Graybill (Reference 18).
[7] In this case, there is no best critical region.

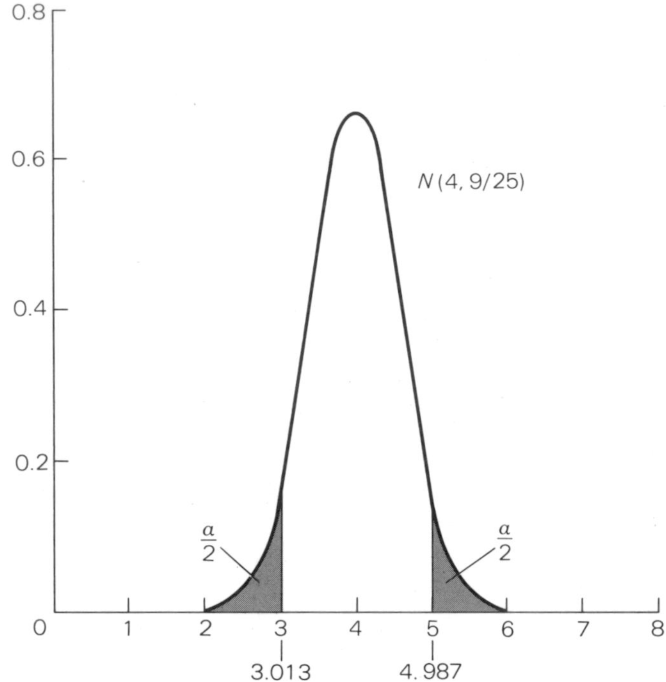

Figure 5.4 Two-Tailed Test

To help see that this is the "correct" critical region, observe that $s(=\sqrt{s^2})$ is an estimate of σ. Consequently, if we replaced S by σ in the formula for W, then W would be a normal random variable with mean 0 and variance 1 and the critical region obtained would be the same as in Case 1, where σ^2 is known.

In order to find c, we must know the distribution of W. It was shown in Theorem 1(b) that W has the Student's-t distribution with $n - 1$ degrees of freedom; that is, "W is τ_{n-1}." Hence, using Table C, we have

$$\alpha = P_{H_0}(W \geq c) = P(T_{n-1} \geq c) = 1 - P(T_{n-1} \leq c),$$

so $c = t_{n-1}(1 - \alpha)$, where $t_{n-1}(1 - \alpha)$ is defined by $P(T_{n-1} \leq t_{n-1}(1 - \alpha)) = 1 - \alpha$. For example, suppose $\mu_0 = 4$, $\bar{x} = 4.78$, $s^2 = 9$, $n = 25$, and $\alpha = .10$. Then $c = t_{24}(.9) = 1.318$ and $w = (\bar{x} - \mu_0)/(s/\sqrt{n}) = 78/60 < c$, so we would accept H_0. (Had we known that $\sigma^2 = 9$ rather than $s^2 = 9$, then we would have rejected H_0. See Case 1.)

In this case, finding the probability of committing a type II error is much beyond our means.

EXAMPLE 2

The manufacturer of the new fiberglass tire claims that fiberglass tires last 40,000 miles on the average, which is twice as long as the ordinary less expensive tires. An impartial consumer's information service has decided to check the manufacturer's claim by testing 5 of these tires. What did the service conclude if they used $\alpha = .05$ and the tires lasted 35,500, 39,350, 41,000, 40,700, and 39,950 miles each?

SOLUTION: There is ample reason to believe that tire life has a normal distribution, so we can assume that X_i is normal, $1 \leq i \leq 5$. The service tested $H_0: \mu = 20{,}000,\ 0 < \sigma^2 < \infty$, against $H_A: \mu = 40{,}000,\ 0 < \sigma^2 < \infty$. Since $\bar{x} = 39{,}300$, $s^2 = (2222)^2$, and $\sqrt{n} = 2.24$, $w \approx 19.3 > 2.132 = t_4(.95)$, we reject H_0. •

Case 3: Comparing the Mean of Two Populations with Known Variances

In many situations it is necessary to compare the means of two distinct normal populations. We now give a method for doing this.

Suppose we have two populations with variates X and Y. We assume X is $N(\mu_X, \sigma_X^2)$, Y is $N(\mu_Y, \sigma_Y^2)$, and X and Y are independent. We wish to test

$$H_0: \mu_X = \mu_Y,\ \sigma_X^2 = \sigma_1^2,\ \sigma_Y^2 = \sigma_2^2$$

against

$$H_A: \mu_X \neq \mu_Y,\ \sigma_X^2 = \sigma_1^2,\ \sigma_Y^2 = \sigma_2^2,$$

where σ_1^2 and σ_2^2 are known (given) numbers. To perform this test we take a random sample $\mathbf{X}$ of size n from the first population and a random sample $\mathbf{Y}$ of size m from the second; $\mathbf{X}$ and $\mathbf{Y}$ are independent.[8] Note that H_0 is not simple, so the Neyman-Pearson lemma is not applicable. However, it seems reasonable to reject $\mu_X = \mu_Y$ if $\bar{X} - \bar{Y}$ is either very large or very small. This is precisely what we shall do. Thus, given α, we reject H_0 if $\bar{X} - \bar{Y} \geq c$ or if $\bar{X} - \bar{Y} \leq -c$—that is,

$$R_\alpha = \{(\mathbf{x}, \mathbf{y}): |\bar{x} - \bar{y}| \geq c\}.$$

As made evident in Figure 5.5, R_α is a two-tailed test. Under H_0, $\bar{X} - \bar{Y}$ is $N(0, \sigma_1^2/n + \sigma_2^2/m)$. Hence,

$$\alpha = P(|\bar{X} - \bar{Y}| \geq c) = P(|Z| \geq c/\sqrt{\sigma_1^2/n + \sigma_2^2/m}),$$

[8] Here, $\mathfrak{F} = \{f_{\mathbf{X},\mathbf{Y}}^{\theta_X,\theta_Y}: \theta_X \in \Theta,\ \theta_Y \in \Theta,\ n, m \in \mathfrak{N}\}$, where

$$\Theta = \{(\mu, \sigma^2): -\infty < \mu < \infty,\ 0 < \sigma^2 < \infty\} \text{ and } f_{\mathbf{X},\mathbf{Y}}^{\theta_X,\theta_Y} = f_{\mathbf{X}}^{\theta_X} \cdot f_{\mathbf{Y}}^{\theta_Y}$$

and $f_{\mathbf{X}}^{\theta_X}$ and $f_{\mathbf{Y}}^{\theta_Y}$ are the joint distributions of independent normal random variables with parameters (μ_X, σ_X^2) and (μ_Y, σ_Y^2).

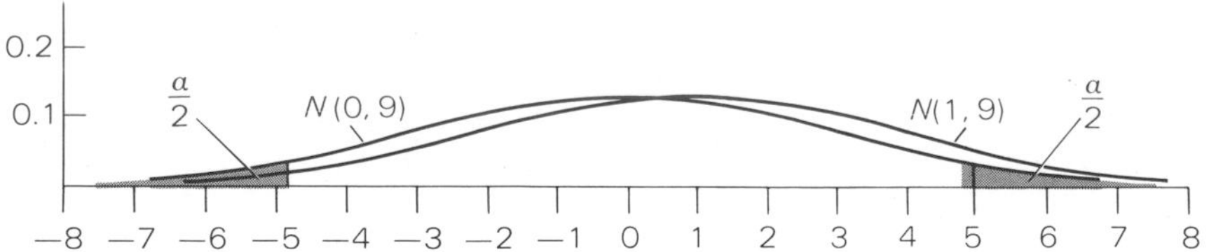

Figure 5.5 Two-Tailed Test

so

$$c = z_{1-(\alpha/2)}\sqrt{\sigma_1^2/n + \sigma_2^2/m}.$$

For example, suppose $\sigma_1^2 = 16$, $\sigma_2^2 = 35$, $n = 4$, $m = 7$, and $\alpha = .10$, then $z_{1-(\alpha/2)} = 1.645$, so $c = 1.645(3)$ since $\sqrt{\frac{16}{4} + \frac{35}{7}} = 3$. Thus,

$$R_{.10} = \{(\mathbf{x}, \mathbf{y})\colon \bar{x} - \bar{y} \geq 4.935 \text{ or } \bar{x} - \bar{y} \leq -4.935\}.$$

The probability of type II error is found as in Case 1, so, for example, if $\mu_X - \mu_Y = 1$ is the alternative hypothesis, then

$$\begin{aligned} P(|\bar{X} - \bar{Y}| \leq 4.935) &= P(|(\bar{X} - \bar{Y} - 1)/3| \leq (4.935 - 1)/3) \\ &= P(|Z| \leq 1.31) = P(Z \leq 1.31) - P(Z \leq -1.31) \\ &= .9049 - .0951 = .8098; \end{aligned}$$

that is, .8098 is the probability of committing a type II error when $\mu_X - \mu_Y = 1$.

EXAMPLE 3

In point of fact, two manufacturers make fiberglass tires and each claims his brand is better than the other. Enough is known about these tire manufacturing companies to assert that the variance of tire life is 700^2 and 750^2 miles for brand X and brand Y, respectively. What would our impartial consumer's service report if it used $\alpha = .05$, the 5 brand-X tires lasted 35,500, 39,350, 41,000, 40,700, and 39,950 miles each, and the 4 brand-Y tires lasted 45,200, 37,200, 39,600, and 40,100 miles each?

SOLUTION: Clearly, we want to test $H_0\colon \mu_X = \mu_Y, \sigma_1^2 = 700^2, \sigma_2^2 = 750^2$ against $H_A\colon \mu_X \neq \mu_Y$, $\sigma_1^2 = 700^2$, $\sigma_2^2 = 750^2$. Since $z_{1-(\alpha/2)} = 1.96$ and $\sigma_{\bar{X}-\bar{Y}} = \sqrt{700^2/5 + 750^2/4} = 10\sqrt{47{,}725/20} \approx 488.5$, $c = 1.96(488.5) = 957.5$. Thus, we reject H_0 as $\bar{x} - \bar{y} = 39{,}300 - 40{,}525 = -1225$ and $-1225 < -957.5$. •

EXAMPLE 4

The dean of admissions wanted to know how much brighter, if at all, the Ph.D. students are than the masters students in a certain department. He used IQ scores as his measure of brightness and found that the IQs of 9 of the masters students were 118, 122, 121, 143, 117, 126, 132, 107, and 128, while the IQs of 4 of the Ph.D. students were 130, 122, 131, and 147. What would you conclude if you were the dean's statistician and you were cognizant of the fact that IQs of college students are normally distributed with variance 225? Use $\alpha = .05$.

Solution: Let us represent the IQs of masters and Ph.D. students by X and Y, respectively. We want to test $H_0: \mu_X = \mu_Y, \sigma_X^2 = \sigma_Y^2 = 225$ against $H_A: \mu_X < \mu_Y, \sigma_X^2 = \sigma_Y^2 = 225$. Note that H_A involves $\mu_X < \mu_Y$ and not $\mu_X \neq \mu_Y$. This fact is revealed in the wording of the problem. The same reasoning that led us to find the appropriate rejection region when H_A involved $\mu_X \neq \mu_Y$ yields $R_\alpha = \{(\mathbf{x}, \mathbf{y}): \bar{x} - \bar{y} \leq -c\}$, where

$$c = z_{1-\alpha}\left(\sqrt{\frac{\sigma_1^2}{n} + \frac{\sigma_2^2}{m}}\right).$$

In this case, $-c = -1.645(9) = -14.8$. Since $\bar{x} - \bar{y} = 123\frac{7}{9} - 132\frac{1}{2} = -8\frac{13}{18} > -c$, we accept H_0. •

Case 4: Comparing the Mean of Two Populations with Unknown Variances

Often we wish to compare the means of two normal populations when their variances are unknown. If we assume that these variances are equal, then we can obtain a relatively simple procedure for testing the equality of the two means; if we do not make this assumption, more complex methods are needed.

Suppose we wish to test

$$H_0: \mu_X = \mu_Y, \sigma_X^2 = \sigma_Y^2 \quad \text{against} \quad H_A: \mu_X \neq \mu_Y, \sigma_X^2 = \sigma_Y^2,$$

at level of significance α. A reasonable rejection region is the one that rejects H_0 when V is either large or small, where the random variable V is defined by

$$V = \frac{\bar{X} - \bar{Y}}{\left[\dfrac{(n-1)S_X^2 + (m-1)S_Y^2}{n+m-2}\right]^{1/2}\left(\dfrac{1}{n} + \dfrac{1}{m}\right)^{1/2}}.$$

It follows from Theorem 1(c) that V has the Student's-t distribution with $n + m - 2$ degrees of freedom, so

$$R_\alpha = \{(\mathbf{x}, \mathbf{y}): v > c \text{ or } v < -c\},$$

where $c = t_{n+m-2}(1 - (\alpha/2))$.

EXAMPLE 5

My little brother likes two kinds of candy bars equally well, so his only concern is to buy the kind that weighs more (on the average). Which kind should I advise him to buy if 4 of the first kind weighed 1.4, 1.5, 1.3, and 1.4 ounces while 3 of the second kind weighed 1.5, 1.7, and 1.6 ounces?

SOLUTION: Let us assume that the weights of both the first and second kinds (X and Y) are normal with equal variances. (In light of the production process for making candy bars, this is a reasonable assumption.) Then we must test $H_0\colon \mu_X = \mu_Y, \sigma_X^2 = \sigma_Y^2$ against $H_A\colon \mu_X \neq \mu_Y, \sigma_X^2 = \sigma_Y^2$. Using $\alpha = .10$, we will accept H_0 if $-2.015 \leq V \leq 2.015$, since V is τ_{4+3-2} (see Table C). We have

$$v = \frac{1.4 - 1.6}{\left(\dfrac{.02 + .02}{5}\right)^{1/2}\left(\dfrac{1}{3} + \dfrac{1}{4}\right)^{1/2}} = -2.9 < -2.015,$$

so we reject H_0. Obviously, the second brand is recommended for purchase. •

EXAMPLE 6. (Social Mobility)

Elder[9] has investigated some factors that appear to determine upward social mobility of women through marriage. He made use of data about 78 girls available from a longitudinal study (1932–1958) called the Oakland Growth Study (initiated by the Institute of Human Development, U.C. Berkeley, and funded by the National Institute of Public Health). Elder defined the girls' original (1929) social status in terms of family social class (measured by Hollingshead's Index of Social Position—a composite of father's educational and occupational status), whereas their later (1958) status was defined in terms of their husbands' occupational status. Thirty-five of the 78 girls had middle-class origins and the other 43 were from working-class families. Perhaps his most interesting finding is the fact that physical attractiveness—which was measured by several experts at the time the girls graduated from high school and included assessments of qualities such as physique, facial features, and sex appeal—directly influenced the likelihood of marriage to a man of higher status, whereas IQ and academic aptitude had relatively little *direct* effect. The reader is referred to Elder's article to learn how this conclusion was statistically validated.

[9] Glen H. Elder, Jr., "Appearance and Education in Marriage Mobility," *American Sociological Review*, Vol. 34, pp. 519–533 (1969).

Appealing to common-sense type arguments involving factors such as the nutritional and clothing advantages of a middle-class background and the experts' bias towards middle-class girls, we would expect the middle-class girls to have been rated as more attractive than their working-class counterparts. In validating this hypothesis at $\alpha = .01$, Elder used the facts that the observed values of $\bar{X}$, $\bar{Y}$, S_X^2, and S_Y^2 are given by $\bar{x} = 55.8$, $\bar{y} = 46.3$, $s_X^2 = 127.69$, and $s_Y^2 = 204.49$, where X_i represents the appearance rating of the ith middle-class girl and Y_j represents the appearance rating of the jth working-class girl. Was Elder's conclusion correct?

Solution: It is reasonable to suppose that the X_i and Y_j are independent while physical appearance, like most biometric, psychometric, and other sociometric characters, is normally distributed. Moreover, the ratings are averages of several experts' opinions, and in turn each expert's opinion is itself an average of his assessment of the many qualities that comprise overall physical appearance. Consequently, we can appeal to the central limit theorem to help justify our assumption of normality. Also, equality of σ_X^2 and σ_Y^2 is not unreasonable. We now must test H_0: $\mu_X = \mu_Y$, $\sigma_X^2 = \sigma_Y^2$ against H_A: $\mu_X > \mu_Y$, $\sigma_X^2 = \sigma_Y^2$. (Note that we use $\mu_X > \mu_Y$ and not $\mu_X \neq \mu_Y$. The wording of the problem makes this choice apparent.) In light of our assumptions, Table C, and the fact that $\alpha = .01$ and $n + m - 2 = 35 + 43 - 2 = 76$, we will reject H_0 if $V \geq 2.38$, since V is τ_{76}. We have

$$v = \frac{55.8 - 46.3}{\left[\dfrac{(34)127.69 + 42(204.49)}{76}\right]^{1/2}\left(\dfrac{1}{35} + \dfrac{1}{43}\right)^{1/2}} = \frac{9.5}{2.97} > 2.38,$$

so we reject H_0 and accept Elder's conclusion. •

Case 5: Comparing Proportions of Two Populations

A recurring situation is the one in which we have two sequences of independent random variables $X_1, X_2, \ldots, X_n$ and $Y_1, Y_2, \ldots, Y_m$, where X_i is $B(1, p_X)$, Y_j is $B(1, p_Y)$, and the X's and Y's are mutually independent. Typically we want to test

$$H_0\colon p_X = p_Y \quad \text{against} \quad H_A\colon p_X \neq p_Y.$$

Common sense tells us that a "good" test of H_0 against H_A consists in rejecting H_0 when $\bar{X} - \bar{Y}$ is too large or too small—that is,

$$R_\alpha = \{(\mathbf{x}, \mathbf{y})\colon \bar{x} - \bar{y} \leq -c \quad \text{or} \quad \bar{x} - \bar{y} \geq c\}.$$

In order to find c, however, we must know the distribution of $\bar{X} - \bar{Y}$. If both n and m are large, then by the central limit theorem $\bar{X} - \bar{Y}$ is approximately normally distributed. Also, when H_0 is true, $p_X = p_Y = p$, so

$$E(\bar{X} - \bar{Y}) = 0 \quad \text{and} \quad \text{Var}(\bar{X} - \bar{Y}) = p(1 - p)\left(\frac{1}{n} + \frac{1}{m}\right).$$

But p is unknown. As in Cases 2 and 4, where we used s^2 as our estimate of σ^2, we use $\hat{p}$ as our estimate of p, where

$$\hat{p} = \frac{\sum_{i=1}^{n} x_i + \sum_{j=1}^{m} y_j}{n + m}.$$

When $n + m$ is large, $\hat{p}$ will be very close to p. Thus, if n and m are large, we can assume that

$$\bar{X} - \bar{Y} \text{ is } N\left[0, \hat{p}(1 - \hat{p})\left(\frac{1}{n} + \frac{1}{m}\right)\right].$$

Consequently,

$$c = z_{1-(\alpha/2)}\sqrt{\hat{p}(1 - \hat{p})\left(\frac{1}{n} + \frac{1}{m}\right)}.$$

EXAMPLE 7. (Chromosomes)

Some geneticists theorize that chromosomal abnormality may predispose a man to antisocial behavior, including crimes of violence. In particular, they claim that an XYY pattern does this. To test their theory, they studied the chromosomes of 197 inmates at Carstairs State Hospital and found 7 of the men to have an XYY pattern, while a random sample of 1500 men yielded but 1 XYY pattern. Using $\alpha = .01$, what do you think of their claim?

SOLUTION: We want to test H_0: $p_1 = p_2$ against H_A: $p_1 > p_2$, where p_1 is the proportion of men in criminal hospitals who have the XYY pattern and p_2 is the percentage of men in the general population having the XYY pattern. Here $n = 197$ and $m = 1500$, so our estimate of p is

$$\hat{p} = \frac{7 + 1}{197 + 1500} = \frac{8}{1697},$$

so $[\hat{p}(1 - \hat{p})(1/n + 1/m)]^{1/2} \approx .00517$. Also, $z_{1-\alpha} = 2.33$, so our rejection region is $\{(\mathbf{x}, \mathbf{y}): \bar{x} - \bar{y} \geq 2.33(.00517)\}$, as the alternative hypothesis is $p_1 > p_2$ and not $p_1 \neq p_2$. We have $\bar{x} - \bar{y} = \frac{7}{197} - \frac{1}{1500} > 2.33(.00517)$, so we reject the null hypothesis; that is, we accept the geneticists' claim. •

EXAMPLE 8. (Tenure)

In the academic year 1966–1967, 14 out of 771 faculty members at UCLA with tenure resigned to take other positions, whereas 22 out of 794 did so in 1967–1968, a year of political unrest on campus. Is this evidence of change at this university? Use $\alpha = .05$.

SOLUTION: We want to test H_0: $p_X = p_Y$ against H_A: $p_X \neq p_Y$, where p_X and p_Y are, respectively, the probabilities that any one faculty member with tenure would leave in 1966–1967 and in 1967–1968. In this case $n = 771$ and $m = 794$, so

$$\hat{p} = \frac{14 + 22}{771 + 794} = .023.$$

Hence, $[.023(1 - .023)(1/771 + 1/794)]^{1/2} \approx .0076$ is our estimate of $\sigma_{\bar{X}-\bar{Y}}$, so that $c = 1.96(.0076) = .014896$. Consequently, we accept H_0, since $\bar{x} - \bar{y} = -.0096$ and $-.014896 < -.0096 < .014896$. •

▶▶

Case 6: Testing the Variance with Known Mean

Suppose we wished to test H_0: $\sigma^2 = \sigma_0^2$, $\mu = \mu_0$ against H_A: $\sigma^2 > \sigma_0^2$, $\mu = \mu_0$, where σ_0^2 and μ_0 are given numbers. Since H_0 is simple, we can apply the Neyman-Pearson lemma. We have

$$L_{\sigma^2}(\mathbf{x}) = \frac{\prod_{i=1}^{n} \frac{1}{\sqrt{2\pi}\,\sigma_0} e^{-(x_i-\mu_0)^2/(2\sigma_0^2)}}{\prod_{i=1}^{n} \frac{1}{\sqrt{2\pi}\,\sigma} e^{-(x_i-\mu_0)^2/(2\sigma^2)}}$$

$$= \left(\frac{\sigma}{\sigma_0}\right)^n \exp\left[\frac{1}{2}\sum_{i=1}^{n} (x_i - \mu_0)^2 \left(\frac{1}{\sigma^2} - \frac{1}{\sigma_0^2}\right)\right].$$

Since $1/\sigma^2 - 1/\sigma_0^2 < 0$, $L_{\sigma^2}(\mathbf{x})$ decreases as $\Sigma_{i=1}^{n}(x_i - \mu_0)^2$ increases. Thus,

$$R_\alpha = \left\{\mathbf{x}\colon \sum_{i=1}^{n} (x_i - \mu_0)^2 \geq c\right\}.$$

To find c, recall the definition of a chi-square random variable. Then under H_0, we see that

$$\sum_{i=1}^{n} \left(\frac{X_i - \mu_0}{\sigma_0}\right)^2 \text{ is } \chi_n^2.$$

Consequently,

$$c = \sigma_0^2 \chi_n^2(1 - \alpha),$$

where $\chi_n^2(1 - \alpha)$ is defined by $P(\chi_n^2 \leq \chi_n^2(1 - \alpha)) = 1 - \alpha$.

EXAMPLE 9

A cheaper process has been discovered for making ball bearings. The mean diameter size is 3 inches, as desired, and the ball-bearing diameters are normally distributed. However, if more than 10 percent of the ball bearings have diameters that are more than $\frac{1}{10}$ inch different from 3 inches, the process will be producing too many "defective" parts. What should we conclude about this process if we observe ball bearings with diameters 3.1, 3.2, 3.0, 3.0, and 2.9 inches and we use $\alpha = .05$?

SOLUTION: Let X be the diameter of a ball bearing. Then X is $N(3, \sigma^2)$. The process is OK if at least 90 percent of the diameters are within $\frac{1}{10}$ inch of 3 inches; that is, the process is OK if X is $N(3, (1/16.45)^2)$, since

$$P(|X - 3| \leq \tfrac{1}{10}) = P\left(\left|\frac{X - 3}{1/16.45}\right| \leq 1.645\right) = P(|Z| \leq 1.645) = .90.$$

Thus, we want to test $H_0: \mu = 3, \sigma^2 = (1/16.45)^2$ against $H_A: \mu = 3, \sigma^2 > (1/16.45)^2$. Since $\alpha = .05$ and $n = 5$, we reject H_0 when

$$\sum_{i=1}^{5} (x_i - 3)^2 \geq (1/16.45)^2 \cdot \chi_5^2(.95) = (1/16.45)^2(11.1) = .041.$$

In fact, $\Sigma_{i=1}^5 (x_i - 3)^2 = .06 > .041$, so we reject H_0. •

Case 7: Testing the Variance with Unknown Mean

Suppose we wanted to test

$$H_0: \sigma^2 = \sigma_0^2, \; -\infty < \mu < \infty \quad \text{against} \quad H_A: \sigma^2 > \sigma_0^2, \; -\infty < \mu < \infty,$$

where σ_0^2 is a given number. Then if we estimate μ by $\bar{X}$, it seems reasonable to reject H_0 only when

$$X \equiv \sum_{i=1}^{n} \left(\frac{X_i - \bar{X}}{\sigma_0}\right)^2$$

is large. It was shown in Theorem 1(a) that X is χ_{n-1}^2. Thus, we reject H_0 only if $X \geq \chi_{n-1}^2(1 - \alpha)$.

Tests for comparing the variance of two normal populations can also be given.[10]

▶

[10] See Mood and Graybill (Reference 18), pp. 307–308.

REMARKS.

1. One-tailed tests are used when testing $\mu = \mu_0$ against either $\mu < \mu_0$ or $\mu > \mu_0$ and when testing $\mu_X = \mu_Y$ against either $\mu_X < \mu_Y$ or $\mu_X > \mu_Y$. Two-tailed tests are used when testing $\mu = \mu_0$ against $\mu \neq \mu_0$ and when testing $\mu_X = \mu_Y$ against $\mu_X \neq \mu_Y$.
2. If a test calls for the use of the Student's-t distribution, and if n is large, say $n \geq 30$, then Table A can be used in place of Table C with only a very small loss in accuracy. If n is not large, this should never be done.

EXERCISES:

1. The five engineering students in our class have the following IQs: 95, 110, 115, 120, and 135, whereas the four mathematics students have IQs of 100, 115, 125, and 140. Realizing that the IQs of both engineering and mathematics students are normally distributed with variance 225, test the hypothesis that, in general, the IQs of engineering and mathematics students are equal against the alternative that the IQs of mathematics students are higher. Use $\alpha = .20$.

2. In a random sample of 5 Barclay bars, the bars weighed 1.7, 1.4, 2.0, 1.6, and 1.8 ounces, respectively. The manufacturer of these delicious chocolates claims that the weight of the bars is, on the average, 1.8 ounces. However, we have reason to believe that the average weight of his chocolate bars (including almonds, of course) is 1.55. As a ravenous consumer of Barclay bars, what do you think about the manufacturer's claim? Use $\alpha = .10$.

3. The principal of Kaneklcy High School (KHS) believes that, partly because of the financial status of their parents, the percentage of his graduates attending college is larger than the citywide average. Over the preceding 5 years, 20 percent of all high-school graduates in the city had attended college, while in the same period 350 of his 1500 graduates attended college. Is the principal justified in saying that the percentage of graduates of KHS attending college is significantly higher than 20 percent, at the 1 percent significance level? Discuss the probability of committing type II error for this example, and perform some calculations.

4. In the past, only 20 percent of people fitted with contact lenses could wear them without undue irritation. A new kind has been put on the market this year, and its manufacturer claims that 25 percent can wear his lenses without undue irritation. Does his claim appear to be justified if 275 out of 1000 people found no irritation in wearing these lenses? Use $\alpha = .08$. What is the probability of type II error?

5. Many years of experience with a university mathematics aptitude test has yielded a mean score of 500 and a variance of 100^2. The 11 students

from Wilmet averaged 570. Is this reason to believe that students from Wilmet perform better on this test? Use $\alpha = .12$. (You may assume that the variance of test scores for Wilmet students is 100^2.)

6. Driving north on the eight-lane Bayshore Freeway, an observer suspected that a disproportionate number of cars were driving in the (supposedly fast) left lane. The next day a count was made to see if there really were more than a fourth of the vehicles in the left lane. Of the 560 cars observed, 160 of them were in the left lane. Is this evidence of a disproportionate number of cars traveling in the left lane? Use $\alpha = .05$.

7. A random sample of 8 attractive 21-year-old fashion models from the world of haute couture in the Chelsea district of London showed that their mini-skirts rose 10.6, 11.3, 11.2, 10.7, 10.9, 12.0, 11.8, and 11.9 inches above the knee, while a random sample of 10 attractive 21-year-old fashion models in Manhattan showed lengths of 11.8, 12.2, 10.9, 11.8, 12.0, 11.2, 11.9, 10.8, 11.4, and 11.0 above the knee. Is this evidence, at a significance level of .05, that the average lengths above the knee differ in London and Manhattan?

8. A psychologist claims that students would perform better on tests if they did not feel the pressure of grades so much. To test his claim, the same test was given to 25 students; 15 were told it did not count and 10 were told that their grades depended upon doing well. The 15 had an $\bar{x} = 111$ and an $s^2 = 225$, whereas the 10 had an $\bar{x} = 103$ and an $s^2 = 280$. As a statistician, what do you think of his claim? Use $\alpha = .10$. You may assume that the variances are equal.

9. In an effort to cut costs and improve efficiency, the GAO (government accounting office) has decided to test a new brand of ballpoint pen, which, incidentally, costs the same as the old brand. The goal is, of course, to see if the new brand lasts longer than the old brand. The old brand lasts 9 hours, on the average. In a random sample of size 4, the new brand lasted 13, 9, 11, and 15 hours. As the GAO's statistician, what would you conclude using $\alpha = .10$?

10. (Continuation of Example 7) The geneticists also claim that men with XYY pattern are taller on the average. In fact, the 7 XYY-pattern inmates average 73 inches in height, whereas the other 190 inmates average 67 inches. Is this difference significant? Use $\alpha = .10$ and assume that the standard deviation of height for both groups is 3 inches.

11. (Continuation of Example 6) Elder was also interested in differences in IQ between middle-class and working-class girls. In fact, he observed the average IQ among the 35 middle-class and 43 working-class girls to be 112.8 and 107.2, respectively. Knowing that IQ scores are normally distributed with variance 225, would you conclude that there is no difference in IQ for working- and middle-class girls? Use $\alpha = .05$.

12. (Continuation) There appeared to be some evidence of differences in the adolescent heterosexual experience of upwardly mobile and nonmobile girls. In particular, data obtained retrospectively in the 1958 interview revealed that 16 out of the 23 nonmobile working-class girls had sexual intercourse during high school while only 2 out of the 13 upwardly mobile girls did. (Only 36 responses were obtained on this question.) Would you conclude, using $\alpha = .10$, that there was a difference in heterosexual experience?

13. Suppose that you are an executive of a large department store that has one outstanding competitor. The two stores have maintained a price balance in the past in that your store A has sold any one item at a higher price than B one-half of the time and at a lower price one-half of the time. However, you suspect that store B is deliberately destroying this balance and is consistently underselling your merchandise. In order to investigate this possibility, you decide to send a "shopper" into store B to check the prices of 250 randomly selected articles. On the basis of the number of these articles in which B undersells A, you wish to decide whether there is evidence of a significant deviation from past policy. Suppose store B undersells on 155 out of the 250 randomly selected articles. Do you accept or reject the null hypothesis at level of significance $\alpha = .05$?

    The level of significance is only one measure of the manner in which a statistical test performs. What is a second characteristic of the test procedure that should be considered? Make some calculations as regards this characteristic. (This exercise is due to Raymond Jessen.)

14. A large manufacturing concern wanted to know whether or not a new plant design would decrease the time it takes to make widgets; at present, it takes 73 seconds, on the average. To find out, they constructed a model of the new plant and simulated the manufacturing process on a computer. The simulated production time for each widget involved the sum of a large number of independent uniform random variables. What should they have concluded if they used $\alpha = .01$ and the 7 simulated widget production times were 62, 69, 61, 60, 58, 63, and 61 seconds?

15. John claims to be a better speller than his younger brother Bruce. When asked to back up this statement with facts, John replied that in the last 10 spelling bees Bruce missed the 1st, 7th, 11th, 3rd, 9th, 4th, 2nd, 5th, 1st, and 2nd words, whereas John missed the 4th, 6th, 11th, 18th, 7th, 3rd, 11th, 2nd, 3rd, and 9th words. Is this substantial proof of his being a better speller than his brother? Use $\alpha = .15$. [*Hint:* Recall Exercise 69 of Chapter 2 and use the central limit theorem.]

16. Test the hypothesis that the means of two independent normal populations differ by no more than 1 if the first population has variance 4, the second has variance 5, and samples of size 4 from the two populations are $(-4.4, 4.0, 2.0, -4.8)$ and $(6.0, 1.0, 3.2, -.6)$. Use $\alpha = .05$.

17. The army wishes to test a manufacturer's claim that, when properly aimed, 95 percent of its missiles land within a 3-mile radius of the target. The kind of missile currently in use lands within this radius only 90 percent of the time. To test this contention the army fired 5 missiles and observed distances of 2, $2\frac{1}{2}$, 1, $1\frac{1}{2}$, and $\frac{1}{2}$ miles from the target. What should the army statistician have concluded if he used $\alpha = .10$? [Of course, he knew that distance (D) from the target satisfies $D^2 = X^2 + Y^2$, where X and Y are the vertical and horizontal components. You may assume that these two components are each normally distributed with mean 0 and variance σ^2 and that they are independent random variables.]

6

CHI-SQUARE TESTS

1. GOODNESS-OF-FIT TESTS

We are often interested in knowing whether or not the particular probabilistic mechanism we have in mind is the appropriate mathematical model for a given random experiment. Consequently, we would like to test the null hypothesis that our probabilistic mechanism is the appropriate one against the alternative hypothesis that it is not. Such tests are called "goodness-of-fit" tests, and, fortunately, we can employ chi-square random variables and Table B in conjunction with a very simple procedure (described below) to construct "good" rejection regions for these tests. Naturally, the empirical evidence employed in performing the test consists of the outcomes obtained by actually performing the random experiment.

The first thing we do is to state (in words rather than in mathematical terms) the null hypothesis and the alternative hypothesis against which it is to be tested. For example, in rolling a die it is natural to suppose that the die is fair. Thus, the null hypothesis is "the die is fair" and the alternative is "the die is not fair." In the simulation experiment (see Section 3 of Chapter 3), our null hypothesis is that the rat has no memory; the alternative is the rat has some memory. As a final example, we might test that examination scores are normally distributed against the alternative that the distribution of examination scores is not a normal random variable but some other kind of random variable.

Second, we classify each possible outcome of the random experiment into one of k classes or *cells;* that is, we partition the sample space Ω of the random experiment into k events. Thus, if the random experiment is performed n times, the sample space $\mathcal{S}$ of the total experiment is taken to be the set of n-tuples $\mathbf{x}$, where each component of each n-tuple is one of the k numbers 1, 2, . . ., k. Often there is a natural way in which to partition Ω. For example, in the case of the die we partition the sample space Ω into $k = 6$ cells; namely, the number appearing on the die is a one, a two, a three, a four, a five, or a six.

Third, we restate the hypotheses in terms of the partition. For example, testing the die is fair against it is not fair is equivalent to testing

$$H_0\colon p_i = \tfrac{1}{6}, 1 \le i \le 6 \quad \text{against} \quad H_A\colon p_i \ne \tfrac{1}{6} \text{ for some } i, 1 \le i \le 6,$$

where p_i is the probability that the number appearing on the die is i. More generally we have

$$H_0\colon p_i = p_i', 1 \le i \le k \quad \text{and} \quad H_A\colon p_i \ne p_i' \text{ for some } i, 1 \le i \le k,$$

where p_i is the probability that the outcome of the random experiment is a member of cell i, and p_i' is this probability when the null hypothesis is true.

Finally, we construct a "good" rejection region. Suppose that we have performed the random experiment n times, and let N_i be the observed number of these n outcomes that were in cell i, $1 \le i \le k$. We shall use n_i to denote the actual value of the random variable N_i, and we refer to the n_i's as the *observed frequencies*. Of course, $e_i \equiv np_i'$ is the expected or theoretical number of outcomes in cell i when the null hypothesis is true. In fact, N_i is $B(n, p_i')$ when the null hypothesis is true as the n trials (experiments) are assumed to be independent. We refer to the e_i's as the *theoretical* or *expected frequencies*. Naturally, we accept H_0 if the observed frequencies and the expected frequencies are close to each other or similar, and we reject H_0 if they are not similar. We shall use the random variable X defined by

$$X = \sum_{i=1}^{k} \frac{(N_i - e_i)^2}{e_i}$$

as our criterion of similarity. (This criterion was first proposed by Karl Pearson.) Since dissimilar means that $N_i - e_i$ is not small for each i, and X is large when $N_i - e_i$ is not small for each i, we will reject H_0 when X is large and accept H_0 when X is small. Thus, we have

$$R_\alpha = \{\mathbf{x} \in \mathcal{S}\colon x \ge c\}.$$

All that remains for us to do is to determine the number c. When n is large and H_0 is true, X has approximately the chi-square distribution with $k - m$ degrees of freedom, where m is (usually) the number of quantities, determined from the observed data, that are used in calculating the expected frequencies.

Determining m is not always easy, but we show how to do so in the next three examples, where the entire testing procedure is also illustrated. Of course, we can now determine c from the following equation:

$$P(\chi^2_{k-m} \geq c) = \alpha.$$

As a rule of thumb, when performing a goodness-of-fit test, the partitioning should be done so that (1) $e_i \geq 1$ for each i, $1 \leq i \leq k$, and (2) at most 20 percent of the e_i's are smaller than 5.

EXAMPLE 1. (Die)

In order to determine whether or not a die is fair, we partition the sample space Ω of the random experiment "rolling a die" into $k = 6$ parts. We say that the outcome is a member of cell i if an i appears when the die is rolled, $1 \leq i \leq 6$. Thus, we want to test[1]

$$H_0\colon p_i = \tfrac{1}{6},\ 1 \leq i \leq 6 \quad \text{against} \quad H_A\colon p_i \neq \tfrac{1}{6} \text{ for some } i,\ 1 \leq i \leq 6,$$

where p_i is the probability that the number appearing on the die is i, $1 \leq i \leq 6$. The die was rolled $n = 150$ times and the number of ones, twos, threes, fours, fives, and sixes that was observed was 25, 28, 26, 29, 24, and 18, respectively; that is, $n_1 = 25$, $n_2 = 28$, $n_3 = 26$, $n_4 = 29$, $n_5 = 24$, and $n_6 = 18$. Since $p_i' = \frac{1}{6}$, $e_i - 150(\frac{1}{6}) = 25$ for each i. Consequently, we obtain the value 3.04 for the random variable X. In this example X is chi-square with 5 degrees of freedom, since $k = 6$ and $m = 1$. The total number n $(= 150)$ of observations was the only quantity obtained from the observed data that we used in calculating the expected frequencies, so we see from the rule above that $m = 1$. Using Table B and $\alpha = .05$ we see that $R_\alpha = \{x\colon x \geq 11.2\}$. Hence, we accept H_0. •

EXAMPLE 2. (Rat in a Maze)

Using the results of the simulation experiment of Chapter 3, decide whether or not the rat running the maze had any memory. Use $\alpha = .05$.

[1] This formulation is, in fact, consistent with that we have been using. This is seen by letting $\mathfrak{F} = \{f_{\mathbf{X}}^{\theta}\colon \theta \in \Theta,\ n \in \mathfrak{N}\}$, where

$$\Theta = \left\{\theta\colon \theta = (p_1, p_2, \ldots, p_6),\ p_i \geq 0,\ 1 \leq i \leq 6, \text{ and } \sum_{i=1}^{6} p_i = 1\right\},$$

$$f_{\mathbf{X}}^{\theta}(\mathbf{x}) = \prod_{i=1}^{n} f_{X_i}^{\theta}(x_i) \quad \text{and} \quad f_{X_i}^{\theta}(x_i) = p_j \quad \text{if } x_i = j,\ j = 1, 2, \ldots, 6.$$

Thus, we have $H_0\colon \theta = \theta_0$ and $H_A\colon \theta \in \Theta \sim \{\theta_0\}$, where $\theta_0 = (\frac{1}{6}, \frac{1}{6}, \frac{1}{6}, \frac{1}{6}, \frac{1}{6}, \frac{1}{6})$.

SOLUTION: The null hypothesis is that the rat has no memory, whereas the alternative is that he does have some memory. One's first thought is to partition the outcomes of the random experiment (running the rat through the maze) according to the number D of decisions he makes until he reaches the cheese. We must, however, follow the rule of thumb given above for partitioning Ω. Since $n = 100$, we see (using footnote 12 of Chapter 3) that the expected frequency is less than 5 for each value of D greater than 9. Consequently, we will partition the outcomes into $k = 8$ cells. Those outcomes for which $D = i + 1$ will be said to be in cell i, $i = 1, 2, \ldots, 7$, and those outcomes for which $D \geq 9$ will be said to be in cell 8. Hence, the null hypothesis can be restated as follows:

$$H_0\colon p_1 = \tfrac{1}{4},\ p_2 = \tfrac{1}{8},\ p_3 = \tfrac{2}{16},\ p_4 = \tfrac{3}{32},\ p_5 = \tfrac{5}{64},\ p_6 = \tfrac{8}{128},\ p_7 = \tfrac{13}{256},\ p_8 = \tfrac{63}{256}.$$

It now follows that $e_i = np'_i = 100p'_i$, so

$$e_1 = \tfrac{6400}{256},\quad e_2 = \tfrac{3200}{256},\quad e_3 = \tfrac{3200}{256},\quad e_4 = \tfrac{2400}{256},$$
$$e_5 = \tfrac{2000}{256},\quad e_6 = \tfrac{1600}{256},\quad e_7 = \tfrac{1300}{256},\quad e_8 = \tfrac{6300}{256}.$$

Table 2 of Chapter 3 reveals that $n_1 = 26$, $n_2 = 7$, $n_3 = 19$, $n_4 = 3$, $n_5 = 12$, $n_6 = 9$, $n_7 = 2$, and $n_8 = 22$, so the actual value of X is 15.62. The same reasoning employed in Example 1 shows that $m = 1$, so X is χ^2_7. Thus, Table B shows that $R_\alpha = \{x\colon x \geq 14.1\}$, so we reject H_0, since the observed value of X is greater than 14.1. •

EXAMPLE 3

The 47 scores listed below are those obtained by students on a final examination. Since the scores represent the total number of points received by each student on the many test problems, one might conjecture that the scores are normally distributed. Test this conjecture using $\alpha = .10$.

241	217	205	192	167	152	151
111	110	105	96	94	91	89
74	51	44	30			
229	175	157	151	149	148	147
141	141	138	132	130	119	119
118	117	117	116	95	86	78
72	66	61	53	53	48	36
16						

SOLUTION: We want to test the null hypothesis "the scores have a normal distribution" against the alternative hypothesis that they do not. In order to find the expected frequencies for a given partition, we need to know n, μ, and σ^2. Since we do not know μ and σ^2, we will estimate them by $\bar{X}$ and S^2. Using the data, we find $\bar{x} = 115.5$ and $s = 53$. Let us then partition the scores into $k = 10$ cells as follows: a score is said to be in the first, second, ..., or tenth cell if it is in the interval[2]

$$\begin{aligned}
&(-\infty, \bar{x} - 2s],\ (\bar{x} - 2s, \bar{x} - s],\ (\bar{x} - s, \bar{x} - .6s],\\
&(\bar{x} - .6s, \bar{x} - .3s],\ (\bar{x} - .3s, \bar{x}],\ (\bar{x}, \bar{x} + .3s],\\
&(\bar{x} + .3s, \bar{x} + .6s],\ (\bar{x} + .6s, x + s],\\
&(\bar{x} + s, \bar{x} + 2s], \quad \text{or} \quad (\bar{x} + 2s, \infty), \text{ respectively.}
\end{aligned}$$

From Table A we see that if we actually have $\mu = \bar{x}$ and $\sigma = s$, then the null hypothesis can be restated as follows:

$$\begin{aligned}
H_0\colon\ &p_1 = p_{10} = .0228,\ p_2 = p_9 = .1359,\ p_3 = p_8 = .1156,\\
&p_4 = p_7 = .1078,\ p_5 = p_6 = .1179.
\end{aligned}$$

It follows that the expected frequencies are all greater than one and that only 20 percent are less than five, as stated in our guideline for partitioning. From the data we see that the observed frequencies are $n_1 = 0$, $n_2 = 7$, $n_3 = 4$, $n_4 = 6$, $n_5 = 3$, $n_6 = 7$, $n_7 = 5$, $n_8 = 7$, $n_9 = 3$, and $n_{10} = 3$, so that the observed value of the random variable X is 9.02. In calculating the expected frequencies we used n, $\bar{x}$, and s, so $m = 3$ and $k - m = 7$. Thus, X is chi-square random variable with 7 degrees of freedom, and hence $R_{.10} = \{x\colon x \geq 12\}$. Since $x < 12$, we accept H_0. •

EXAMPLE 4. (Draft Lottery)

On December 1, 1969, the Selective Service people placed 366 allegedly identical blue capsules into an urn. Each capsule contained a slip of paper with one of the days of the year written on it. Then the capsules were withdrawn one by one from the urn. The order of the days drawn determined the vulnerability to the draft of men who were between 19 and 26 years old as of January 1, 1970. Moreover, it was announced that the men whose birthdays were among the first 120 days drawn were almost certain to be drafted.

As it turned out, the number of days from the 12 months of the year among the first 120 days drawn by the Selective Service people was, starting with January:

8 7 4 8 9 11 12 13 10 9 12 17.

Use these data and $\alpha = .05$ to test the null hypothesis that the drawing was random.

[2] The interval $(a, b]$ is the set of those numbers x such that $a < x \leq b$.

SOLUTION: Naturally, we partition the outcome of each draw into $k = 12$ cells corresponding to the 12 months of the year. Realizing that January has 31 days, February has 29 (including February 29th), and so on, we see that the null hypothesis is

$$H_0: p_1 = \tfrac{31}{366}, p_2 = \tfrac{29}{366}, p_3 = \tfrac{31}{366}, p_4 = \tfrac{30}{366}, p_5 = \tfrac{31}{366}, p_6 = \tfrac{30}{366},$$
$$p_7 = \tfrac{31}{366}, p_8 = \tfrac{31}{366}, p_9 = \tfrac{30}{366}, p_{10} = \tfrac{31}{366}, p_{11} = \tfrac{30}{366}, p_{12} = \tfrac{31}{366}.$$

We continue to use

$$X \equiv \sum_{i=1}^{12} \frac{(N_i - e_i)^2}{e_i}$$

to measure the "goodness-of-fit" of the empirical evidence to the expected frequencies. Here, $n = 120$ so $e_i = 120p_i'$ and $m = 1$. The observed value of X is found to be $x = 11.949$. Of course, we will reject the null hypothesis if x is too large. In this example, however, it is not true that X is χ^2_{11}. The reason for this is that the 120 drawings are done without replacement so they are not independent. [N_i is $H(120, 366p_i', 366)$.]

As before, the random variable X is a reasonable measure of the discrepancy between the observed and the expected outcomes, but we now must find the distribution of X under the null hypothesis. As is typical of many such problems, it is next to impossible to actually find the distribution of X. Consequently, we must simulate it (see Example 11 of Chapter 3).

In 399 simulated draws from the distribution of X, we found that 42 of the values of X obtained exceeded 11.949. Thus, $P_{H_0}(X \geq 11.949)$ is close to $.1075 > \alpha$, so we must accept the null hypothesis that the drawing was random. •

The use of simulation is an exceedingly important method in hypothesis testing.

EXERCISES:

1. If a coin were flipped 1000 times and it landed heads only 420 times, would you still believe the coin to be fair? Use $\alpha = .05$. How would your *procedure* be different if the coin were flipped but 5 times and 2 heads appeared?

2. A close inspection of the table of random digits found in Chapter 3 reveals that there are 108, 106, 88, 92, 101, 81, 103, 95, 108 and 118 0's, 1's, 2's, . . . , and 9's, respectively. Do you still think that this table has the properties it is supposed to have? Use $\alpha = .05$. (Also see Exercise 34 of Chapter 9.)

3. A certain tavern orders 4 times as much tap beer (by volume) as it orders bottled beer. What would you say about the ordering policy if last week 436 mugs of tap beer and 64 bottles of beer were sold? Use $\alpha = .10$ and assume that a mug and a bottle contain equal volumes of beer.

4. It is an unwritten rule of the university that in upper-division courses approximately 20, 30, 40, and 10 percent of the students should receive course grades of A, B, C, or below C, respectively. Do you think the instructor was following the university guideline if he gave 18 A's, 7 B's, 15 C's, and 10 grades lower than C to his class of 50 students? Use $\alpha = .05$.

5. According to a genetic model (proposed by Mendel), offspring of certain crossings should fall into three categories in the ratios of 9:3:4. If an experiment gave 72, 35, and 38 offspring in these categories, would you feel this model is appropriate? Use $\alpha = .05$.

6. One basketball fan claims that guards, centers, and forwards do not, on the average, score the same number of field goals. On one particular team, the two guards, the two forwards, and the center scored 140, 190, and 95 field goals over the entire season, respectively. Test his contention using $\alpha = .025$.

7. Come to a conclusion in Exercise 6 of Chapter 5 using $\alpha = .10$. [*Hint:* Use $k = 2$.]

8. Test the null hypothesis that the test scores given in Example 3 are uniformly distributed. Use $k = 9$, intervals of length 25, and $\alpha = .05$. [*Hint:* Notice that $m = 3$. Why?]

9. Suppose it were known in Example 3 that $\mu = \bar{x}$. How would this change the testing procedure? What if it were also known that $\sigma = s$?

10. Suppose that n is the only quantity used in computing the expected frequencies and that $H_0: p_1 = p_1'$, $p_2 = 1 - p_1'$, so that $k = 2$.
   (a) Show that $X = (N_1 - np_1')^2/[np_1'(1 - p_1')]$.
   (b) Use part (a) and the central limit theorem to show that

$$Y \equiv \frac{N_1 - np_1'}{\sqrt{np_1'(1 - p_1')}}$$

   is (approximately) normally distributed with mean 0 and variance 1 when n is large, and, hence, X is χ_1^2. [*Hint:* N_1 is $B(n, p_1')$.]

11. Show that $E(X) = k - 1$, where n is the only quantity used in computing the expected frequencies. [*Hint:* N_i is $B(n, p_i')$ for each i.]

12. In Example 2, we rejected the null hypothesis that the rat has no memory. But the evidence used in reaching this conclusion was generated by our

table of random digits in a manner that is in agreement with the null hypothesis. Consequently, we should have accepted the null hypothesis. What has "gone wrong" here?

2. TESTS OF INDEPENDENCE IN TWO-WAY CONTINGENCY TABLES

In this section we consider problems in which our observations are classified according to two characteristics. For example, the girls at a certain sorority might be classified according to hair color and eye color (see Table 6.1) or according to year in college and number of dates per week (see Table 6.2).

Table 6.1 is called a 3-by-2 contingency table, as there are 3 categories within the first characteristic and 2 categories within the second charac-

Table 6.1

	Eye color	
Hair color	*Blue*	*Brown*
Blonde	9	4
Brunette	11	23
Red	2	1

Table 6.2

	Number of dates per week		
Year	0 *or* 1	2	*More than* 2
Freshman	3	9	3
Sophomore	3	7	4
Junior	3	4	4
Senior	4	1	5

teristic. Each distinct classification is called a cell, and the classification corresponding to the ith row and jth column is called cell i, j. Thus, Table 6.1 has 6 cells and there is 1 observation in cell 3, 2. Similarly, Table 6.2 is a 4-by-3 contingency table with 12 cells, and there are 7 observations in cell 2, 2.

In general, we shall suppose that there are n individuals classified according to two characteristics R and C and that there are r categories, labeled $R_1, R_2, \ldots, R_r$, in R, and c categories, labeled $C_1, C_3, \ldots, C_c$, in C. We will denote by n_{ij} the number of individuals observed to be in both categories R_i and C_j. Thus, an *r-by-c contingency table* appears as follows:

R	C			
	C_1	C_2	...	C_c
R_1	n_{11}	n_{12}	...	n_{1c}
R_2	n_{21}	n_{22}	...	n_{2c}
. . .				
R_r	n_{r1}	n_{r2}	...	n_{rc}

We wish to test the null hypothesis that the two classifications are independent that the probability that an individual belongs to the category C_j is not affected by the R category to which the individual belongs. In other words, the null hypothesis states that

$$P(R_i \cap C_j) = P(R_i)P(C_j), \qquad \text{for } i = 1, 2, \ldots, r \text{ and } j = 1, 2, \ldots, c.$$

Denoting $P(R_i \cap C_j)$ by p_{ij}, $P(R_i)$ by p_i, and $P(C_j)$ by q_j, we have

$$H_0\colon \; p_{ij} = p_i q_j, \; i = 1, 2, \ldots, r, \; j = 1, 2, \ldots, c, \; \sum_{i=1}^{r} p_i = \sum_{j=1}^{c} q_j = 1.$$

The alternative hypothesis states that the two classifications are not independent—that is,

$$H_A\colon \; p_{ij} \neq p_i q_j \quad \text{for some pair } i, j.$$

Naturally we accept H_0 if the observed frequencies n_{ij} are close to the expected frequencies e_{ij}. We will use the random variable X defined by

$$X = \sum_{i=1}^{r} \sum_{j=1}^{c} \frac{(N_{ij} - e_{ij})^2}{e_{ij}}$$

as our criterion of closeness, where the random variable N_{ij} is the number of individuals in cell i, j. Since $e_{ij} = np_i q_j$ and we do not know p_i and q_j, we must estimate them. Natural estimates—that is, "guesses"—of p_i and q_j are

$$\hat{p}_i = \frac{\sum_{j=1}^{c} n_{ij}}{n} \quad \text{and} \quad \hat{q}_j = \frac{\sum_{i=1}^{r} n_{ij}}{n}.$$

Thus, $e_{ij} = n\hat{p}_i\hat{q}_j$. When n is relatively large ($n \geq 10rc$ is usually sufficiently large), X is approximately chi-square with $(r - 1)(c - 1)$ degrees of freedom, so that $R_\alpha = \{\mathbf{x} \in \mathcal{S}: x \geq s\}$, where $P(\chi^2_{(r-1)(c-1)} \geq s) = \alpha$.

EXAMPLE 5

Using $\alpha = .10$ and the information contained in Table 6.1, test whether or not hair and eye color are independent.

SOLUTION: We have

$$H_0: p_{ij} = p_i q_j, \; i = 1, 2, 3, \; j = 1, 2.$$

Since $n = 50$, we have

$$\hat{p}_1 = \tfrac{13}{50}, \quad \hat{p}_2 = \tfrac{34}{50}, \quad \hat{p}_3 = \tfrac{3}{50}, \quad \hat{q}_1 = \tfrac{22}{50}, \quad \hat{q}_2 = \tfrac{28}{50}.$$

Thus,

$$e_{11} = \tfrac{13 \cdot 22}{50}, \; e_{12} = \tfrac{13 \cdot 28}{50}, \; e_{21} = \tfrac{34 \cdot 22}{50}, \; e_{22} = \tfrac{34 \cdot 28}{50}, \; e_{31} = \tfrac{3 \cdot 22}{50}, \; e_{32} = \tfrac{3 \cdot 28}{50},$$

so

$$x = \frac{(9 - 5.72)^2}{5.72} + \frac{(4 - 7.28)^2}{7.28} + \frac{(11 - 14.96)^2}{14.96} + \frac{(23 - 19.04)^2}{19.04} + \frac{(2 - 1.32)^2}{1.32} + \frac{(1 - 1.68)^2}{1.68}$$

$$= 6.48.$$

As $r = 3$ and $c = 2$, $(r - 1)(c - 1) = 2$, so we see from Table B that $R_{.10} = \{\mathbf{x}: x \geq 4.6\}$; consequently, we reject H_0. •

EXAMPLE 6. (Visibility of Auditor's Disclosure)

The purpose of the study "The Visibility of the Auditor's Disclosure of Deviation from APB Opinion" by Purdy, Smith, and Gray[3] was "to provide one piece of empirical evidence impinging upon the broad question of what constitutes adequate disclosure in published financial statements." More specifically, one of the questions they asked was whether the disclosure of a deviation from standard accounting practices will be equally visible to the readers of financial statements if the disclosure is placed in the auditor's report or in a footnote to the financial statements. Rather than give the entire elaborate procedure they devised for testing the null hypothesis of no difference in visibility against the alternative hypothesis of some difference in visibility, we have distilled the pertinent information. An annual report was given to 133 subjects. Of these, 44 were chosen at random, and the annual report presented to this group contained the auditor's disclosure in a footnote, whereas the annual report presented to the other 89 subjects contained the auditor's disclosure in the financial statements. After having read the annual report, the subjects were asked a series of questions of which 6 related to the auditor's disclosure.

Using the data in Table 6.3, test the null hypothesis of "no difference in visibility." Use $\alpha = .10$.

Table 6.3

	Number of correct answers					
Warning placed in	1	2	3	4	5	6
Footnote	3	3	4	10	18	6
Auditor's Report	1	10	16	20	28	14

SOLUTION: We wish to test H_0: $p_{ij} = p_i q_j$, $i = 1, 2$, $j = 1, 2, \ldots, 6$. Since $n = 143$, we have

$$\hat{p}_1 = \tfrac{44}{133},\quad \hat{p}_2 = \tfrac{89}{133},\quad \hat{q}_1 = \tfrac{4}{133},\quad \hat{q}_2 = \tfrac{13}{133},$$
$$\hat{q}_3 = \tfrac{20}{133},\quad \hat{q}_4 = \tfrac{30}{133},\quad \hat{q}_5 = \tfrac{46}{133},\quad \hat{q}_6 = \tfrac{20}{133}.$$

[3] Purdy, Smith and Gray, "The Visibility of the Auditor's Disclosure of Deviation from APB Opinion," *Empirical Research in Accounting: Selected Studies*, 1969, *Journal of Accounting Research*.

Thus,

$$e_{11} = \frac{44\cdot 4}{133}, \quad e_{12} = \frac{44\cdot 13}{133}, \quad e_{13} = \frac{44\cdot 20}{133}, \quad e_{14} = \frac{44\cdot 30}{133},$$

$$e_{15} = \frac{44\cdot 46}{133}, \quad e_{16} = \frac{44\cdot 20}{133}, \quad e_{21} = \frac{89\cdot 4}{133}, \quad e_{22} = \frac{89\cdot 13}{133},$$

$$e_{13} = \frac{89\cdot 20}{133}, \quad e_{14} = \frac{89\cdot 30}{133}, \quad e_{15} = \frac{89\cdot 46}{133}, \quad e_{16} = \frac{89\cdot 20}{133},$$

so

$$\begin{aligned} x = &\frac{(3-1.32)^2}{1.32} + \frac{(3-4.30)^2}{4.30} \\ &+ \frac{(4-6.62)^2}{6.62} + \frac{(10-9.92)^2}{9.92} + \frac{(18-15.22)^2}{15.22} + \frac{(6-6.62)^2}{6.62} \\ &+ \frac{(1-2.68)^2}{2.68} + \frac{(10-8.70)^2}{8.70} + \frac{(16-13.38)^2}{13.38} \\ &+ \frac{(20-20.08)^2}{20.08} + \frac{(28-30.78)^2}{30.78} + \frac{(14-13.38)^2}{13.38} \\ = &\ 6.174. \end{aligned}$$

As $r = 2$ and $c = 6$, $(r-1)(c-1) = 5$, so we see from Table B that $R_{.10} = \{\mathbf{x} \in \mathcal{S}: x \geq 9.24\}$; consequently, we accept H_0. •

EXAMPLE 7. (Tenure)

A 2-by-2 contingency table may also be used to test the null hypothesis that two proportions are equal, as in Case 5 of Chapter 5. Consider Example 8 of Chapter 5. In this case we have H_0: $p_{ij} = p_i q_j$, $i = 1, 2$, $j = 1, 2$, and Table 6.4 is the appropriate 2-by-2 contingency table. •

Table 6.4

	Status	
Year	*Resign*	*Remain*
1966–1967	14	757
1967–1968	22	772

EXAMPLE 8. (Voter Perception)

In mid-October 1960, shortly after the famed TV debates in which the candidates John F. Kennedy and Richard M. Nixon appeared side by side, a marketing research organization conducted face-to-face interviews of 3018 registered voters in California. Among the many questions posed were "If the presidential elections were being held today, which one of these men would you vote for—Nixon or Kennedy?" and "In your opinion who is taller—Nixon or Kennedy?" The results of the survey are given in Table 6.5. Kassarjian[4] hypothesized that a voter's preference would tend to prejudice his perception of the candidates' relative height. In particular, Nixon voters would perceive Nixon as taller, and Kennedy voters would perceive Kennedy as taller.

Using a chi-square test and ignoring the responses of the respondents who answered "don't know" to either question, Kassarjian found the observed value of χ_1^2 to be 132.93. This value was large enough to reject the null hypothesis of no relationship between political beliefs and perceived relative height of the candidates in favor of the alternative hypothesis of a positive relationship. It is interesting to note that phenomena such as the one in Kassarjian's study are encountered frequently. (For further details and references see Kassarjian, p. 85.) •

Table 6.5

	Respondent intending to vote for		
	Nixon	*Kennedy*	*Don't know*
Size of sample	1505	1325	188
Percent saying the taller man is			
Nixon	42.9	23.4	29.9
Kennedy	47.3	68.1	46.8
Don't Know	9.8	8.5	23.9

The example above was chosen to illustrate a major pitfall in the use of χ^2 tests for goodness-of-fit and for independence.[5] Note that the observed value of χ_1^2 was exceedingly large. Thus, we can reject the null hypothesis, even at extremely small values of α.[6] That is, we can wholeheartedly say our conclusion is *statistically significant*. This is not, however, the same as saying that our conclusion has *practical significance*. On the one

[4] Harold H. Kassarjian, "Voting Intentions and Political Perceptions," *The Journal of Psychology*, Vol. 56, pp. 85–88 (1963).

[5] The author is indebted to Professor Harold H. Kassarjian for having suggested this example.

[6] If Z is $N(0, 1)$, then Z^2 is χ_1^2, so (using Mill's ratio)

$$\begin{aligned} P(\chi_1^2 \geq 132.93) &= P(Z^2 \geq 132.93) = 2P(Z \geq \sqrt{132.93}) \\ &= 2P(Z \geq 11.53) \approx 2\left[\frac{1}{\sqrt{2\pi}} e^{-(11.53)^2/2}\right] \frac{1}{11.53} < \left(\frac{1}{10}\right)^{30}. \end{aligned}$$

hand statistical significance tells us that such differences as were observed in the proportions in the first two columns would be most unlikely[7] to occur if in the population [of voters] there were, in fact, no differences whatsoever [in the proportions in the two columns]. But on the other hand, statistical significance tells us nothing at all about the *magnitude* of these differences. Thus, with a very large sample such as the one used in the voter survey, we are likely to detect differences that are real but minute and unimportant. The point is, we are not really interested in knowing whether there is a difference. Instead, we are interested in knowing whether this difference is large enough to be of practical importance.

EXERCISES

13. Use $\alpha = .10$ and Table 6.2 to determine if there is any relationship between the number of dates per week a sorority girl has and her year in school.

14. A poll of 300 students showed that 15 out of 100 humanities, 14 out of 100 science, and 8 out of 100 social science students wear sandals. Does this constitute evidence of difference in dress of these three groups? Use $\alpha = .10$.

15. A serum supposed to have some effect in preventing colds was tested on 500 individuals, and that year 252 were found to have no colds, 144 one cold, and 104 more than one cold. Among 500 people not being treated that year we found 224 with no colds, 136 with one cold, and 140 with more than one cold. Test at $\alpha = .05$ whether or not the serum has any effect on preventing colds.

16. On the basis of Table 6.2, do you think there is a significant difference in the dating habits of freshman and senior girls? Use $\alpha = .01$.

17. Solve Example 7 using $\alpha = .05$.

18. To find out whether a new film had equal appeal for different age groups, a random sample was taken at a sneak preview: 73 out of 100 persons under 21 liked it, 43 out of 65 persons between 21 and 25 liked it, and 43 out of 150 persons over 25 liked it; all the others didn't like it. What should be concluded from this sample? Use $\alpha = .01$.

19. The results of a May 1969 student referendum on the question of ROTC at UCLA were as follows:

	Undergraduates	*Graduates*
Yes—with academic credit	3217	653
Yes—with no academic credit	2416	602
No—under any circumstances	1882	665
Don't know	68	15

[7] See footnote 6.

Does it appear that undergraduate and graduate students at UCLA hold the same attitude towards ROTC? Use $\alpha = .10$.

20. The following information has been taken from the Annual Report of the Committee on Athletic Policy, 1968–69, Los Angeles Division, Academic Senate, University of California.

"Do athletes select an undue proportion of "snap" or "Mickey Mouse" courses?

The Committee did notice the recurrence with unusual frequency of certain courses and also of languages chosen by athletes in its perusing of their academic records. Some of these courses obviously are easy to get good grades in or at least to get passing grades in. We discounted in this study courses in Physical Education in which athletes might be expected to earn good grades.

Although we did find evidence of "Mickey" courses, two points should be made. (1) Athletes in different sports tended to choose different "Mickey" courses which suggests that the information about the easy courses gets around by word of mouth among the students rather than by dissemination by the Athletic Department. In any event, athletes are not unique in choosing easy courses. (2) We found no evidence for preferential grading of students. As a case in point, we present data on one course we identified as a "Mickey." For obvious reasons, we decline to identify the course itself.

Course "Mickey Mouse 999"

	Number receiving grade				
Sport	*A*	*B*	*C*	*P*	*Total*
Football	35	2	1	—	38 (38% of team)
Swimming	4	—	—	—	4
Tennis	1	—	—	—	1
Volleyball	—	1	—	—	1
Baseball	4	—	—	1	5
Basketball	5	1	1	—	7
Crew	1	—	—	—	1
Total	50	4	2	1	57

Of 56 students taking the course for a grade, 50, or 89%, received a grade of A. For comparison, we obtained the grade distribution of *all* students taking the course during the past two years. The data follow:

Number receiving:	*A*	*B*	*C*	*P*	*Total*
	99	14	3	11	127

Of 116 taking "Mickey Mouse 999" for a grade, 99, or 85%, received A's."

Do you agree with the Committee's statement "We found no evidence for preferential grading of (athletic) students"? Use $\alpha = .05$. (*Note:* Perhaps the most important part of this problem is deciding which of the many numbers are relevant and which are irrelevant!)

21. A young psychiatrist suggested to his doctor friends that half of all headaches are psychosomatic. To verify his claim the doctor told 50 patients complaining of headaches that he had a new pill that might help. He administered aspirin—the usual headache remedy—to 25 of the patients and a placebo (water, flour, and sugar) to the others. The next day he obtained the following responses from the aspirin group: better than aspirin, 8; the same as aspirin, 13; worthless, 4. From the placebo group there were 10, 9, and 6 in each of the three categories. Does this evidence lend credence to his theory? Use $\alpha = .10$.

22. Show that H_0 in Case 5 of Chapter 5 is the same as that for a 2-by-2 contingency table.

23. Show that for a 2-by-2 contingency table the value of X is the square of $Z \equiv (\bar{X} - \bar{Y})/\hat{\sigma}_{\bar{X}-\bar{Y}}$ in Case 5 of Chapter 5, so that the two methods for testing the equality of proportions are exactly the same.

7

ESTIMATION

The art of estimation
Comes with the revelation
That perfect information
Would require application
And result in frustration
Far beyond consideration,

And that equal compensation
Comes with due deliberation
And careful consideration
Of a randomly sampled population
(An unbiased illustration)
Of what's under observation.

1. INTRODUCTION

The use of the term *estimation* in statistics is quite similar to its everyday (nontechnical) usage. While the doctor might be interested in estimating how long it will take his heart patient to recover or the city planner in estimating how large the city's population will be next year, the statistician is interested in estimating one of the parameters in his probability model. For example, the statistician might be interested in estimating the average height of males in the United States, the total number of fish in the lake, or the probability that it will be the youngest of the three sisters who breaks the next dish.

If it were physically possible and economically feasible for the statistician to examine the entire population being studied, then the true values of the parameters could be precisely determined—and known with certainty. Hence, there would be no need to make estimates. As the poem indicates, however, it is seldom possible or economically feasible to examine the entire population. Consequently, as in all problems of statistical inference, we as statisticians must base our conclusions on the information contained in a sample from the population.

When making estimates, the doctor and the city planner, like the statistician, make use of two types of information: (1) the data specific to this particular heart patient or to this particular city, and (2) the theory, such as physiology or urban land economics, generic to all heart patients or to all cities. For the statistician, the specific information corresponds to the empirical evidence or sample he obtains, whereas the generic information corresponds to his probabilistic mechanism.

Conceptually, the problem of estimation is quite closely related to that of hypothesis testing. The main problem of hypothesis testing can be stated roughly as follows: Given the probabilistic mechanism and given the empirical evidence that has been gathered, determine in which of two disjoint sets the true value of the unknown parameter lies. The problem of estimation differs in that we seek to determine exactly the true value of the unknown parameter rather than in which of two disjoint sets it lies. For example, suppose we are interested in knowing μ, the average height of males in the United States. In hypothesis testing we would pose and answer the question of whether or not this average differs from, say 5 ft 8 in. In estimation our job is to find a single number that represents our best guess of the true value of μ.

At first glance, it might appear that estimation is descriptive, in contrast to the decision-making orientation of hypothesis testing. Admittedly, the greatest use of estimation may be for description; however, it is frequently used in decision making.

In developing the theory of estimation, we shall proceed along the lines of Chapter 4. In Sections 2, 3, and 4 we define what we mean by an estimator, list some desirable properties of estimators (that is, define a good estimator), and describe how to obtain a good estimator. Interval estimation is the topic of Section 5.

EXERCISES:

1. A number of expensive electronic components of a certain kind have been produced. One use for these components is in a million-dollar rocket. For all intents and purposes, the rocket malfunctions if and only if this component

does. If the likelihood of this component's malfunctioning is too high, then a different kind of component should be used. Thus, we are interested in knowing this likelihood. Give two reasons why we must be content with obtaining only an estimate of this likelihood.

2. A government official was charged with the responsibility of verifying a manufacturer's claim that his light bulbs have a mean lifetime of 1000 hours. Briefly discuss some of the difficulties the official might encounter. In particular, why can't he precisely determine the mean lifetime?

3. Reformulate Examples 1, 2, and 3 of Chapter 4 so that the problems posed are of estimation and not of hypothesis testing.

2. ESTIMATORS

The introduction mentioned two types of information that we use in making estimates. The first type is the empirical evidence or the sample from the population under study. Naturally, we want our sample to be representative of the population. Roughly speaking, this will be attained if each member of the population is as likely to be in our sample as any other member. That is to say, we want our sample to be a random sample. Using slightly different words, we now restate the definition of a random sample given on page 110.

DEFINITION

Let the random variables $X_1, X_2, \ldots, X_n$ have the joint distribution $g_{\mathbf{X}}$ given by

$$g_{\mathbf{X}}(\mathbf{x}) = \prod_{i=1}^{n} f_X(x_i),$$

where $\mathbf{X} = (X_1, X_2, \ldots, X_n)$, $\mathbf{x} = (x_1, x_2, \ldots, x_n)$, and the distribution of each X_i is f_X. Then $\mathbf{X}$ is said to be a *random sample of size n* from the population with distribution f_X. We refer to $\mathbf{X}$ as the *observable random variables*, while we refer to $\mathbf{x}$ (the observed values of $\mathbf{X}$) as the *empirical evidence*.

Sometimes the population from which we are sampling is finite, and we must draw our sample without replacement. In this case, a representative sample of size n will be obtained only if each set of n members of the population is as likely to constitute our sample as any other set of n members. This idea is expressed more formally in the following definition.

DEFINITION

Let $a_1, a_2, \ldots, a_N$ be a population with N members, and let X_i be the ith member drawn. Then $\mathbf{X} = (X_1, X_2, \ldots, X_N)$ is said to be a *random sample of size n without replacement* from our population of size N[1] if

$$P(\mathbf{X} = \mathbf{x}) = \frac{1}{N!/(N-n)!}$$

for each of the $N!/(N-n)!$ ordered samples of size n without replacement.

EXAMPLE 1

Let 1, 2, . . . , 10 be the 10 members of our population. If

$$P(X_1 = i, X_2 = j, X_3 = k) = \frac{1}{10!/7!} = \frac{1}{10 \cdot 9 \cdot 8} \qquad \text{for } i \neq j \neq k \neq i \text{ and } 1 \leq i, j, k \leq 10,$$

then $\mathbf{X} = (X_1, X_2, X_3)$ is a random sample of size 3 without replacement from our population of size 10. •

The second type of information we use in making estimates is the probabilistic mechanism. We repeat the definition given in Chapter 4.

DEFINITION

A *probabilistic mechanism* simply (1) specifies the (functional) form of the distribution of each of the random variables in which we are interested; (2) specifies the interdependence relationships between these random variables—that is, specifies their joint distribution; and (3) specifies the set Θ of possible values that the unknown parameter(s) can assume.

We now illustrate the definition of a probabilistic mechanism with a few familiar examples from Chapter 4.

EXAMPLE 2. (Three Sisters)

In the case of the three sisters, mother was interested in knowing the probability θ that the youngest of the three sisters would be responsible for breaking the next dish. In Chapter 4 we decided it was reasonable to assume that there is a probability θ that she breaks the ith dish that is the same for each i—no matter which of the first $i - 1$ dishes she broke.

[1] Although the random variables $X_1, X_2, \ldots, X_n$ are not independent, they each have the same distribution, namely, $P(X_i = a_j) = 1/N$ for $j = 1, 2, \ldots, N$. Also, we remark that while it is possible to sample with replacement from a finite population, this should not ordinarily be done (see Exercise 31).

Let $X_i = 1$ if she breaks the ith dish and $X_i = 0$ if she does not break the ith dish. Then we have assumed that $\mathbf{X} = (X_1, X_2, \ldots, X_n)$ is a random sample of size n and that X_i is $B(1, \theta)$ for each i. Also, we have $\Theta = \{\theta: 0 \leq \theta \leq 1\}$. •

EXAMPLE 3. (Spinach)

Consider Example 1 of Chapter 4, where school officials wanted to estimate the proportion θ of students who would eat the frozen spinach. They assumed that each student would make his decision to eat or not to eat the spinach independently of all the other students' decisions, and that one student was as likely as another to eat the spinach.

Let $X_i = 1$ if the ith student eats the spinach and zero otherwise. It follows that $\mathbf{X} = (X_1, X_2, \ldots, X_n)$ is a random sample of size n and that X_i is $B(1, \theta)$. Since θ is the *proportion* of students who would eat the frozen spinach, we have $\Theta = \{\theta: 0 \leq \theta \leq 1\}$. •

EXAMPLE 4. (Fish)

In Example 2 of Chapter 4 the game warden wanted to estimate the number θ of fish in the lake. As explained in that example, there were 100 tagged fish in the lake, and the game warden caught a random sample without replacement of size 120.

Let X be the number of tagged fish that the game warden caught. Then X is $H(120, 100, \theta)$. The parameter set Θ that lists the possible values for the number of fish in the lake is $\Theta = \{120, 121, \ldots\}$. •

EXAMPLE 5. (Tanks)

The problem posed in Example 3 of Chapter 4 is that of estimating the number θ of tanks produced in a single month at a German factory. The tanks produced at the factory were numbered serially from 1 to θ inclusive. It was assumed that the numbers on the n captured tanks constituted a random sample of size n without replacement.

Let X_i be the number on the ith tank captured, $i = 1, 2, \ldots, n$. Then $\mathbf{X} = (X_1, X_2, \ldots, X_n)$ is a random sample of size n without replacement from the population whose members are $1, 2, \ldots, \theta$. The set of possible values for the number θ of tanks produced in that month is $\Theta = \{n, n + 1, \ldots\}$. •

EXAMPLE 6. (Plant Foreman)

In Exercise 21 of Chapter 4 the plant foreman wanted to estimate μ, the number of parts a customer would have to inspect until he found a defective part. It was reasonable to assume that each part had probability θ of being defective and that defective parts were produced independently of one another.

Let X_i be the number of parts that must be inspected in order to find the ith defective after the $(i-1)$st defective has already been found. Then $X_1, X_2, \ldots, X_n$ are independent random variables and X_i is $G(\theta)$ for each i. The set Θ of possible values of θ is easily seen to be $\Theta = \{\theta: 0 \leq \theta \leq 1\}$. Also, $\mu = 1/\theta$. •

EXAMPLE 7. (Height)

The problem in Example 1 of Chapter 5 is that of estimating the average height θ of men from California. It is known that heights of males are approximately normally distributed with a standard deviation of 3 inches, and a random sample from the population is easily obtained.

Let X_i be the height of the ith man in the sample. Then $X_1, X_2, \ldots, X_n$ are independent random variables and X_i is $N(\theta, 9)$ for each i. Since heights cannot be negative, θ must be positive, so we take as our parameter set $\Theta = \{\theta: 0 < \theta < \infty\}$. •

The principal goal of this chapter is to provide procedures—referred to as estimators—that will yield "good" estimates. As we have pointed out earlier, our estimate of the unknown parameter should depend upon both the empirical evidence and the probabilistic mechanism. It is clear upon reflection, however, that the procedure for obtaining our estimate must not depend upon *knowing* the value of an unknown parameter. This leads us to the next definition.

DEFINITION

Any real-valued function of the observable random variables that does not depend upon any *unknown* parameter is called a *point estimator* or simply an *estimator*.[2] Given the values of the observable random variables (that is, the empirical evidence), the value of the estimator is called the *estimate*.

It is common to denote an estimator of θ by $\hat{\theta}$, read "theta hat." Often, however, we use the more explicit functional form of the estimator, and we write $\hat{\theta}_n(\mathbf{X})$ or $\hat{\theta}(X_1, X_2, \ldots, X_n)$ instead of simply $\hat{\theta}$. [The subscript n on $\hat{\theta}_n(\mathbf{X})$ indicates that there are n observable random variables.] Thus, if the empirical evidence is $\mathbf{x}$, then $\hat{\theta}(\mathbf{x})$ is our estimate of θ. Moreover, we need not always use the letter θ to denote an unknown parameter. For example, if the unknown parameter is the expectation or the variance, we will denote it by μ or σ^2, respectively. But we will *always* denote the parameter set by Θ.

[2] Many texts use the term statistic instead of the term estimator.

EXAMPLE 8

Let $\mathbf{X} = (X_1, X_2, \ldots, X_n)$ be a random sample, where X_i is $N(\mu, 9)$ for each i. Then we may consider the following definitions for $\hat{u}(X_1, \ldots, X_n)$:

(a) $\hat{\mu}(X_1, X_2, \ldots, X_n) = \frac{1}{2}(X_1 + X_2)$,

(b) $\hat{\mu}(X_1, X_2, \ldots, X_n) = \dfrac{X_2}{1 + X_4}$,

(c) $\hat{\mu}(X_1, X_2, \ldots, X_n) = \frac{1}{2}[\max(X_1, X_2, \ldots, X_n) + \min(X_1, X_2, \ldots, X_n)]$,

(d) $\hat{\mu}(X_1, X_2, \ldots, X_n) = \dfrac{1}{n}\displaystyle\sum_{i=1}^{n} X_i^2$,

(e) $\hat{\mu}(X_1, X_2, \ldots, X_n) = \bar{X}_n \equiv \dfrac{1}{n}\displaystyle\sum_{i=1}^{n} X_i$, and

(f) $\hat{\mu}(X_1, X_2, \ldots, X_n) = \bar{X} + 2/n$.

They are each estimators of μ. [Of course, the $\hat{\mu}$ given in (b) and (d) are patently ridiculous.] But

(g) $\hat{\mu}(X_1, X_2, \ldots, X_n) = \mu + \dfrac{1}{n^2}\displaystyle\sum_{i=1}^{n} X_i$ and

(h) $\hat{\mu}(X_1, X_2, \ldots, X_n) = \begin{cases} \bar{X} & \text{if } \bar{X} \geq \mu, \\ \bar{X} + 2/n & \text{if } \bar{X} < \mu, \end{cases}$

are not estimators, since they both depend upon the unknown parameter μ.

If $n = 4$ and the observed values were 67, 74, $69\frac{1}{2}$, and 71 inches as in Example 1 of Chapter 5, then our estimates would be $70\frac{1}{2}$, $\frac{74}{72}$, $70\frac{1}{2}$, $4959\frac{1}{16}$, $70\frac{3}{8}$, and $70\frac{7}{8}$ inches, respectively. •

It follows from the definition that an estimator is a random variable. Therefore, every estimator has a distribution. Also, the estimator's distribution depends upon the true value of θ. Furthermore, it is clear from the definition and Example 8 that our estimate will depend upon the empirical evidence, but it is not clear how it depends upon the probabilistic mechanism. In the next section we shall show that "good" estimators do, in fact, take the probabilistic mechanism into account.

EXERCISES:

4. The accounting department has asked the production manager to make a one-month estimate of the materials expense involved in making a new injection-molded plastic part. He knows that 10,000 nondefective plastic parts will be needed this month. Also, the materials expense is \$.25 per part molded—regardless of whether or not the molded part is defective. Unfortunately, injection molding frequently produces defective parts. In fact, 15 out of the last 100 parts molded were defective. What is a reasonable estimate of his materials expense? What assumptions did you make in arriving at your estimate?

5. Let p_X and p_Y be the probabilities that a person with malaria survives if he uses medication X and Y, respectively. Let $\mathbf{X}$ and $\mathbf{Y}$ be independent random samples of size n and m, where $X_i = 1$ if the ith person given medication X survives and $X_i = 0$ otherwise; similarly for medication Y. Suggest an estimator for $\theta = p_X - p_Y$. What estimate does it yield if $\mathbf{x} = (1, 0, 0, 0, 1, 1)$ and $\mathbf{y} = (1, 1, 0)$?

6. Suppose that $p_X = p_Y$ in the problem above. Suggest three different estimators for p_X. Using the data given above, what estimates do they yield? Which of the three estimators do you prefer? Why?

7. In Example 24 of Chapter 1 we showed that the probability of getting a winning ticket in Union Oil's "Beat the Dealer" promotion game is $\frac{1}{560}$. Moreover, you can win, 1, 5, 10, 20, or 100 dollars. Let μ be the expected amount you will win on any given ticket. Suggest an estimator for μ. What estimate would it yield if we inspected 300 tickets and found 1 \$100 ticket, 0 \$20 tickets, 2 \$10 tickets, 15 \$5 tickets, and 282 \$1 tickets?

In Exercises 8 through 18 describe the probabilistic mechanism and the sample, suggest an estimator for the unknown parameter, and use the data to find the estimate produced by this estimator.

8. Grandpa claims extraordinary tactile abilities. In particular, he can often tell without looking whether or not a penny has been placed "heads up" in his flat open hand. The extent of his ability is unknown. To get more information, a coin was flipped and placed in his hand 6 times. Grandpa guessed correctly on the 1st, 2nd, 3rd, and 5th time. (See Exercise 5 of Chapter 4.)

9. What can you say about John's preference for blondes if 5 of his last 8 dates were blondes? Is the usual assumption of independence really appropriate here? (See Exercise 6 of Chapter 4.)

10. In a certain multiple-choice examination there are 5 questions. If a student knows the subject material on the test, he will be able to give the correct answer to any given question on the test with probability $\frac{3}{4}$. If he does not know the material, he will have a probability of $\frac{1}{5}$ of answering any question. Jack answered 3 of the questions correctly. What is a good guess as

to the probability that Jack knows the material? (Remember, either $\hat{\theta}(\mathbf{x}) = \frac{1}{5}$ or $\hat{\theta}(\mathbf{x}) = \frac{3}{4}$! See Exercise 9 of Chapter 4.)

11. Today, John spelled 5 words correctly before giving an incorrect answer. What percentage of words can he spell correctly? (See Exercise 10 of Chapter 4.)

12. Fred agreed to return from his fishing trip after catching his 3rd fish. He returned home on the 16th day. If Fred never catches more than one fish a day, what is the probability of his catching a fish on any given day? (See Exercise 12 of Chapter 4.)

13. A large jar contains 20 marbles. How many red marbles do you think there are in the jar if 4 marbles were withdrawn and only 1 of them was red? (See Exercise 13 of Chapter 4.)

14. (Continuation) What would you say if the marbles had been drawn with replacement instead of without replacement?

15. (Continuation) Suppose we knew that there were 7 red marbles in the jar, but we didn't know how many marbles there were in all. How many would you say there were if 4 marbles were withdrawn and only 1 was red?

16. (Continuation) What if the marbles were drawn with replacement?

17. A stack of cards were numbered serially starting with number 1. After being well shuffled, two cards were removed from the deck. How many cards would you guess there to be in the deck if the numbers on the two cards draw were 2 and 19?

18. (Continuation) What if the two cards had been drawn with replacement?

19. Show that the problems of estimation and hypothesis testing are equivalent when the parameter set Θ contains two points—that is, $\Theta = \{\theta_1, \theta_2\}$. In particular, let $\mathbf{X}$ and $\mathcal{S}$ be the observable random variables and the set of possible values of $\mathbf{X}$, and suppose that H_0: $\theta = \theta_1$. Show that (i) for each rejection region R there is exactly one estimator $\hat{\theta}$ such that $\{\mathbf{x} \in \mathcal{S}: \hat{\theta}(\mathbf{x}) = \theta_2\} = R$ and (ii) for each estimator $\hat{\theta}$ there is exactly one rejection region R such that $R = \{\mathbf{x} \in \mathcal{S}: \hat{\theta}(\mathbf{x}) = \theta_2\}$.

3. DESIRABLE PROPERTIES OF ESTIMATORS

Potentially, we have an unlimited number of estimators to choose from. But by properly taking into account the probabilistic mechanism, we can often single out one as being the best.

3.1 Unbiased Estimators

As pointed out earlier, an estimator is a random variable, so we cannot be sure that the estimates it produces will be equal to the true value of the unknown parameter. It seems reasonable, however, to ask that the average estimate it produces be equal to θ. Thus, we ask that the expected value of the estimator equal the true value of the unknown parameter, so the estimator's distribution is centered at θ. When this property holds, we say that our estimator is unbiased.

DEFINITION

An estimator $\hat{\theta}$ is said to be an *unbiased* estimator of θ if for each θ in Θ we have

$$E(\hat{\theta}) = \theta.$$

EXAMPLE 9

Suppose we have taken a random sample of size n with or without replacement from a population with distribution f_X, and we want to estimate $\mu = E(X)$. Then, as shown in Theorem 6 of Chapter 2,

$$\bar{X}_n \equiv \frac{1}{n} \sum_{i=1}^{n} X_i$$

is an unbiased estimator of μ. Furthermore, if $E(X)$ is known, but $\operatorname{Var}(X) = \sigma^2$ is unknown, then Theorem 2 of Chapter 2 may be used to show that

$$\frac{1}{n} \sum_{i=1}^{n} (X_i - \mu)^2$$

is an unbiased estimator of σ^2. •

EXAMPLE 10. (Three Sisters)

In Example 2 we wanted to estimate the probability p that the youngest of three sisters will be responsible for breaking a dish. We let $X_i = 1$ if she breaks the ith dish and $X_i = 0$ if she did not break the ith dish. An estimator of p that seems reasonable and immediately comes to mind is $\bar{X}_n$. Since each X_i is $B(1, p)$ and the expected value of a binomial random variable with parameters 1 and p is p, we have for each $0 \le p \le 1$ that

$$E(\bar{X}_n) = E\left(\frac{1}{n} \sum_{i=1}^{n} X_i\right) = \frac{1}{n} \sum_{i=1}^{n} E(X_i) = \frac{1}{n} \sum_{i=1}^{n} p = p,$$

so

$$\bar{X}_n$$

is unbiased. Recalling that $n = 4$, $x_1 = x_3 = x_4 = 1$, and $x_2 = 0$, our estimate of p is $\frac{3}{4}$. •

EXAMPLE 11. (Fish)

Let N be the number of fish in the lake and let X be the number of tagged fish that the game warden caught. Then X is $H(120, 100, N)$. It is clear upon reflection that large values of X indicate that N is small, for we are likely to observe a large proportion of tagged fish in our sample when the proportion of tagged fish in the lake is large. In fact, it is reasonable to *guess* that on the average these two proportions will be equal. Thus, we suggest

$$\hat{N} = \frac{120 \cdot 100}{X}$$

as our estimator for N. [Naturally if X is $H(n, r, N)$, then we would suggest $\hat{N} = n \cdot r/X$.] Unfortunately, $\hat{N}$ is not unbiased. Worse yet, there is no unbiased estimator of N—a situation we had not anticipated![3] Nevertheless, $\hat{N} = 120 \cdot 100/X$ still appears to be the most reasonable estimator of N, and we recommend its use. In light of the fact (Example 2 of Chapter 4) that the game warden caught 5 tagged fish, $120 \cdot 100/5 = 2400$ is our estimate of N. •

EXAMPLE 12. (Tanks)

Let N be the number of tanks produced in the month, and let X_i be the number on the ith tank captured, so that $\mathbf{X} = (X_1, X_2, \ldots, X_n)$ is a random sample of size n without replacement from a population whose members are $1, 2, \ldots, N$. Since $E(X_i) = \sum_{j=1}^{N} j \frac{1}{N} = (N + 1)/2$, it follows from Theorem 2 of Chapter 2 that

$$\hat{N} = 2\bar{X}_n - 1$$

is an unbiased estimator of N. In Example 3 of Chapter 4 we had $n = 3$ and the empirical evidence was $\mathbf{x} = (2, 13, 99)$, so our estimate of N is 75. Although $\hat{N}$ is unbiased, the empirical evidence reveals that $\hat{N}$ is not a totally satisfactory estimator of N. Why? •

EXAMPLE 13. (Plant Foreman)

Let θ be the probability that a part is defective, let μ be the expected number of parts we must inspect in order to find a defective, and let X_i be the number of parts inspected in order to find the ith defective subsequent to finding the $(i - 1)$st defective. Then $\mathbf{X} = (X_1, X_2, \ldots, X_n)$ is a ran-

[3] Suppose X is $H(n, r, N)$, and $\hat{N}$ is an estimator of N. Then there is a number K such that $\hat{N}(x) \leq K$ for $x = 0, 1, 2, \ldots, n$, so $E(\hat{N}) \leq K$ for each $N \in \Theta$. Hence, if $N > K$, then $E(\hat{N}) \neq N$, so $\hat{N}$ is not unbiased.

dom sample and X_i is $G(\theta)$. As pointed out in Example 9, $E(\bar{X}_n) = \mu$, so

$$\bar{X}_n$$

is an unbiased estimator of μ. This suggests using $\hat{\theta} = 1/\bar{X}_n$ as an estimator of θ, since $\mu = 1/\theta$.[4] In Exercise 21 of Chapter 4 we had $n = 1$ and $x_1 = 5$, so 5 is our estimate of μ. •

EXAMPLE 14. (Height)

In Example 7 we wanted to estimate the average height μ of men from California. Once again, the most obvious choice for $\hat{\mu}$ is $\bar{X}_n$, and, as shown in Example 9, $\hat{\mu}$ is unbiased. Using $\bar{X}_n$ as our estimator and the data in Example 1 of Chapter 5, we arrive at $70\frac{3}{8}$ inches as our estimate of μ. •

The next example is so useful that we label it a theorem.

THEOREM 1

Let $\mathbf{X} = (X_1, X_2, \ldots, X_n)$ be a random sample from a population with unknown mean μ and unknown variance σ^2, then

$$S^2 \equiv \frac{1}{n-1} \sum_{i=1}^{n} (X_i - \bar{X})^2$$

is an unbiased estimator of σ^2.

PROOF. Fix i, $1 \le i \le n$. Then using Theorems 2 and 6 of Chapter 2 and equation (2.5), we have

$$\begin{aligned} E[(X_i - \bar{X})^2] &= E[(X_i - \mu + \mu - \bar{X})^2] = E[(X_i - \mu)^2] + E[(\mu - \bar{X})^2] \\ &\qquad + 2E[(X_i - \mu)(\mu - \bar{X})] \\ &= \sigma^2 + \frac{\sigma^2}{n} + 2\{E(X_i\mu) - E(\mu^2) + E(\mu\bar{X}) - E(X_i\bar{X})\} \\ &= \sigma^2 + \frac{\sigma^2}{n} + 2\left\{\mu^2 - \mu^2 + \mu^2 - \frac{1}{n}\sum_{j=1}^{n} E(X_iX_j)\right\} \\ &= \sigma^2 + \frac{\sigma^2}{n} + 2\left\{\mu^2 - \frac{1}{n}E(X_i^2) - \frac{1}{n}\sum_{j\neq i} E(X_iX_j)\right\} \\ &= \sigma^2 + \frac{\sigma^2}{n} + 2\left\{\mu^2 - \frac{1}{n}E(X_i^2) - \frac{n-1}{n}\mu^2\right\} \\ &= \sigma^2 + \frac{\sigma^2}{n} - 2\frac{\sigma^2}{n} = \frac{n-1}{n}\sigma^2. \end{aligned}$$

[4] In general, it is not true that $E(1/Y) = 1/E(Y)$ for a random variable Y. We can show that for $n = 1$, $E(1/\bar{X}_n) = [-\log \theta/(1 - \theta)]\theta$, so $1/\bar{X}_n$ is not an unbiased estimator of θ.

Thus,

$$E(S^2) = \frac{1}{n-1}\sum_{i=1}^{n} E(X_i - \bar{X})^2 = \frac{1}{n-1}\sum_{i=1}^{n}\frac{n-1}{n}\sigma^2 = \sigma^2. \blacklozenge$$

3.2 Consistent Estimators

Instead of—or perhaps in addition to—being unbiased, we might ask that the estimates produced by our estimator be close or at least have a high probability of being close to θ. In particular, as the sample size increases, our estimates should get better more often. This property is referred to as consistency.

DEFINITION

Let $\hat{\theta}_1, \hat{\theta}_2, \ldots, \hat{\theta}_n, \ldots$ be a sequence of estimators, where $\hat{\theta}_n$ is the estimator when there are n observable random variables. This sequence of estimators is said to be *consistent* if for each θ in Θ and each number $\epsilon > 0$ we have

$$P(|\hat{\theta}_n - \theta| > \epsilon) \to 0 \qquad \text{as } n \to \infty.$$

Loosely speaking, a sequence of estimators is consistent if the estimates it produces are likely to be close to θ for large samples. Typical of what we have in mind when we speak of a sequence of estimators are the following three sequences:

$$\bar{X}_1, \bar{X}_2, \bar{X}_3, \ldots, \qquad \text{where } \bar{X}_n = \frac{1}{n}\sum_{i=1}^{n} X_i;$$

$$S_2^2, S_3^2, S_4^2, \ldots, \qquad \text{where } S_n^2 = \frac{1}{n-1}\sum_{i=1}^{n}(X_i - \bar{X}_n)^2;$$

and

$$X_1, \ \max(X_1, X_2), \ \max(X_1, X_2, X_3), \ \ldots.$$

(These are sequences of estimators for μ, σ^2, and the population size N, respectively.) For convenience, we sometimes say $\tilde{\theta}_n$ is consistent if the sequence $\tilde{\theta}_1, \tilde{\theta}_2, \ldots$ is.

The next theorem provides us with an easily applied criterion for determining consistency.

THEOREM 2

Let $\hat{\theta}_1, \hat{\theta}_2, \ldots$ be a sequence of estimators. If for each θ in Θ we have

$$E(\hat{\theta}_n) \to \theta \qquad \text{as } n \to \infty \tag{7.1}$$

and

$$\operatorname{Var}(\hat{\theta}_n) \to 0 \qquad \text{as } n \to \infty, \tag{7.2}$$

then the sequence is consistent.

▶▶

PROOF. Let $\epsilon > 0$ be given. By (7.1) we can choose M such that

$$|E(\hat{\theta}_n) - \theta| < \frac{\epsilon}{2} \qquad \text{whenever } n \geq M.$$

Then

$$\{\mathbf{x}: |\hat{\theta}_n(\mathbf{x}) - \theta| \geq \epsilon\} \subset \{\mathbf{x}: |\hat{\theta}_n(\mathbf{x}) - E(\hat{\theta}_n)| \geq \frac{\epsilon}{2}\} \qquad \text{whenever } n \geq M,$$

since

$$\begin{aligned} |\hat{\theta}_n - \theta| &= |\hat{\theta}_n - E(\hat{\theta}_n) + E(\hat{\theta}_n) - \theta| \\ &\leq |\hat{\theta}_n - E(\hat{\theta}_n)| + |E(\hat{\theta}_n) - \theta| \leq |\hat{\theta}_n - E(\hat{\theta}_n)| + \frac{\epsilon}{2}. \end{aligned}$$

Therefore, we can use the Chebychev inequality and (7.2) to obtain

$$\begin{aligned} P(|\hat{\theta}_n - \theta| > \epsilon) &\leq P\,(|\hat{\theta}_n - E(\hat{\theta}_n)| > \epsilon/2 \\ &\leq \frac{\operatorname{Var}(\hat{\theta}_n)}{(\epsilon/2)^2} \to 0 \qquad \text{as } n \to \infty. \end{aligned}$$ ♦

EXAMPLE 15

If $\mathbf{X}$ is a random sample of size n from a population with mean μ and variance σ^2, then $\bar{X}_n$ is a consistent estimator of μ. Theorem 6 of Chapter 2 shows that $E(\bar{X}_n) = \mu$ and $\operatorname{Var}(\bar{X}_n) = \sigma^2/n$ for each n, so both (7.1) and (7.2) hold. Hence, $\bar{X}_n$ is consistent. This establishes the consistency of the estimators suggested in Examples 10, 13, and 14. •

EXAMPLE 16

If $\mathbf{X}$ is a random sample of size n without replacement from a population of size N with mean μ and variance σ^2, then $\bar{X}_n$ is a consistent estimator of μ. As shown in Theorem 2 of Chapter 2, $E(\bar{X}_n) = \mu$ for each n, so

(7.1) holds, while Theorem 7 of Chapter 2 shows that

$$\operatorname{Var}(\bar{X}_n) = \frac{N - n}{N - 1}\frac{\sigma^2}{n} \qquad \text{for each } n \leq N,$$

so (7.2) also holds. Hence, $\bar{X}_n$ is a consistent estimator of μ. Our estimator of N in Example 12 (tanks) is also consistent, since $\hat{N}_n$ is unbiased and $\operatorname{Var}(\hat{N}_n) = 4(N - n)\sigma^2/[(N - 1)n]$. •

EXAMPLE 17

If **X** is a random sample of size n from a population with mean μ and variance σ^2 and if $E(X^4)$ exists, then S_n^2 is a consistent estimator of σ^2. This follows from Theorem 2 if we use the facts that S_n^2 is unbiased and

$$\operatorname{Var}(S_n^2) = \frac{1}{n}\left[E(X^4) - \frac{n-3}{n-1}\sigma^4\right]$$

(see Exercise 27). Note too that $[(n - 1)/n]S_n^2$ is also consistent (see Exercise 29), so we see that consistent estimators need not be unbiased. •

3.3 Minimum-Variance Unbiased Estimators

First we asked that our estimator be unbiased—that is, have a distribution that has mean θ. Next, we asked for consistency—that is, that our estimator's distribution be such that there is a high probability that our estimate will be close to θ. Although both of these conditions permitted us to drastically reduce the number of estimators under consideration, there may still be many unbiased, consistent estimators. We will now propose another criterion that will enable us to choose the best from among these.

Consistency is closely related to the variance of our estimator. In view of Theorem 2, we can think of consistency as asking, in addition to (asymptotic) unbiasedness, that the variance of our estimator get closer to zero as the sample size gets larger. Analogously, the criterion we are about to propose asks, in addition to unbiasedness, that the variance of our estimator be as small as possible so that the variance of $\hat{\theta}_n$ goes to zero as rapidly as possible.

DEFINITION

Let **X** be a random sample of size n (with or without replacement), and let $\hat{\theta}_n$ be an estimator of θ such that

$$\hat{\theta}_n \text{ is an unbiased estimator of } \theta \tag{7.3}$$

and

$$\operatorname{Var}(\hat{\theta}_n) < \operatorname{Var}(\tilde{\theta}_n), \qquad \text{where } \tilde{\theta}_n \text{ is any other unbiased estimator of } \theta. \tag{7.4}$$

Then $\hat{\theta}_n$ is the *minimum-variance unbiased estimator* of θ, abbreviated MVU.

Having defined an MVU estimator, we must now find such estimators for the various cases (that is, probabilistic mechanisms) of interest. We must remember (see Example 11) that sometimes MVU estimators do not exist.

We hasten to add that restricting attention to the class of unbiased estimators is not always appropriate. Instead, it is sometimes preferable to consider the class of consistent estimators (which includes some biased as well as unbiased estimators), and then pick one estimator from among this class. For instance, we might seek a consistent estimator $\hat{\theta}$ with minimum *mean-squared error* where

$$\text{mean-squared error of } \hat{\theta} \equiv E\{(\hat{\theta} - \theta)^2\} = \text{Var}(\hat{\theta}) + [E(\hat{\theta}) - \theta]^2.$$

(Verify this equality! The second term on the right-hand side is, of course, equal to zero for unbiased estimators.) We might, quite possibly, prefer a slightly biased estimator $\hat{\theta}$ over any unbiased estimator if its variance is much smaller than that of any unbiased estimator.

We state without proof that the estimators suggested in Examples 10, 13, and 14 and in Theorem 1 are MVU estimators. The mathematics involved in showing this is somewhat advanced and is taken up in the next subsection. Luckily, in most of the examples we are likely to encounter, an estimator of the form

$$\hat{\theta}_n(\mathbf{x}) = a + \sum_{i=1}^{n} b_i X_i, \quad S^2, \quad \text{or} \quad \frac{1}{n}\sum_{i=1}^{n} (X_i - \mu)^2$$

is MVU. (In Examples 10, 13, and 14 we had $a = 0$ and $b_i = 1/n$, whereas $a = -1$ and $b_i = 2/n$ in Example 12.) However, this is not always the case, as demonstrated in the following example.

EXAMPLE 18. (Tanks)

In Example 12 we showed that $\hat{N} = 2\bar{X} - 1$ is an unbiased estimator of N, while it follows from Theorems 4 and 7 of Chapter 2 and the fact that

$$\text{Var}(X_i) = \sum_{i=1}^{N} j^2 \frac{1}{N} - \left(\sum_{i=1}^{N} j \frac{1}{N}\right)^2$$

$$= \frac{(N+1)(2N+1)}{6} - \left(\frac{N+1}{2}\right)^2 = \frac{N^2 - 1}{12}$$

that

$$\operatorname{Var}(\hat{N}) = 4\frac{N-n}{N-1}\frac{1}{n}\frac{N^2-1}{12} = \frac{(N-n)(N+1)}{3n}.$$

Next, consider the estimator

$$N^* \equiv \frac{n+1}{n}\max(X_1, X_2, \ldots, X_n) - 1.$$

It can be shown that N^* is unbiased by employing Corollary 1 and Exercise 14 of Chapter 2, while it follows from Theorem 4 and Exercise 41 of that chapter that

$$\operatorname{Var}(N^*) = \frac{(N-n)(N+1)}{n(n+2)}.$$

Thus, $\hat{N}$ is not MVU. In fact, N^* is MVU. The reason that $\hat{N} = 2\bar{X} - 1$ is not MVU is that it fails to use all of the available information. We saw this in Example 12, where its estimate of N was 75 yet one of the observations was 99, which shows that $N \geq 99$. •

In the next subsection we take up the (rather difficult) study of estimators that use all of the available information in the sample. These estimators are called *sufficient*, and they provide us with MVU estimators.

In concluding this subsection we point out that we have not used the sample median as an estimator of μ. The reason is as follows. If X is a random variable with a symmetric distribution—that is, $f_X^\mu(\mu + t) = f_X^\mu(\mu - t)$ for all t—then the median and the mean of X are equal. Thus, the sample median $M(\mathbf{X})$ is a reasonable estimator of μ, and, of course, $M(\mathbf{X})$ is unbiased. Nonetheless, $\bar{X}$ is preferred to $M(\mathbf{X})$, as $\operatorname{Var}(\bar{X})$ is usually smaller than $\operatorname{Var}(M(\mathbf{X}))$. In fact, if the distribution of X is nearly normal, then $\operatorname{Var}(M(\mathbf{X})) \approx 1.25 \operatorname{Var}(\bar{X})$. [There are, however, very complicated examples for which $\operatorname{Var}(M(\mathbf{X})) < \operatorname{Var}(\bar{X})$.]

*3.4 Sufficient Estimators

In this subsection we shall present a criterion for determining whether or not we have found an MVU estimator. (A second criterion is presented in Exercise 40.) As illustrated in Example 18, it seems—and is true—that the MVU estimator of θ must necessarily utilize all of the information in the sample. Fortunately, the converse is, in general, true: any unbiased estimator that utilizes all the available information is the MVU estimator.

* This subsection can be omitted without loss of continuity.

In using an estimator, we have condensed all of the information about θ that is available in the random sample $\mathbf{X}$ into a single random variable. In this condensing process, we might have lost some of the information about θ. For example, it is clear that we have lost information about θ if $\hat{\theta}_n(\mathbf{X}) = X_1$. This raises the questions of when we can accomplish this condensation without loss of information and what we mean by not losing information. We now answer the latter question.

DEFINITION

Let $\mathbf{X}$ be a random sample (with or without replacement) from the distribution $f_{\mathbf{X}}^{\theta}$, let $\hat{\theta}(\mathbf{X})$ be an estimator of θ, and let $\tilde{\theta}(\mathbf{X})$ be any other estimator of θ that is not a function of $\hat{\theta}$. If for each of the estimators $\tilde{\theta}$, the conditional distribution of $\tilde{\theta}$ given $\hat{\theta}$ does not depend on θ—that is, if $f_{\tilde{\theta}|\hat{\theta}}^{\theta} = f_{\tilde{\theta}|\hat{\theta}}$ for each θ in Θ—then $\hat{\theta}$ is a *sufficient estimator* of θ.

It may be quite difficult and tedious to show that an estimator is sufficient. With the aid of the next theorem, however, this task is often simplified.

THEOREM 3

Let $\mathbf{X}$ be a random sample (with or without replacement) from the distribution f_X^{θ}. Then $\hat{\theta}$ is a sufficient estimator for θ if and only if we can find two nonnegative functions h and k such that

$$f_{\mathbf{X}}^{\theta}(\mathbf{x}) = h(\hat{\theta}(\mathbf{x}), \theta) \cdot k(\mathbf{x}),$$

where $k(\mathbf{x})$ does not depend on θ.

This theorem is due to R. A. Fisher and Jerzy Neyman.[5]

▶▶

Sometimes there is more than one unknown parameter, as is the case if X is $N(\mu, \sigma^2)$ and both μ and σ^2 are unknown. This situation also arises in Chapter 8. More generally, suppose there are m unknown parameters $\theta_1, \theta_2, \ldots, \theta_m$ and write $\boldsymbol{\theta} = (\theta_1, \theta_2, \ldots, \theta_m)$ and $\hat{\boldsymbol{\theta}} = (\hat{\theta}_1, \hat{\theta}_2, \ldots, \hat{\theta}_m)$; then the following generalization of Theorem 3 holds: If $\mathbf{X}$ is a random sample with joint distribution $f_{\mathbf{X}}^{\boldsymbol{\theta}}$, then $\hat{\boldsymbol{\theta}}$ is sufficient for $\boldsymbol{\theta}$ if and only if we can find two nonnegative functions h and k such that

$$f_{\mathbf{X}}^{\theta}(\mathbf{x}) = h(\hat{\boldsymbol{\theta}}(\mathbf{x}), \boldsymbol{\theta}) \cdot k(\mathbf{x}).$$

Using this generalization of Theorem 3, we can show (see Exercise 37) that $(\bar{X}, S^2)$ is sufficient for (μ, σ^2) when X is $N(\mu, \sigma^2)$.

▶

[5] The proof of Theorem 3 can be found in Hogg and Craig (Reference 17), p. 213.

EXAMPLE 19

Let $\mathbf{X}$ be a random sample from a distribution that is $N(\mu, \sigma^2)$, where σ^2 is known and $-\infty < \mu < \infty$. Then $\bar{X}_n$ is a sufficient estimator for μ. We have for each $\mathbf{x}$ that

$$\begin{aligned} f_{\mathbf{X}}^{\mu}(\mathbf{x}) &= \prod_{i=1}^{n} \frac{1}{\sqrt{2\pi}\,\sigma} \exp\left[-\frac{1}{2\sigma^2}(x_i-\mu)^2\right] \\ &= \left(\frac{1}{\sqrt{2\pi}\,\sigma}\right)^n \exp\left[-\frac{1}{2\sigma^2}\sum_{i=1}^{n}(x_i-\mu)^2\right] \\ &= \left(\frac{1}{\sqrt{2\pi}\,\sigma}\right)^n \exp\left(-\frac{1}{2\sigma^2}\cdot n(\bar{x}-\mu)^2\right)\cdot\exp\left(-\frac{1}{2\sigma^2}\sum_{i=1}^{n}(x_i-\bar{x})^2\right) \end{aligned}$$

since

$$\Sigma(x_i-\mu)^2 = \Sigma(x_i-\bar{x}+\bar{x}-\mu)^2 = \Sigma(x_i-\bar{x})^2 + \Sigma(\bar{x}-\mu)^2. \bullet$$

EXAMPLE 20

Let $\mathbf{X}$ be a random sample of size n from a distribution that is $B(m, p)$, where m is known and $0 \le p \le 1$. Then $\bar{X}$ is a sufficient estimator for p. We have for each $\mathbf{x}$ that

$$\begin{aligned} f_{\mathbf{X}}^{p}(\mathbf{x}) &= \prod_{i=1}^{n} \binom{m}{x_i} p^{x_i}(1-p)^{m-x_i} \\ &= \prod_{i=1}^{n} \exp\left\{[\log p - \log(1-p)]x_i + \log\binom{m}{x_i} + m\log(1-p)\right\} \\ &= \exp\left\{[\log p - \log(1-p)]\sum_{i=1}^{n} x_i + \sum_{i=1}^{n}\log\binom{m}{x_i}\right. \\ &\qquad \left. + nm\log(1-p)\right\} \\ &= \exp\{[\log p - \log(1-p)]n\bar{x} + nm\log(1-p)\} \\ &\qquad \cdot\exp\left\{\sum_{i=1}^{n}\log\binom{m}{x_i}\right\}. \bullet \end{aligned}$$

We note the useful fact that if $\hat{\theta}$ is a sufficient estimator for θ and if $b \neq 0$, then $\theta^* = a + b\hat{\theta}$ is a sufficient estimator for $a + b\theta$. In fact, if u

is a real-valued 1-1 (one-to-one) function,[6] then $u(\hat{\theta})$ is a sufficient estimator for $u(\theta)$.

Next, a well-known and very useful theorem known as the Rao-Blackwell theorem[7] states that an unbiased estimator that is a function of a sufficient estimator has smaller variance than an unbiased estimator that is not a function of a sufficient estimator. Consequently, in searching for an MVU estimator we need only consider unbiased estimators that are functions of a sufficient estimator.

Finally, if our probabilistic mechanism satisfies a condition known as *completeness* [see, for example, Hogg and Craig (Reference 17), p. 220], then we can show that there is at most one unbiased estimator that is a function of a sufficient estimator. It then follows from the Rao-Blackwell theorem that this estimator is the MVU estimator. We hasten to add that *all* of the distributions we have studied so far satisfy the condition of completeness (see Exercises 38 and 39).

EXERCISES for Subsection 3.1

20. Decide whether or not the estimators you suggested in the exercises named are unbiased.
    (a) Exercise 4.
    (b) Exercise 5.
    (c) Exercise 6.
    (d) Exercise 7.
    (e) Exercise 11.
    (f) Exercise 13.
    (g) Exercise 14.
    (h) Exercise 15.
    (i) Exercise 16.
    (j) Exercise 17.
    (k) Exercise 18.

21. At a recent engineering conference, the lecturer said that a certain random variable X took the values 10, 15, and 20 with probabilities $\frac{1}{4}$, $\frac{1}{2}$, and $\frac{1}{4}$. He then said that $\frac{1}{4}(10 - 15)^2 + \frac{1}{2}(15 - 15)^2 + \frac{1}{4}(15 - 20)^2$ was the variance of X. However, he didn't know what to say when one of the engineers asked why he didn't divide by $(n - 1)$ in order to find the variance, as is indicated in Theorem 1. What was the engineer's confusion, and how would you resolve it?

22. In a recent election, N_i people voted in precinct i, and N_i is know for each i, $i = 1, 2, \ldots, 100$. Thus, $N \equiv \sum_{i=1}^{100} N_i$ votes were cast. A random sample of voters of size n_i was taken and Y_i of these n_i voters were observed to have voted in favor of the incumbent, $i = 1, 2, \ldots, 100$. Find a "good" unbiased estimator for the number I of votes received by the incumbent.

[6] A real-valued function h is 1-1 if we have $h(a) \neq h(b)$ whenever $a \neq b$.

[7] See Mood and Graybill (Reference 18), p. 176, for a formal statement and proof of this theorem.

23. Let $\tilde{\theta}$ and θ^* be unbiased estimators of θ_1 and θ_2, respectively. Find unbiased estimators of (i) $\theta_1 + \theta_2$, (ii) $(\theta_1 + \theta_2)/2$, (iii) $a\theta_1 + b\theta_2$.

24. Let $\mathbf{X}$ be a random sample of size n from a population with distribution f_X^p, where X is $B(1, p)$. Show that

$$\hat{p}^2 = \frac{n}{n-1}\bar{X}^2 - \frac{1}{n-1}\bar{X}$$

is an unbiased estimator of p^2. Find an unbiased estimator of $p(1 - p)$.

25. Let $\mathbf{X}$ be a random sample of size n from a population with distribution whose variance is σ^2 and $0 < \sigma^2 < \infty$. Show that

$$S \equiv \sqrt{\frac{1}{n-1}\sum_{i=1}^{n}(X_i - \bar{X})}$$

is not an unbiased estimator of $\sigma \equiv \sqrt{\sigma^2}$. [*Hint:* Use Exercise 44 of Chapter 2.] This is illustrative of the fact that, in general, it is not necessarily true that $E(u(\hat{\theta})) = u(\theta)$, even though $E(\hat{\theta}) = \theta$ and $u(\cdot)$ is 1-1.

26. Suppose X is $B(1, p)$. Sometimes one is interested in the ratio $p/(1 - p)$ rather than in p itself. For example, when studying human populations we are interested in the ratio of males to females. As we have seen before, unbiased estimators do not always exist. Let $\mathbf{X}$ be a random sample of size n. Show that there is no unbiased estimator for $p/(1 - p)$. [*Hint:* See footnote 3.]

27. Let $\mathbf{X}$ be a random sample of size n. Show that

$$\mathrm{Var}(S^2) = \frac{1}{n}\left[E(X^4) - \frac{n-3}{n-1}\sigma^4\right].$$

28. Let $\mathbf{X}$ be a random sample of size n without replacement from the population $1, 2, \ldots, N$. Show that

$$\hat{\sigma}^2 = \frac{n(N-1)}{N(n-1)}\left(\frac{1}{n}\sum_{i=1}^{n}X_i^2 - \bar{X}^2\right)$$

is an unbiased estimator of σ^2.

EXERCISES for Subsections 3.2 and 3.3

29. Use Exercise 27 and Theorem 2 to show that

$$\frac{1}{n}\sum_{i=1}^{n}(X_i - \bar{X})^2$$

is consistent.

30. Let $\mathbf{X}$ be a random sample of size n from the distribution f_X^θ, where $f_X^\theta(x) = 1/\theta$ for $0 < x \leq \theta$ and $f_X^\theta(x) = 0$ otherwise; $\theta \in \Theta = \{\theta\colon 0 < \theta < \infty\}$. Show that

$$\hat{\theta} = \frac{n+1}{n} \max(X_1, X_2, \ldots, X_n)$$

is unbiased, find $\text{Var}(\hat{\theta})$, and verify that $\text{Var}(\hat{\theta}) < \text{Var}(2\bar{X})$. Also, show that $\hat{\theta}_1, \hat{\theta}_2, \ldots$ is consistent.

31. Show that a random sample without replacement contains more information about a finite population than does a random sample with replacement. In particular, consider samples of size n from the population $a_1, a_2, \ldots, a_N$ of size N with

$$\mu = \Sigma a_i/N \quad \text{and} \quad \sigma^2 = \Sigma(a_i - \mu)^2/N.$$

Show that $E(\bar{X}_w) = E(\bar{X}) = \mu$ and $\text{Var}(\bar{X}_w) < \text{Var}(\bar{X})$, where $\bar{X}_w$ is the mean of a random sample without replacement and $\bar{X}$ is the mean of a random sample with replacement.

32. Let $\mathbf{X}$ be a random sample of size n. Suppose we consider only those estimators of θ of the form

$$\hat{\theta}(\mathbf{X}) = a + \sum_{i=1}^{n} b_i X_i.$$

Show that the MVU estimator of this form satisfies $b_i = b$ for each i.

33. Let $\mathbf{X}$ be a random sample of size n from a population with distribution f_X^θ, where $-\infty < E(X), E(X^2), E(X^3), E(X^4) < \infty$. Show that $\hat{\sigma}_1^2, \hat{\sigma}_2^2, \hat{\sigma}_3^2, \ldots$ is consistent for σ^2 if (for n even)

$$\hat{\sigma}_n^2 = \frac{1}{n} \sum_{i=1}^{n/2} (X_{2i} - X_{2i-1})^2.$$

Is the variance of $\hat{\sigma}_n^2$ smaller than that of S_n^2? Assume that $E(X^3) = 0$.

34. Let $X_1, X_2, \ldots$ be independent random variables with

$$P(X_i = j) = 1/N \qquad \text{for } j = 1, 2, \ldots, N$$

for each i, and define

$$\hat{N}_n = \text{maximum}\{X_1, X_2, X_3, \ldots, X_n\}.$$

Here, $N \in \Theta = \{1, 2, \ldots\}$. Is the sequence $\hat{N}_1, \hat{N}_2, \ldots$ consistent? Does $E(\hat{N}_n) \to N$ as $n \to \infty$? Are there constants K_n such that $E(K_n\hat{N}_n) = N$ for each $N \in \Theta$?

EXERCISES for Subsection 3.4

35. Are the estimators suggested in the following exercises sufficient? (a) 6, (b) 11, (c) 14, (d) 15, (e) 16, (f) 17.

36. Show that the estimator in Exercise 30 is sufficient.

37. Show that $(\bar{X}, S^2)$ is sufficient for (μ, σ^2) when X is $N(\mu, \sigma^2)$. [*Hint:* Choose $k(\mathbf{x}) \equiv 1$.]

38. (Exponential family) A large family of random variables have distributions of the form

$$f_X^\theta(x) = \begin{cases} \exp\{p(\theta)K(x) + S(x) + q(\theta)\}, & a < x < b, \\ 0 & \text{otherwise} \end{cases}$$

with $-\infty \le a, b \le +\infty$ for each $\theta \in \Theta$, where $p(\theta)$ is not a constant and $dK(x)/dx$ is a continuous function that is not identically zero. A random variable X that satisfies the above conditions is said to be a member of the *exponential family*. Now if $\mathbf{X}$ is a random sample of size n, then

$$\begin{aligned} f_{\mathbf{X}}^\theta(\mathbf{x}) &= \exp\left\{p(\theta) \sum_{i=1}^n K(x_i) + \sum_{i=1}^n S(x_i) + nq(\theta)\right\} \\ &= \exp\left\{p(\theta) \sum_{i=1}^n K(x_i) + nq(\theta)\right\} \cdot \exp\left\{\sum_{i=1}^n S(x_i)\right\}. \end{aligned}$$

Hence, $\hat{\theta}_n = \sum_{i=1}^n K(x_i)$ is sufficient for θ. Moreover, the condition of completeness is satisfied. Thus it is useful to know if a random variable is in the exponential family. Members of this family also possess other good properties. In Examples 20 and 19 we demonstrated that if X is $B(n, p)$ with n known and if Y is $N(\mu, \sigma^2)$ with σ^2 known, then X and Y are in the exponential family. Show that the following random variables are in the exponential family and find a sufficient estimator for the unknown parameter when the sample size is n:

(a) X is $G(\theta)$.
(b) X is χ_m^2.
(c) $f_X^\theta(x) = \theta x^{\theta-1}$, $0 < \theta < 1$, $0 < x < \infty$, zero elsewhere. (Here X is an exponential random variable. See Exercise 6 of Chapter 3.)
(d) $f_X^\theta(x) = (e^{-\theta}\theta^x/x!)$, $x = 0, 1, 2, \ldots$, $0 < \theta < \infty$. (Here X is a Poisson random variable.)
(e) $f_X^\theta(x) = [1/(\theta!\beta^{\theta+1})]x^\theta e^{-x/\beta}$, $0 < x < \infty$, $-1 < \theta < \infty$, zero elsewhere (β is a given positive number; X is known as a *gamma random variable*).
(f) $f_X^\theta(x) = [(\theta + \beta + 1)!/(\theta!\beta!)]x^\theta(1 - x)^\beta$, $0 < x < 1$, $-1 < \theta < \infty$, zero elsewhere ($\beta > -1$ is a given number; X is known as a *beta random variable*).

39. (Range family) Another large family of random variables has the form

$$f_X^\theta(x) = \begin{cases} M(x)Q(\theta), & 0 < x < \theta, \\ 0 & \text{elsewhere,} \end{cases}$$

where $\theta \in \Theta = \{\theta\colon 0 < \theta < \infty\}$. Of course the uniform random variables are seen to be of this form if we set $M(x) = 1$ and $Q(\theta) = 1/\theta$. These random variables are members not of the exponential family but of another large family known as the *range family*. In this case, the condition of completeness is also satisfied, so the MVU estimator will be easy to find if we can find a sufficient estimator.

Let **X** be a random sample of size n, and use the factorization theorem Theorem 3) to show that

$$\hat{\theta}_n(\mathbf{X}) \equiv \max\{X_1, X_2, \ldots, X_n\}$$

is a sufficient estimator for θ.

40. (Cramér-Rao inequality). Let **X** be a random sample of size n from a population with distribution f_X^θ such that the set $\{x\colon f_X^\theta(x) > 0\}$ does not depend upon θ. Then if $\tilde{\theta}_n$ is an unbiased estimator of θ, we have

$$\operatorname{Var}(\tilde{\theta}_n) \geq \frac{1}{nE\left\{\left[\dfrac{d}{d\theta}\log f_X^\theta\right]^2\right\}}.$$

Hence, if we have equality for an unbiased estimator $\hat{\theta}$, then $\operatorname{Var}(\tilde{\theta}) \geq \operatorname{Var}(\hat{\theta})$ for all unbiased estimators $\tilde{\theta}$, so $\hat{\theta}$ is MVU. Show that the estimator $\bar{X}$ achieves the lower bound if (a) X is $N(\mu, 1)$ or (b) X is $B(1, p)$.

4. METHODS FOR FINDING GOOD ESTIMATORS

Using the technique presented in Section 3.4, we can (usually) determine whether or not a given estimator is MVU. However, this technique does not tell us how to generate potential candidates; we still face the problem of how to obtain the MVU estimator. In this section we present two general methods for generating estimators. It is very often the case (see Theorem 4) that one of the estimators generated by these two methods is indeed MVU.

4.1 The Method-of-Moments

Our first technique for generating estimators, first introduced by Karl Pearson in 1894, is known as the method of moments.

This simple and intuitive method works as follows. Let **X** be a random sample of size n (with or without replacement) from a population with distribution f_X^θ, $\theta \in \Theta$. Often, we can write $E(X)$, the first population moment of X, in the form $E(X) = g(\theta)$, where g is simply some function; for example, $E(X) = 2\theta + 1$ or $E(X) = 3\theta^2$. The method of moments tells us to equate $\bar{X}$, the first sample moment, with $E(X)$ and then solve this equation for θ. That is, solve $\bar{X} = g(\theta)$. The resulting random variable is called the *method-of-moments estimator*, and we denote it by $\tilde{\theta}$.

More generally, given a random sample $\mathbf{X}$ of size n, we say that $E(X^k)$ is the *kth population moment* and

$$M_k \equiv \frac{1}{n} \sum_{i=1}^{n} X_i^k$$

is the *kth sample moment*. Then the *method-of-moments estimator* $\tilde{\theta}$ of the unknown parameter(s) θ is obtained by solving as many of the equations

$$E(X^k) = M_k$$

as is necessary to permit a solution in terms of θ. Of course, we must be able to write $E(X^k)$ as a function of θ in order to solve the equations for θ. The number of these equations that must be solved varies with the specific example under consideration, but this seldom presents a problem. We illustrate this method in the next few examples.

EXAMPLE 21. (Three Sisters)

In the case of the three sisters, we have a random sample $\mathbf{X}$ of size n, where each X_i is $B(1, \theta)$. Here, θ is the probability that the youngest sister breaks a dish and $\Theta = \{\theta : 0 \leq \theta \leq 1\}$ First, we must find a function g such that $E(X) = g(\theta)$ for each θ in Θ. Since $E(X) = \theta$ for each θ in Θ, we choose g to be the identity function—that is, $g(\theta) = \theta$. Next, we equate $\bar{X}$ with $g(\theta)$ and solve for θ. This yields $\bar{X} = \theta$, so

$$\tilde{\theta} = \bar{X}.$$

That is, $\bar{X}$ is the method-of-moments estimator of θ. •

EXAMPLE 22. (Fish)

In Example 2 of Chapter 4 the game warden wanted to estimate the number N of fish in the lake. In Example 4 of the present chapter we showed that X is $H(120, 100, N)$, where X is the number of tagged fish that the game warden caught. First, we must find a function g such that $E(X) = g(N)$ for each N in $\Theta = \{120, 121, \ldots\}$. Since $E(X) = 120 \cdot 100/N$, we choose $g(N) = 120(100/N)$. Next, we equate X $(= \bar{X}_1)$

with $g(N)$ and solve for N. This yields $120 \cdot 100/X = N$, so

$$\tilde{N} = \frac{120 \cdot 100}{X}. \bullet$$

EXAMPLE 23. (Tanks)

In Example 3 of Chapter 4 we wanted to estimate the number N of tanks produced in a single month. We have a random sample $\mathbf{X}$ of size n without replacement from the population $1, 2, 3, \ldots, N$. Since we have $E(X_i) = (N + 1)/2$ for each i, we equate $\bar{X}$ with $(N + 1)/2$ and solve for N. This yields $2\bar{X} - 1 = N$, so

$$\tilde{N} = 2\bar{X} - 1. \bullet$$

EXAMPLE 24. (Plant Foreman)

In Exercise 21 of Chapter 4 the plant foreman wanted to estimate the probability θ of a part's being defective. We have a random sample $\mathbf{X}$ of size n, where each X_i is $G(\theta)$. First, note that $E(X) = 1/\theta$ for each θ in $\Theta = \{\theta : 0 < \theta \leq 1\}$. Next, we equate $\bar{X}$ with $1/\theta$ and solve for θ. This yields $\theta = 1/\bar{X}$, so

$$\tilde{\theta} = \frac{1}{\bar{X}}. \bullet$$

EXAMPLE 25

Suppose we want to estimate both μ and σ^2 when we have a random sample $\mathbf{X}$ of size n, where each X_i is $N(\mu, \sigma^2)$. Since $E(X) = \mu$ and $E(X^2) = \sigma^2 + \mu^2$, we solve the equations

$$\bar{X} = \mu \quad \text{and} \quad \frac{1}{n}\sum_{i=1}^{n} X_i^2 = \sigma^2 + \mu^2$$

for μ and σ^2. This yields

$$\bar{X} = \mu \quad \text{and} \quad \frac{1}{n}\sum_{i=1}^{n} X_i^2 - \bar{X}^2 = \sigma^2,$$

so

$$\tilde{\mu} = \bar{X} \quad \text{and} \quad \tilde{\sigma}^2 = \frac{1}{n}\sum_{i=1}^{n} X_i^2 - \bar{X}^2. \bullet$$

In general, the method-of-moments yields estimators that are good, but it does occasionally give estimators that are not good as in the tank example. We shall now consider another method of generating estimators.

4.2 The Method of Maximum-Likelihood

To motivate the method of maximum-likelihood, we shall consider the following simple estimation problem. A certain jar contains a total of 5000 Lincoln-head and Indian-head pennies, but we do not know the true proportion of Lincoln-head pennies. For the sake of simplicity, let us suppose that the true proportion of Lincoln-head pennies is either $\frac{1}{5}$, $\frac{1}{4}$, $\frac{1}{2}$, or $\frac{3}{5}$. Consequently, if we draw a single coin from the jar, the probability p that it is a Lincoln-head is either $\frac{1}{5}$, $\frac{1}{4}$, $\frac{1}{2}$, or $\frac{3}{5}$. Therefore, if 3 coins are drawn with replacement from the jar and if X is the number of Lincoln-heads that are drawn, then X is $B(3, p)$ with $p = \frac{1}{5}$, $p = \frac{1}{4}$, $p = \frac{1}{2}$, or $p = \frac{3}{5}$, depending on the true value of p. Our estimation problem then is to decide whether $p = \frac{1}{5}$, $p = \frac{1}{4}$, $p = \frac{1}{2}$, or $p = \frac{3}{5}$.

An intuitively appealing way to choose between these four values of p is to *choose p so as to make the probability of obtaining the particular value of X that we observed as large as possible*. The probabilities associated with the four values of p and the various values of X are displayed in Table 7.1.

Table 7.1

	Values of X			
Values of p	0	1	2	3
$\frac{1}{5}$	$\frac{64}{125}$	$\frac{48}{125}$	$\frac{12}{125}$	$\frac{1}{125}$
$\frac{1}{4}$	$\frac{27}{64}$	$\frac{27}{64}$	$\frac{9}{64}$	$\frac{1}{64}$
$\frac{1}{2}$	$\frac{1}{8}$	$\frac{3}{8}$	$\frac{3}{8}$	$\frac{1}{8}$
$\frac{3}{5}$	$\frac{8}{125}$	$\frac{36}{125}$	$\frac{54}{125}$	$\frac{27}{125}$

Thus, in accordance with the procedure described above, our estimator p^* is defined by

$$p^*(x) = \begin{cases} \frac{1}{5}, & \text{if } x = 0, \\ \frac{1}{4}, & \text{if } x = 1, \\ \frac{3}{5}, & \text{if } x = 2 \text{ or } 3. \end{cases}$$

This reflects the fact that large values of X are more likely to arise from a population with $p = \frac{3}{5}$, whereas the smallest values of X are more likely to arise from a population with $p = \frac{1}{5}$.

We now generalize and formalize these ideas.

DEFINITION

Let $\mathbf{X}$ be a random sample (with or without replacement) from a population with distribution f_X^θ, and let $f_{\mathbf{X}}^\theta$ be the joint distribution of $\mathbf{X}$ [so $f_{\mathbf{X}}^\theta(\mathbf{x}) = \Pi f_X^\theta(x_i)$ if we are sampling with replacement]. Then for each θ in

Θ we define the *likelihood function* L_θ at the point $\mathbf{x}$ by

$$L_\theta(\mathbf{x}) = f^\theta_{\mathbf{X}}(\mathbf{x}).$$

The likelihood function evaluated at the point $\mathbf{x}$ gives the relative likelihood that the observable random variables $\mathbf{X}$ assume the particular values $\mathbf{x}$. As proposed above, we should choose our estimate of θ so as to make this relative likelihood for the particular values of $\mathbf{x}$ that we observed as large as possible.

DEFINITION

Let L_θ be the likelihood function for the observable random variables $\mathbf{X}$. For each $\mathbf{x}$, let $\theta^*(\mathbf{x})$ be the value of θ in Θ that maximizes L_θ. Then θ^* is the *maximum-likelihood estimator* of θ.

Thus, if $\mathbf{x}$ is the empirical evidence (the observed value of $\mathbf{X}$), then $\theta^*(\mathbf{x})$ satisfies

$$L_{\theta^*(\mathbf{x})}(\mathbf{x}) \geq L_\theta(\mathbf{x}) \qquad \text{for each } \theta \text{ in } \Theta.$$

We denote maximum-likelihood estimators by θ^* in order to distinguish them from the estimators obtained by the method of moments.

EXAMPLE 26. (Three Sisters)

In this example we have a random sample of size n from a population whose distribution is $B(1, p)$ with $p \in \Theta = \{p: 0 \leq p \leq 1\}$. Hence,

$$L_p(\mathbf{x}) = \prod_{i=1}^{n} p^{x_i}(1-p)^{1-x_i} = p^{\Sigma x_i}(1-p)^{n-\Sigma x_i}.$$

We now use a very common technique for finding a maximum-likelihood estimator.[8] Denote the logarithm of L_p by K_p, so

$$K_p(\mathbf{x}) = \Sigma x_i \log p + (n - \Sigma x_i) \log (1-p).$$

Observe that

$$\frac{d}{dp} K_p(x) = \Sigma x_i/p + (n - \Sigma x_i)(-1/(1-p)).$$

So, upon setting this expression equal to 0 and solving for p, we find that $K_p(\mathbf{x})$ is maximized at $p = \bar{X}$. But logarithm is an increasing function, so

[8] Students unfamiliar with calculus should skip this derivation—and others that use calculus—and merely take note of the conclusion.

this value of p also maximizes L_p. Hence,

$$p^* = \bar{X}.$$

That is, $\bar{X}$ is the maximum-likelihood estimator of θ. •

EXAMPLE 27. (Fish)

In this example the game warden wants to estimate the number of fish in the lake. Here X, the number of tagged fish he caught, is $H(120, 100, N)$. That is, for each N in $\Theta = \{120, 121, \ldots\}$

$$L_N(x) = \frac{\binom{100}{x}\binom{N-100}{120-x}}{\binom{N}{120}} \qquad \text{for } x = 0, 1, 2, \ldots, 100.$$

We can show (see Exercise 53) that for fixed values of x, L_N increases and then decreases as N increases. Thus N^* will satisfy (a) $L_{N^*} \geq L_{N^*-1}$ and (b) $L_{N^*} \geq L_{N^*+1}$. From (a) it follows that $N^* \leq 100 \cdot 120/x$, whereas it follows from (b) that $N^* \geq (100 \cdot 120/x) - 1$ (see Exercise 54). Hence, N^* is the integer that satisfies

$$\frac{100 \cdot 120}{X} - 1 \leq N^* \leq \frac{100 \cdot 120}{X}. \bullet$$

EXAMPLE 28. (Tanks)

In this example we want to estimate the number N of tanks produced. We have a random sample of size n without replacement from a population with distribution f_X^N, where $f_X^N(i) = 1/N$ for $i = 1, 2, \ldots, N$. Thus, for each N in $\Theta = \{n, n+1, \ldots\}$ we have

$$L_N(\mathbf{x}) = \begin{cases} \dfrac{(N-n)!}{N!} & \text{if } 1 \leq x_i \leq N \quad \text{and} \quad x_i \neq x_j \text{ for } i \neq j, \\ 0 & \text{otherwise.} \end{cases}$$

If we choose N less than $\max(x_1, x_2, \ldots, x_n)$, then $L_N(\mathbf{x}) = 0$. Also, $(N-n)!/N!$ decreases as N increases. Hence,

$$N^* = \max(X_1, X_2, \ldots, X_n). \bullet$$

Note that in Example 28 the maximum-likelihood estimator is not unbiased. Moreover, maximum-likelihood estimators are generally not unbiased.

As the examples indicate, the method of moments frequently generates the same estimator as the method of maximum-likelihood. But the maximum-likelihood estimator is generally preferable when the two methods do not generate the same estimator. This last fact is made apparent by the following theorem. The proof is left as an exercise (see Exercise 55).

THEOREM 4

Let $\mathbf{X}$ be a random sample of size n (with or without replacement) from a population with distribution f_X^θ. If $\hat{\theta}$ is a sufficient estimator for θ and if θ^* is a maximum-likelihood estimator of θ, then θ^* is a function of $\hat{\theta}$.

Although maximum-likelihood estimators are not always unique (see Exercise 52), we remark that in general (1) maximum-likelihood estimators are functions of the sufficient estimator and are consistent, and (2) the distribution of θ_n^* approaches the normal, $E(\theta_n^*)$ approaches θ, and $\mathrm{Var}(\theta_n^*)$ approaches

$$\frac{1}{n} E\left[\left(\frac{d}{d\theta} \log f_X^\theta(X)\right)^2\right]^{-1}$$

as n gets large.

Finally, we mention another nice property of maximum-likelihood estimators: they are invariant. That is, if u is a 1-1 function,[9] then the maximum-likelihood estimator of $u(\theta)$ is $u(\theta_n^*)$. For example, suppose that $X_1, \ldots, X_n$ are independent and each X_i is $N(\mu, \sigma^2)$ and both μ and σ^2 are unknown. Then it is easy to show that the maximum-likelihood estimators of μ and σ^2 are

$$\mu^* = \bar{X} \quad \text{and} \quad \sigma^{2*} = \frac{1}{n} \sum_{i=1}^{n} (X_i - \bar{X})^2.$$

It now follows directly from the invariance property that σ^*, the maximum-likelihood estimator of σ, is given by

$$\sigma^* = \sqrt{\sigma^{2*}} = \sqrt{\frac{1}{n} \sum_{i=1}^{n} (X_i - \bar{X})^2}.$$

Similarly, the maximum-likelihood estimators of μ^2 and μ/σ are $\mu^{*2} = \bar{X}^2$ and μ^*/σ^*.

[9] A real-valued function h is 1-1 if we have $h(a) \neq h(b)$ whenever $a \neq b$. That is, each element in the range is associated with a unique element in the domain.

EXERCISES:

41. Find the maximum-likelihood estimator for the parameter in Exercise 11.

42. A statistics professor is trying to determine the number of undergraduates relative to the number of graduate students in his class. He has 30 students in all and has been told the ratio, but he has forgotten whether there are 20 undergraduates and 10 graduates, or vice versa. He decides to take a sample of 4 students and ask them their status. Furthermore, he decides to *sample without replacement*, since it will give him more information. Provide him with a reasonable estimator of the number of graduate students in his class.

43. Find the maximum-likelihood estimator and the method-of-moments estimator for the parameter in Exercise 5.

44. Find the maximum-likelihood estimator and the method-of-moments estimator for the parameters in Exercises (a) 14, (b) 15, (c) 16, and (d) 17.

45. Let X_i and θ be as in Example 24. Find the maximum-likelihood estimator of θ.

46. Let X, μ, and σ^2 be as in Example 25. Find the maximum-likelihood estimator of μ and of σ^2.

47. Let $\mathbf{X}$ be a random sample of size n from a population with distribution f_X, where X is $N(\mu, \sigma^2)$ and μ is known. Find the method-of-moments estimator of σ^2 and the maximum-likelihood estimator of σ^2.

48. Assume that the age at death of Americans is normally distributed with variance 144. A sample of 10,000 histories yielded $\bar{x} = 72.1$. Compute the maximum-likelihood estimate of the probability that an American will live to be (a) at least 50, (b) no older than 90.

49. Using nontechnical language, explain why Theorem 4 tells us that maximum-likelihood estimators are preferable to method-of-moments estimators.

50. Find the maximum-likelihood estimator for θ when

$$f_X^\theta(x) = \begin{cases} 1/\theta, & 0 < x \le \theta, \\ 0 & \text{otherwise.} \end{cases}$$

and $\theta \in \Theta = \{\theta\colon 0 < \theta < \infty\}$.

51. Use the invariance property of maximum-likelihood estimators to find the maximum-likelihood estimator of
    (a) σ when X is $N(\mu, \sigma^2)$ and μ is unknown.
    (b) $E(X)$ when $f_X^\theta(x)$ is as in Exercise 50.
    (c) $p/(1 - p)$ when X is $B(1, p)$.

52. Sometimes there may be more than one maximum-likelihood estimator. To see this, let

$$f_X^\theta(x) = \begin{cases} 1 & \text{if } \theta - \frac{1}{2} \leq x \leq \theta + \frac{1}{2}, \\ 0 & \text{otherwise,} \end{cases}$$

where $\Theta = \{\theta\colon -\infty < \theta < \infty\}$, and let $\mathbf{X}$ be a random sample of size n from the population with distribution f_X^θ. Show that every estimator θ^* such that $\max(X_1, X_2, \ldots, X_n) - \frac{1}{2} \leq \theta^* \leq \min(X_1, X_2, \ldots, X_n) + \frac{1}{2}$ is a maximum-likelihood estimator for θ. Hence, you can also show that for each number b with $0 < b < 1$, the estimator

$$b \min(X_1, X_2, \ldots, X_n) + (1 - b) \max(X_1, X_2, \ldots, X_n) + b - \tfrac{1}{2}$$

is a maximum-likelihood estimator.

53. Suppose X is $H(n, r, N)$ and n and r are known; then

$$L_N(x) = \frac{\binom{r}{x}\binom{N-r}{n-x}}{\binom{N}{n}} \qquad \text{for } x = 0, 1, 2, \ldots, \min(n,r).$$

Show that for each fixed value of x, L_N first increases as N increases and then decreases as N increases.

54. Suppose X is $H(n, r, N)$ and n and r are known. Show that the maximum-likelihood estimator N^* of N satisfies

$$\frac{n \cdot r}{x} - 1 \leq N^* \leq \frac{n \cdot r}{x}.$$

[*Hint:* Use Exercise 53 and consider the inequalities (a) $L_{N^*}/L_{N^*-1} \geq 1$ and (b) $L_{N^*}/L_{N^*+1} \geq 1$.]

55. Prove Theorem 4 when $X_1, X_2, \ldots, X_n$ are independent and identically distributed random variables. [*Hint:* Use the Fisher-Neyman factorization criterion—that is, Theorem 3.]

5. INTERVAL ESTIMATION

Now we are well equipped to find a good (point) estimator. Given the empirical evidence $\mathbf{x}$, the estimator $\hat{\theta}$ provides us with the estimate $\hat{\theta}(\mathbf{x})$. This estimate is a single number representing our best guess of the true value of the unknown parameter θ. But the estimate does not relay any information about how close it is to θ, nor does it make known the likelihood of its being close to θ. It behooves us then to consider confidence intervals—a second kind of estimator.

Confidence intervals provide a method for stating both how close the estimate is to θ and the likelihood of its being that close to θ. A confidence interval tells us "how close" by giving us an interval—called the confidence interval—within which the true value of θ is likely to be. It tells us "the likelihood" of being in that interval by giving the probability that this interval contains the true value of θ. We now make all this precise.

An interval with endpoints a and b, written "$[a, b]$," is the set of all numbers between a and b. Thus, $[a, b] \equiv \{\theta : a \leq \theta \leq b\}$.

DEFINITION

Let $\mathbf{X}$ be a random sample of size n (with or without replacement) from a population with distribution $f_X{}^\theta$. Then the random interval $[L, U]$ is a $100(1 - \alpha)$ percent *random confidence interval* for θ if for each θ in Θ we have

$$P[L(\mathbf{X}) \leq \theta \leq U(\mathbf{X})] \geq 1 - \alpha.$$

Given the observed value of $\mathbf{X}$, and hence of the two random variables L and U, we refer to $[L(\mathbf{x}), U(\mathbf{x})]$ as a $100(1 - \alpha)$ percent *confidence interval*[10] for θ, and we refer to $100(1 - \alpha)$ as the *confidence coefficient*.

Extreme care must be used in interpreting this definition. Given the empirical evidence $\mathbf{x}$, the 90 percent confidence interval $[L(\mathbf{x}), U(\mathbf{x})]$ either does or does not contain the fixed but unknown number θ. Furthermore, we will never know whether the confidence interval $[L(\mathbf{x}), U(\mathbf{x})]$ contains θ. On the other hand, we do know that 90 percent of the confidence intervals $[L(\mathbf{x}), U(\mathbf{x})]$ provided by the random confidence interval $[L(\mathbf{X}), U(\mathbf{X})]$ will contain θ. This is illustrated in Figure 7.1.

Thus, the statement "the probability that θ is in the interval $[L(\mathbf{X}), U(\mathbf{X})]$ is .9" is correct, but the statement "the probability that θ is in the interval $[L(\mathbf{x}), U(\mathbf{x})]$ is .9" is incorrect. (Note that the interval $[L(\mathbf{X}), U(\mathbf{X})]$ is a random interval, whereas the interval $[L(\mathbf{x}), U(\mathbf{x})]$ is not.)

Naturally, we would like the expected length of our random confidence interval to be as short as possible. Unfortunately, in most problems it is not possible to construct a random confidence interval with minimal expected length. We propose the following method for choosing among the infinite number of random confidence intervals. Choose L and U to be functions of the maximum-likelihood estimator (MLE) θ^* such that

$$P(L > \theta) \leq \frac{\alpha}{2} \quad \text{and} \quad P(U < \theta) \leq \frac{\alpha}{2}.$$

[10] The terms interval estimator and interval estimate would be more in keeping with earlier terminology, but the term confidence interval is more widely used.

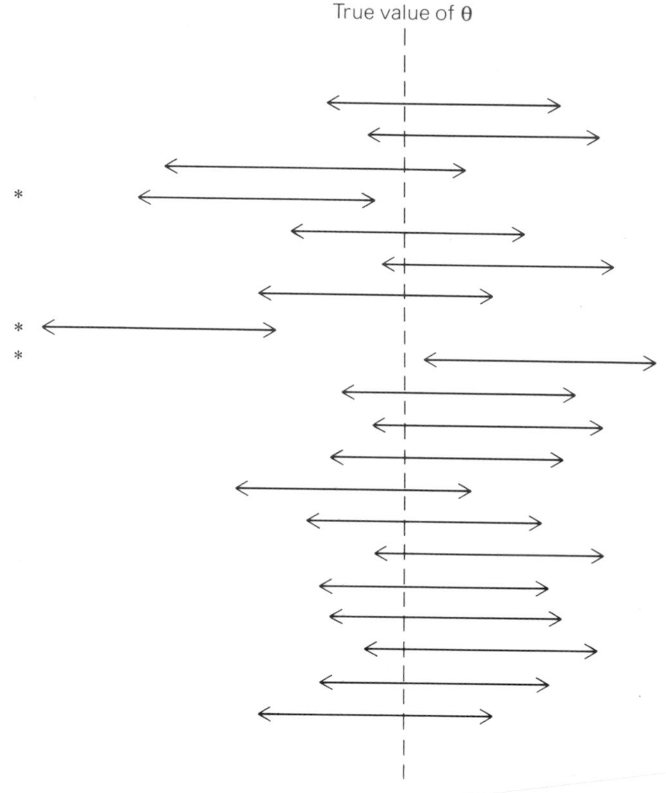

Figure 7.1 20 Typical 90 Percent Confidence Intervals for θ (notice that the starred intervals—4, 8, and 9—do not contain θ)

[This last restriction is innocuous and results in no increase in length if the distribution of $u(\theta^*)$ is symmetric about θ, where $E(u(\theta^*)) = \theta$.]

We now present seven commonly encountered situations in which the relationship between hypothesis testing and interval estimation is made apparent. (These seven cases coincide with the seven cases presented in Chapter 5 except for Case 5. A brief review of those cases would be useful for most readers at this point.)

Throughout the remainder of this section we shall assume that $\mathbf{X}$ is a random sample of size n from a population with distribution f_X^θ.

Case 1:

Suppose X is $N(\mu, \sigma^2)$ and σ^2 is known. Then the MLE of μ is $\bar{X}$ and

$$\left[\bar{X} - \frac{\sigma}{\sqrt{n}} z_{1-(\alpha/2)},\ \bar{X} + \frac{\sigma}{\sqrt{n}} z_{1-(\alpha/2)}\right]$$

is a $100(1 - \alpha)$ percent random confidence interval for μ, since $\bar{X}$ is $N(\mu, \sigma^2/n)$ and from Table A we have

$$1 - \alpha = P\left(-z_{1-(\alpha/2)} \le \frac{\bar{X} - \mu}{\sigma/\sqrt{n}} \le z_{1-(\alpha/2)}\right)$$
$$= P\left(-z_{1-(\alpha/2)} \frac{\sigma}{\sqrt{n}} \le \bar{X} - \mu \le z_{1-(\alpha/2)} \frac{\sigma}{\sqrt{n}}\right)$$
$$= P\left(\bar{X} - z_{1-(\alpha/2)} \frac{\sigma}{\sqrt{n}} \le \mu \le \bar{X} + z_{1-(\alpha/2)} \frac{\sigma}{\sqrt{n}}\right).$$

Suppose $\sigma^2 = 9$, $n = 25$, and α equals .10, .05, or .01. Then $[\bar{X} - .987, \bar{X} + .987]$, $[\bar{X} - 1.176, \bar{X} + 1.176]$, and $[\bar{X} - 1.545, \bar{X} + 1.545]$ are $100(1 - \alpha)$ percent random confidence intervals for μ when $\alpha = .10$, .05, or .01, respectively. If the observed value of $\bar{X}$ is $\bar{x} = 4.61$, then $100(1 - \alpha)$ percent confidence intervals are [3.623, 5.597], [3.434, 5.786], and [3.065, 6.155]. As shown in Figure 7.2, and as is always true, the length of the confidence interval increases as α decreases. (Is this obvious?)

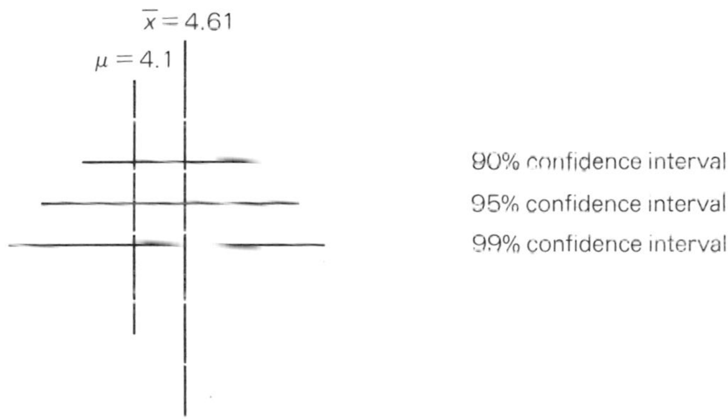

Figure 7.2 Confidence Intervals for μ

Case 2:

Suppose X is $N(\mu, \sigma^2)$ and σ^2 is unknown. Then the MLE of μ is $\bar{X}$, and $\bar{X}$ is $N(\mu, \sigma^2/n)$. Since σ^2 is unknown, we will use S^2 to estimate it. Now $T \equiv (\bar{X} - \mu)/(S/\sqrt{n})$ is τ_{n-1}, so

$$1 - \alpha = P(-t_{n-1}(1 - \alpha/2) \le T \le t_{n-1}(1 - \alpha/2))$$
$$= P(\bar{X} - t_{n-1}(1 - \alpha/2)S/\sqrt{n} \le \mu \le \bar{X} + t_{n-1}(1 - \alpha/2)S/\sqrt{n}).$$

Thus, the interval above is a $100(1 - \alpha)$ percent random confidence interval for μ.

Case 3:

Suppose X is $N(\mu_X, \sigma_X^2)$, Y is $N(\mu_Y, \sigma_Y^2)$, σ_X^2 and σ_Y^2 are known, and $\mathbf{X}$ and $\mathbf{Y}$ are independent random samples of size n and m. Then the MLE of $\mu_X - \mu_Y$ is $\bar{X} - \bar{Y}$, and $\bar{X} - \bar{Y}$ is $N(\mu_X - \mu_Y, \sigma_X^2/n + \sigma_Y^2/m)$. Therefore, it follows that

$$\left[(\bar{X} - \bar{Y}) - z_{1-(\alpha/2)}\sqrt{\frac{\sigma_X^2}{n} + \frac{\sigma_Y^2}{m}},\ (\bar{X} - \bar{Y}) + z_{1-(\alpha/2)}\sqrt{\frac{\sigma_X^2}{n} + \frac{\sigma_Y^2}{m}}\right]$$

is a $100(1 - \alpha)$ percent random confidence interval for $\mu_X - \mu_Y$.

Case 4:

Suppose X is $N(\mu_X, \sigma^2)$, Y is $N(\mu_Y, \sigma^2)$, σ^2 is unknown, and $\mathbf{X}$ and $\mathbf{Y}$ are independent random samples of size n and m. Then the MLE for $\mu_X - \mu_Y$ is $\bar{X} - \bar{Y}$, and $\bar{X} - \bar{Y}$ is $N(\mu_X - \mu_Y, \sigma^2/n + \sigma^2/m)$. As σ^2 is unknown, we will use $(n - 1)S_X^2 + (m - 1)S_Y^2$ to estimate it. Since the variable T is τ_{n+m-2}, where

$$T = \frac{(\bar{X} - \bar{Y}) - (\mu_X - \mu_Y)}{\sqrt{\dfrac{(n-1)S_X^2 + (m-1)S_Y^2}{n + m - 2}\left(\dfrac{1}{n} + \dfrac{1}{m}\right)}},$$

we can conclude that the interval with endpoints

$$(\bar{X} - \bar{Y}) \pm t_{n+m-2}(1 - \alpha/2)\sqrt{\frac{(n-1)S_X^2 + (m-1)S_Y^2}{n + m - 2}\left(\frac{1}{n} + \frac{1}{m}\right)}$$

is a $100(1 - \alpha)$ percent random confidence interval for $\mu_X - \mu_Y$.

Case 5:

Suppose X is $B(1, p)$. Then the MLE of p is $\bar{X}$. The central limit theorem tells us that $\bar{X}$ is $N(p, p(1 - p)/n)$ if n is large. However, the distribution of $\bar{X} - p$ still depends on p, as its variance is $p(1 - p)/n$. Thus, we must estimate $\sigma_{\bar{X}}^2$. Using the invariance property of maximum-likelihood estimators, we have $\sigma_{\bar{X}}^* = \sqrt{\bar{X}(1 - \bar{X})/n}$. When n is large, the normal approximation still suffices to assert that $(\bar{X} - p)/\sqrt{\bar{X}(1 - \bar{X})/n}$ is $N(0, 1)$, so

$$[\bar{X} - z_{1-(\alpha/2)}\sqrt{\bar{X}(1 - \bar{X})/n},\ \bar{X} + z_{1-(\alpha/2)}\sqrt{\bar{X}(1 - \bar{X})/n}]$$

is a $100(1 - \alpha)$ percent random confidence interval for p.

Case 6:

Suppose X is $N(\mu, \sigma^2)$ and μ is known. Then the MLE of σ^2 is $\Sigma_{i=1}^n (X_i - \mu)^2$, and $\Sigma_{i=1}^n (X_i - \mu)^2/\sigma^2$ is χ_n^2, so

$$1 - \alpha = P\left[\chi_2^n(\alpha/2) \le \frac{\Sigma_{i=1}^n (X_i - \mu)^2}{\sigma^2} \le \chi_n^2(1 - \alpha/2)\right]$$
$$= P\left[\frac{\Sigma_{i=1}^n (X_i - \mu)^2}{\chi_n^2(1 - \alpha/2)} \le \sigma^2 \le \frac{\Sigma_{i=1}^n (X_i - \mu)^2}{\chi_n^2(\alpha/2)}\right].$$

Consequently, the interval above is a $100(1 - \alpha)$ percent random confidence interval for σ^2.

Case 7:

Suppose X is $N(\mu, \sigma^2)$ and μ is unknown. Then the MLE of σ^2 is $\frac{1}{n}\Sigma_{i=1}^n (X_i - \bar{X})^2$, and $\Sigma_{i=1}^n (X_i - \bar{X})^2/\sigma^2$ is χ_{n-1}^2, so

$$\left[\frac{\Sigma_{i=1}^n (X_i - \bar{X})^2}{\chi_{n-1}^2(1 - \alpha/2)}, \frac{\Sigma_{i=1}^n (X_i - \bar{X})^2}{\chi_{n-1}^2\alpha/2}\right]$$

is a $100(1 - \alpha)$ percent random confidence interval for σ^2.

We are now familiar with both point estimation and interval estimation, and it is appropriate to inquire at this time as to their relative merits. This is analogous to the problem we faced when we condensed all the information contained in the probability mass function of a random variable into a single number—the expected value On one hand, it is much easier to use the expected value. On the other hand, the probability mass function conveys more information. Thus, point estimation is preferable when the gain in simplicity outweighs the loss of information and vice versa.

We conclude this chapter with one final observation. If we increase the sample size n, then the expected length of our $100(1 - \alpha)$ percent random confidence interval decreases. In fact, as n increases to ∞, the expected length of the random confidence interval decreases. Of particular importance is the problem of determining the sample size needed to obtain a confidence interval of prescribed length. We consider this problem in the next example.

EXAMPLE 29. (Three Sisters)

Let p be the probability that it is the youngest of the three sisters who is responsible for breaking a dish. Then how large must we choose n in order to ensure that the probability that $\bar{X}$ (our estimator of p) differs from p by more than .01 is less than .05?

SOLUTION: From Case 5 we see that ($z_{1-(\alpha/2)} = 1.96$ for $\alpha = .05$)

$$\left[\bar{X} - 1.96\sqrt{\frac{\bar{X}(1-\bar{X})}{n}},\ \bar{X} + 1.96\sqrt{\frac{\bar{X}(1-\bar{X})}{n}}\right]$$

is a 95 percent random confidence interval for p. Moreover, whatever the true value of p, we know that $\sqrt{\bar{X}(1-\bar{X})} \leq \sqrt{\frac{1}{4}} = \frac{1}{2}$, so the length of our random confidence interval for p will never exceed

$$2(1.96)(\tfrac{1}{2})/\sqrt{n} = 1.96/\sqrt{n}.$$

Asking that $\bar{X}$ differ from p by no more than .01 is equivalent to asking that the length of the random confidence interval not exceed .02. This certainly occurs whenever

$$\frac{1.96}{\sqrt{n}} \leq .02,$$

or equivalently whenever

$$n \geq 98^2 = 9604. \bullet$$

EXERCISES:

56. An anthropologist measured the height (in inches) of a random sample of 100 men from a certain isolated island. He found $\bar{x} = 71$ and $s = 3$. Find a 95 and a 99 percent random confidence interval and the concomitant confidence intervals for μ.

57. The admissions office needed to know the average SAT score of this year's entering freshmen. Unfortunately, they had no electronic data-processing equipment, so inspection of all student folders was seen as too burdensome a task. Using the fact that these test scores are normally distributed with $\sigma = 100$, how many folders must be inspected if they are to obtain a 90 percent random confidence interval of length 10?

58. Reformulate the examples from Chapter 5 named below so that they are problems of interval estimation. Choose the confidence coefficient to be $100(1-\alpha)$, where α is the level of significance in the example. Find the $100(1-\alpha)$ percent random confidence interval, and use the data to find the $100(1-\alpha)$ percent confidence interval:
    (a) Example 1.
    (b) Example 2.
    (c) Example 4.
    (d) Example 5.
    (e) Example 8.
    (f) Example 9.

59. A survey of family income in a town of 1000 families produced the following information. A random sample of size 200 without replacement was taken, and the statistician found that $\bar{x} = 9870$ and $s^2 = 1250^2$. Now the dis-

tribution of family income is not normally distributed. Use the Chebychev inequality and Theorem 7 of Chapter 2 to find a 90 percent confidence interval for the town's average family income.

60. Consider the random confidence interval in Case 5. Give an explanation for the fact that the expected length of this random confidence interval increases as $|p - \frac{1}{2}|$ decreases.

61. Occasionally we might want to construct a *one sided random confidence interval*—that is, an interval of the form $[L, \infty]$. Here is an example. A manufacturer of cigarettes is quite sure that the length of his cigarettes is normally distributed with mean μ and standard deviation of .05 millimeters. He has an idea that μ is close to 100.1, but for advertising purposes he would like to find a length that he is quite sure is exceeded by μ. He decides that a 99 percent confidence interval will serve this purpose.
    (a) Find a 99 percent one-sided random confidence interval for μ.
    (b) How large a sample is needed if the interval is to be $[100, \infty]$ when $\bar{x} = 100.1$?

62. Comment on the relationship between interval estimation and hypothesis testing. (Take note of Exercise 61.)

8

LINEAR REGRESSION

1. INTRODUCTION

One of the principal objectives of the natural, physical, and social sciences is to formulate "laws of interaction" that concisely describe the system being studied. Typically, these laws describe the effects that some variables exert (or seem to exert) on other variables. Furthermore, these laws can often be expressed as simple functional relationships between the variables of the system. We illustrate this last point with two examples:

(1) Suppose a projectile is shot straight up into the air, and we are interested in knowing the height h it will reach. Physicists tell us that a good approximation of h is $\frac{1}{2}v^2/g$, where g is the acceleration of gravity and v is the initial velocity of the projectile. That is, the functional relationship between the two variables h and v—remember, g is a constant and not a variable—can be expressed by the equation

$$h = \frac{\frac{1}{2}v^2}{g}.$$

(2) For some firms, the cost y of producing x units of a product is given by the functional relationship

$$y = K + cx,$$

where the constants K and c represent the set-up cost for the entire production run and the marginal cost of producing each unit.

The second of these two functional relationships is called linear. Thus, we say that there is a linear relationship between the variables x and y if there are constants α and β such that $y = \alpha + \beta x$. Linear relationships are useful and important, not only because there are many systems wherein the relationships are actually of this form, but also because many complicated nonlinear relationships, which would be difficult to describe, can often be closely approximated by a linear relationship. Consequently, it behooves us, as scientists, to first focus our attention on linear relationships whose simplicity makes it easier to gain an understanding of the role of the important variables in the system we are studying. Moreover, we shall restrict our study to linear relationships involving only two variables. (The multivariable case is very important, but it requires a knowledge of linear algebra.)

Although the variables in the examples presented above were not random, we shall be concerned with systems in which one of the two variables is random and the other is not. For example, a farmer can control the number x of acres he plants in wheat, but the number Y of bushels of wheat that he will be able to harvest is a random variable (which depends on x). In this case, the linear relationship is expressed by

$$E(Y \mid x) = \alpha + \beta x,$$

or equivalently by

$$Y = \alpha + \beta x + \epsilon,$$

where the random variable ϵ is an unobservable error term with mean zero.

Throughout this chapter, the symbol x will denote the mathematical (*nonrandom*) variable, usually referred to as the *controllable* or *independent variable*; $x_1, x_2, \ldots, x_n$ will denote the particular values of the independent variable used in an experiment. The symbol Y will denote the random variable, usually referred to as the *dependent variable*. The observations on the dependent variable corresponding to x_i will be denoted by Y_i, so $E(Y_i) = \alpha + \beta x_i$. The observed value of Y_i will be labeled y_i, while the actual (though unobserved) value of the random variable ϵ_i will be labeled e_i. The line $y = \alpha + \beta x$ is called the *regression line*. The data for a typical regression problem are presented in Figure 8.1 in what is known as a *scatter diagram*.

EXAMPLE 1. (Insurance)

In order to protect policyholders against the possibility of an insurance company's not being able to pay their claims, governmental regulatory agencies require insurance companies to carry a surplus in addition to other contingency reserves (such as unearned premium reserves and claim reserves). This surplus is needed as protection against unusually large

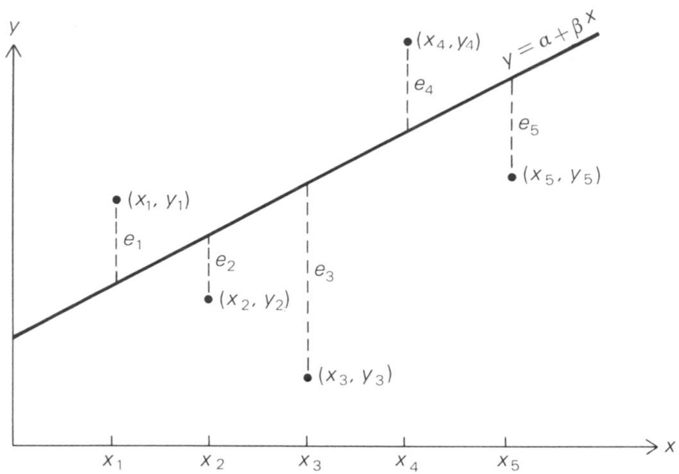

Figure 8.1 Scatter Diagram

claims and against declines in the value of the company's assets. While large surpluses provide extra protection to policyholders, they also bring the policyholders higher premium costs. An intriguing problem then is to decide how large a surplus is adequate.

In discussing this question, Hofflander[1] brings in the notion of a company's combined *loss and expense ratio*. This ratio is simply the total cost in cents incurred per dollar of earned premium. Thus, if for every dollar of premium earned, 18 cents is spent on overhead and 76 cents is spent on paying claims, then the loss and expense ratio is 94. Other things being equal, a low ratio is desirable. Consequently, one might be led to ask whether there are any (easily measured) variables that appear to influence the magnitude of this ratio. In particular, we shall investigate the relationship between annual premium volume and loss and expense ratio in the field of health insurance. We will label these two variables[2]

[1] Alfred E. Hofflander, "Minimum Capital and Surplus Requirements for Multiple Line Insurance Companies: A New Approach," Chapter 6 in Kimball and Denenberg, eds., *Insurance, Government, and Social Policy: Studies in Insurance Regulation* (Homewood, Ill.: Richard D. Irwin, Inc. 1969).

[2] Although an argument can be given to assert that x is a controllable variable, it really appears that x is a random variable. But as we will see in Section 4, the fact that x is not really controllable is of no consequence in our analysis. We shall, therefore, proceed as if x were controllable.

x and Y. And because the relationship between these two variables is in fact linear for other lines of insurance (such as auto, credit, and life), it is reasonable to believe that this relationship is also linear in the health line. The data for 1960 are given in Table 8.1 and depicted in Figure 8.2. •

Table 8.1 Data for Health Insurance Companies, 1960

Company	1	2	3	4	5	6
Annual Premium Volume (in \$1,000's)	1084	828	349	699	657	271
Loss and Expense Ratio (rounded to nearest whole number)	91	85	81	90	71	69

Since the error terms ϵ_i are unobservable, the first problem we encounter is that of estimating the unknown parameters α and β that determine the regression line. Another problem is to determine whether we are justified in assuming a linear relationship between the two variables.

We now precisely spell out the conditions under which we shall answer these and other questions of statistical inference concerning our linear model.

DEFINITION

Let $Y_1, Y_2, \ldots, Y_n$ be independent, observable random variables such that

$$Y_i = \alpha + \beta x_i + \epsilon_i,$$

where α and β are unknown parameters, x_i are observable mathematical (nonrandom) variables, and ϵ_i are uncorrelated, unobservable random variables with mean 0 and variance σ^2, so

$$E(\epsilon_i) = 0,$$
$$E(\epsilon_i\epsilon_j) = 0 \qquad \text{if } i \neq j,$$

and

$$E(\epsilon_i^2) = \sigma^2.$$

Then these conditions define a *simple linear model*. If, in addition, we assume that the random variables ϵ_i are each normally distributed, then these conditions define a *normal simple linear model*.

In Section 2 we will study the normal simple linear model and then in Section 3 the simple linear model. In the final section we offer a few words of warning against the misuse of the methods and models presented in this chapter.

EXERCISES:

Make a scatter diagram for the data in Exercises 1, 2, 3.

1. In deciding on the profitability of hiring an additional salesman, the sales manager made use of the data below in a linear regression model.

Number of salesmen employed	2	3	4	5	6
Average sales in dollars	2000	2700	3300	3800	4350

2. There is reason to believe that there is a linear relationship between the number of pounds per acre of fertilizer and the yield of wheat per acre. Six one-acre plots were planted in wheat with the following results:

Pounds of fertilizer	50	60	70	80	90	100
Yield (in 1000's of bushels)	7	17	23	26	28	31

3. A study of car depreciation revealed the following pairs of car ages (in years) and car prices (in dollars):

(0, 2500), (0, 2450), (0, 2285);
(1, 1950), (1, 2000), (1, 1925), (1, 1850);
(2, 1650), (2, 1600);
(3, 1300), (3, 1350), (3, 1325);
(4, 1050), (4, 925), (4, 1000);
(5, 875), (5, 825);
(6, 625), (6, 675), (6, 600), (6, 650), (6, 625);
(7, 475), (7, 525);
(8, 450), (8, 425), (8, 350), (8, 375).

All 28 cars in the sample had similar characteristics (for example, they all had power steering, automatic transmissions, no air conditioning, and were the same make and style) and cars in the same age group had undergone similar use (as measured by the number of miles on the odometer).

2. THE NORMAL SIMPLE LINEAR MODEL

Throughout this section we shall assume[3] that the unobservable error terms ϵ_i in our simple linear model are independent and normally distributed, each with mean 0 and the same variance σ^2. In the next three subsections we consider problems of estimation, hypothesis testing, and prediction. In the fourth and last subsection we investigate the assumption of linearity.

2.1 Estimating α, β, and σ^2

We use the method of maximum-likelihood to estimate α, β, and σ^2. Since the ϵ_i are independent normal random variables, it follows that

$$L \equiv \underset{\alpha,\beta,\sigma^2}{L(\mathbf{e})} = \left(\frac{1}{\sqrt{2\pi}\,\sigma}\right)^n \exp\left(-\frac{1}{2\sigma^2}\sum_{i=1}^{n} e_i^2\right) \tag{8.1}$$

$$= \left(\frac{1}{\sqrt{2\pi}\,\sigma}\right)^n \exp\left\{-\frac{1}{2\sigma^2}\sum_{i=1}^{n} [y_i - (\alpha + \beta x_i)]^2\right\}. \tag{8.2}$$

On setting $\partial \log L/\partial\sigma^2 = 0$, $\partial \log L/\partial\alpha = 0$, and $\partial \log L/\partial\beta = 0$, we obtain

$$-\frac{n}{2}\frac{1}{\hat{\sigma}^2} + \frac{1}{2}\left[\frac{1}{\hat{\sigma}^2}\right]^2 \sum_{i=1}^{n} (y_i - \hat{\alpha} - \hat{\beta}x_i)^2 = 0, \tag{8.3}$$

$$\sum_{i=1}^{n} (y_i - \hat{\alpha} - \hat{\beta}x_i) = 0, \tag{8.4}$$

and

$$\sum_{i=1}^{n} x_i(y_i - \hat{\alpha} - \hat{\beta}x_i) = 0. \tag{8.5}$$

[3] This assumption is equivalent to the independence and normality of the Y_i.

[Equations (8.4) and (8.5) are called the normal equations.] Solving for $\hat{\alpha}$, $\hat{\beta}$, and $\hat{\sigma}^2$, we see that the maximum-likelihood estimators are

$$\hat{\alpha} = \bar{Y} - \hat{\beta}\bar{x}, \tag{8.6}$$

$$\hat{\beta} = \frac{\sum_{i=1}^{n} (x_i - \bar{x})(Y_i - \bar{Y})}{\sum_{i=1}^{n} (x_i - \bar{x})^2} \quad \left(= \frac{\sum_{i=1}^{n} (x_i - \bar{x})Y_i}{\sum_{i=1}^{n} (x_i - \bar{x})^2} \right), \tag{8.7}$$

and

$$\hat{\sigma}^2 = \frac{1}{n} \sum_{i=1}^{n} (Y_i - \hat{\alpha} - \hat{\beta}x_i)^2. \tag{8.8}$$

A moment's reflection reveals that equation (8.8) is, in fact, intuitive. Here's why. If we knew the true values of the e_i's, then our estimate of σ^2 would have been $(1/n)\Sigma e_i^2$ [as $E(\epsilon_i) = 0$]. But we do not know e_i, so we estimate it by $d_i \equiv Y_i - (\hat{\alpha} + \hat{\beta}x_i)$, and hence $(1/n)\Sigma d_i^2$ becomes our estimate of σ^2. Also, notice that $\hat{\beta}$ would be undefined—or equivalently that the normal equations would have infinitely many solutions—if all the values of x_i were the same. Hence, we will *always* require (or assume) that there be at least two distinct x values.

EXAMPLE 2. (Insurance)

As the total loss and expense incurred on an amount x of premiums earned is the sum of a great many small effects, we can invoke the central limit theorem to justify our assumption that the total loss and expense incurred is normally distributed. Hence, by Theorem 1 of Chapter 4, the loss and expense ratio is normally distributed for each fixed value of x. Using the data in Table 8.1, we find $\bar{x} = 648$, $\bar{y} = 81.17$,

$$\Sigma_{i=1}^{6}(x_i - \bar{x})y_i = 39{,}676 + 15{,}300 - 24{,}219 + 4590 + 639 - 26{,}013 = 9973,$$

and

$$\Sigma_{i=1}^{6}(x_i - \bar{x})^2 = 190{,}096 + 32{,}400 + 89{,}401 + 2601 + 81 + 142{,}129 = 456{,}708,$$

so the maximum-likelihood estimates are

$$\hat{\beta} = \frac{9973}{456{,}708} = .0218 \quad \text{and} \quad \hat{\alpha} = 81.17 - .0218(648) = 67.044.$$

Also,

$$\hat{\sigma}^2 = \tfrac{1}{6}[(.325)^2 + (-.094)^2 + (6.348)^2 + (7.718)^2 + (-10.367)^2 + (-3.952)^2] = 37.179.$$

These results and the corresponding estimated regression line are depicted in Figure 8.2. •

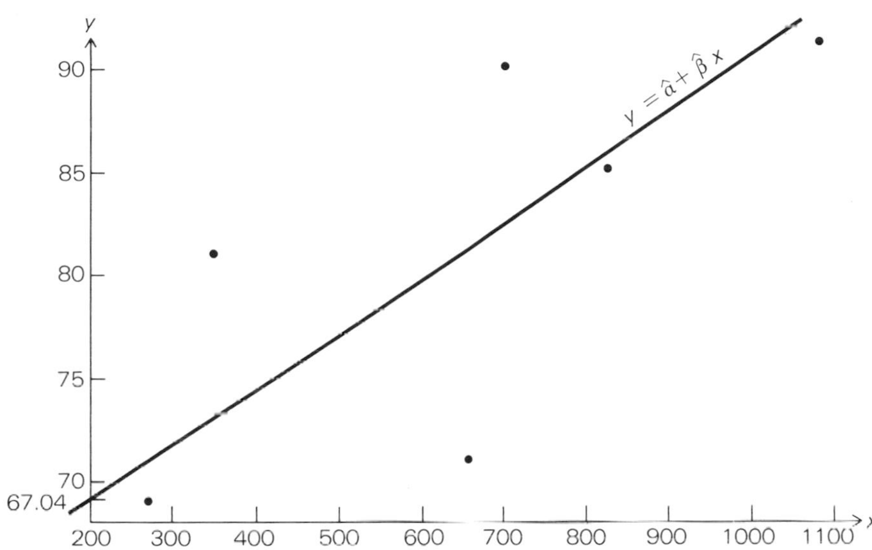

Figure 8.2 Scatter Diagram for Health Insurance Companies, 1960

Next, we seek the distributions of $\hat{\alpha}$, $\hat{\beta}$, and $\hat{\sigma}^2$. Since the Y_i are independent normal random variables, we see from the second form of $\hat{\beta}$ given in (8.7) that $\hat{\beta}$ can be written as a sum of independent normal random variables, Therefore, it follows from Theorem 2 of Chapter 3 that $\hat{\beta}$ is normally distributed. Furthermore, straightforward calculations show (see Exercise 6) that

$$E(\hat{\beta}) = \beta \quad \text{and} \quad \operatorname{Var}(\hat{\beta}) = \frac{\sigma^2}{\sum_{i=1}^{n} (x_i - \bar{x})^2}. \tag{8.9}$$

Similarly we can show that $\hat{\alpha}$ is normally distributed with

$$E(\hat{\alpha}) = \alpha \quad \text{and} \quad \operatorname{Var}(\hat{\alpha}) = \frac{\sigma^2 \sum_{i=1}^{n} x_i^2}{n \sum_{i=1}^{n} (x_i - \bar{x})^2}. \tag{8.10}$$

It can also be shown[4] that

$$\frac{n\hat{\sigma}^2}{\sigma^2} \text{ is } \chi^2_{n-2}, \tag{8.11}$$

$$\hat{\alpha} \text{ and } \hat{\beta} \text{ are distributed independently of } \hat{\sigma}^2, \text{ and} \tag{8.12}$$

$$\operatorname{Cov}(\hat{\alpha}, \hat{\beta}) = -\frac{\hat{\sigma}^2\bar{x}}{\sum_{i=1}^{n}(x_i - \bar{x})^2}. \tag{8.13}$$

Finally, we can use Theorem 3 of Chapter 7 and equation (8.2) to show that $\hat{\alpha}$, $\hat{\beta}$, and $\hat{\sigma}^2$ are sufficient, and it follows from Exercsie 38 of Chapter 7 that they are complete so they are minimum-variance unbiased estimators. Consistency follows from equations (8.9), (8.10), and (8.11). We summarize these results in the next theorem.

THEOREM 1

The maximum-likelihood estimators $\hat{\alpha}$, $\hat{\beta}$, $[n/(n-2)]\hat{\sigma}^2$ defined in (8.6)–(8.8) are minimum-variance unbiased, consistent, and sufficient;

$$\hat{\alpha} \text{ is } N\left(\alpha, \frac{\sigma^2\Sigma x_i^2}{n\Sigma(x_i - \bar{x})^2}\right), \qquad \hat{\beta} \text{ is } N\left(\beta, \frac{\sigma^2}{\Sigma(x_i - \bar{x})^2}\right),$$

and $n\hat{\sigma}^2/\sigma^2$ is χ^2_{n-2}; $\hat{\alpha}$ and $\hat{\beta}$ are independent of $\hat{\sigma}^2$.

2.2 Confidence Intervals and Tests of Hypotheses for α, β, and σ^2

For simplicity in notation we define the sum of squares of the x's and the sum of squared deviations about the average of the x's by SS and SD. That is,

$$SS \equiv \sum_{i=1}^{n} x_i^2$$

and

$$SD \equiv \sum_{i=1}^{n} (x_i - \bar{x})^2.$$

Of course, neither SS nor SD is a random variable.

To begin, recall from (8.12) that $\hat{\alpha}$ and $\hat{\sigma}^2$ are independent and from

[4] See Mood and Graybill (Reference 18), pp. 331–333.

(8.10) and (8.11) that

$$U \equiv \frac{\hat{\alpha} - \alpha}{\sigma} \sqrt{\frac{nSD}{SS}} \quad \text{is} \quad N(0, 1) \qquad \text{and} \qquad \frac{n\hat{\sigma}^2}{\sigma^2} \quad \text{is} \quad \chi^2_{n-2},$$

so from our definition of Student's-t random variable we see that

$$A \equiv \frac{U}{\sqrt{n\hat{\sigma}^2/\sigma^2}} \sqrt{n-2} = (\hat{\alpha} - \alpha) \sqrt{\frac{(n-2)SD}{SS\hat{\sigma}^2}} \qquad \text{is } \tau_{n-2}.$$

Similarly,

$$B \equiv (\hat{\beta} - \beta) \sqrt{\frac{(n-2)SD}{n\hat{\sigma}^2}} \qquad \text{is } \tau_{n-2}.$$

Hence, as shown in Case 2 of Section 5, Chapter 7, the intervals with endpoints

$$\hat{\alpha} \pm t_{n-2}(1 - \delta/2) \sqrt{SS\hat{\sigma}^2/(n-2)SD},$$

and

$$\hat{\beta} \pm t_{n-2}(1 - \delta/2) \sqrt{n\hat{\sigma}^2/(n-2)SD},$$

are $100(1 - \delta)$ percent random confidence intervals[5] for α and β, respectively. As shown in Case 7 of Section 5, Chapter 7,

$$\left[\frac{n\hat{\sigma}^2}{\chi^2_{n-2}(1 - \delta/2)}, \frac{n\hat{\sigma}^2}{\chi^2_{n-2}(\delta/2)}\right]$$

is a $100(1 - \delta)$ percent random confidence interval for σ^2.

Next, suppose we wish to test $H_0: \alpha = \alpha_0$ against $H_A: \alpha > \alpha_0$ at level of significance δ. Then as shown in Case 2 of Section 2, Chapter 5, we should reject H_0 whenever $A \geq t_{n-2}(1 - \delta)$, with α_0 replacing α in the definition of A. Of course, if the alternative hypothesis is $H_A: \alpha \neq \alpha_0$, then we reject H_0 whenever $|A| \geq t_{n-2}(1 - (\delta/2))$—that is, whenever α_0 is not in the $100(1 - \delta)$ percent random confidence interval for α given above.

Tests for β and σ^2 are made in a similar manner. Of particular interest is the test of $H_0: \beta = 0$ against $H_A: \beta \neq 0$, for $\beta = 0$ means that there is no relation between the expected value of Y and the mathematical variable x. That is, the null hypothesis $\beta = 0$ tests whether the means of the family of normal distributions under consideration are independent of the variable x.

[5] We use the letter δ in the expression "$100(1 - \delta)$ percent" because we are already using the letter α to represent one of the unknown parameters.

EXAMPLE 3. (Insurance)

Let us test $H_0: \beta = 0$ against $H_A: \beta \neq 0$ in our insurance example at level of significance .05. As shown above, we will reject H_0 only if the observed value of B exceeds $t_4(.975) = 2.776$ in absolute value, where $\beta = 0$ in the definition of B. From our calculations in Examples 1 and 2 we have

$$\left| .0218 \left[\frac{4(456{,}708)}{6(37.179)} \right]^{1/2} \right| = |1.973| < 2.776.$$

Hence, we accept the null hypothesis that $\beta = 0$, and we conclude that there is no meaningful relationship between the ratio and annual premium volume. •

2.3 Prediction

One of the most important uses of the estimated regression line is in prediction. Often we are interested in predicting the value of Y for some specified value of x, say x_0. For example, suppose a farmer has taken n observations to estimate α and β, where the variables x and Y in the model $E(Y) = \alpha + \beta x$ represent the number of acres he plants in wheat and the number of bushels of wheat he will be able to harvest. We would *predict* a harvest of $\hat{\alpha} + \hat{\beta} x_0$ bushels of wheat if x_0 acres were planted.

Just as we were able to construct a $100(1 - \delta)$ percent random confidence interval for an unknown parameter, we can also construct an interval, called a $100(1 - \delta)$ percent *random prediction interval,* such that the probability that Y_0 will be in the interval is $1 - \delta$. If we knew the true values of α, β, and σ^2, then the desired $100(1 - \delta)$ percent prediction interval would simply be

$$[\alpha + \beta x_0 - z_{1-(\delta/2)}\sigma,\ \alpha + \beta x_0 + z_{1-(\delta/2)}\sigma].$$

This would seem to suggest that we use

$$[\hat{\alpha} + \hat{\beta} x_0 - z_{1-(\delta/2)}\hat{\sigma},\ \hat{\alpha} + \hat{\beta} x_0 + z_{1-(\delta/2)}\hat{\sigma}]$$

as our prediction interval. While this last interval accounts for the variation in Y_0 about its expected value, it does not account for the variation in our estimates of $\alpha + \beta x_0$ and of σ^2. Thus, we must obtain a different—and wider—interval.

Now the variable $U \equiv Y_0 - \hat{\alpha} - \hat{\beta} x_0$ is normally distributed, since it is a weighted sum of the independent normal random variables $Y_0, Y_1, Y_2, \ldots, Y_n$. Also, $E(U) = 0$, since $\hat{\alpha}$ and $\hat{\beta}$ are unbiased and $E(Y_0) = \alpha + \beta x_0$. In view of this and of the fact that Y_0 is independent of $\hat{\alpha}$ and $\hat{\beta}$, we have from (8.9), (8.10), and (8.13) that

$$\begin{aligned}
\sigma_U^2 &= E(U^2) = E\{(Y_0 - \hat{\alpha} - \hat{\beta}x_0)^2\} \\
&= E\{[(Y_0 - \alpha - \beta x_0) - (\hat{\alpha} + \hat{\beta}x_0 - \alpha - \beta x_0)]^2\} \\
&= \mathrm{Var}(Y_0) + \mathrm{Var}(\hat{\alpha} + \hat{\beta}x_0) \\
&= \sigma^2 + \mathrm{Var}(\hat{\alpha}) + x_0^2\,\mathrm{Var}(\hat{\beta}) + 2x_0\,\mathrm{Cov}(\hat{\alpha}, \hat{\beta}) \\
&= \sigma^2 + \frac{\sigma^2\Sigma x_i^2}{n\Sigma(x_i - \bar{x})^2} + \frac{x_0^2\sigma^2}{\Sigma(x_i - \bar{x})^2} - \frac{2x_0\bar{x}\sigma^2}{\Sigma(x_i - \bar{x})^2} \\
&= \sigma^2\left[1 + \left(\frac{1}{n}\Sigma x_i^2 + x_0^2 - 2x_0\bar{x} + \bar{x}^2 - \bar{x}^2\right)\Big/\Sigma(x_i - \bar{x})^2\right] \\
&= \sigma^2\left[1 + \frac{1}{n} + (x_0 - \bar{x})^2/\Sigma(x_i - \bar{x})^2\right].
\end{aligned}$$

Next, recall that $n\hat{\sigma}^2/\sigma^2$ is χ^2_{n-2} and that $\hat{\sigma}^2$ is independent of $\hat{\alpha}$ and $\hat{\beta}$, so $\hat{\sigma}^2$ and U are independent. Hence,

$$\frac{U/\sigma_U}{\sqrt{n\hat{\sigma}^2/\sigma^2}}\sqrt{n - 2} \text{ is } \tau_{n-2}.$$

Consequently, the desired $100(1 - \delta)$ percent prediction interval has end points

$$\hat{\alpha} + \hat{\beta}x_0 \pm \hat{\sigma}t_{n-2}(1 - \delta/2)\sqrt{\frac{n}{n - 2}\left[\frac{n + 1}{n} + \frac{(x_0 - \bar{x})^2}{\Sigma\,(x_i - \bar{x})^2}\right]}.$$

EXAMPLE 4. (Insurance)

Suppose the directors of a fire insurance company decide to start underwriting health insurance, and on the basis of past history and their current sales force they figure that their annual premium volume will be \$525,000. Then, using our previous calculations, a 90 percent prediction interval for their loss and expense ratio is

$$[78.49 - 17.40, 78.49 + 17.40] = [61.09, 95.89],$$

as

$$x_0 = 525, \quad \hat{\alpha} = 67.04, \quad \hat{\beta} = .0218, \quad \hat{\sigma} = 6.097, \quad t_4(.95) = 2.13,$$

and the quantity under the square-root sign is

$$\frac{6}{4}\left[\frac{7}{6} + \frac{15{,}129}{456{,}708}\right] = [1.34]^2. \bullet$$

Note the following three facts: the quantity under the square-root sign is greater than $\sqrt{n/(n-2)}$, $t_{n-2}(1-(\delta/2)) \geq z_{1-(\delta/2)}$, and

$$E(\hat{\sigma}^2) = \frac{n}{n-2}\sigma^2.$$

Together they imply that the expected length of the prediction interval is greater than $2\sigma z_{1-(\delta/2)}$, which would have been the length of our prediction interval had we known α, β, and σ^2. The difference, however, between the expected lengths of these two intervals decreases to zero as n increases to infinity (see Exercise 18). Also note that in forming prediction intervals for additional values of x_0, our estimates of α, β, σ^2, and σ_U^2 must be revised as additional observations are made (see Exercise 17).

*2.4 Testing the Assumption of Linearity

So far, our analysis has proceeded on the assumption that a linear relationship does, in fact, exist between x and the expected value of Y given x; it is important, then, to check whether this assumption is valid. We now present a method for testing this assumption.

In order to test the hypothesis of linearity, we must have several observations for each distinct value of x_i in our sample. That is, we require for each i, $1 \leq i \leq n$, that there be an integer j such that $1 \leq j \leq n$, $j \neq i$, and $x_i = x_j$. This requirement should present no problem, as the x variable is controllable. (If it is not—that is, if x is a random variable—then we are not dealing with a regression problem but rather with a correlation problem. See Section 4.)

For ease in exposition, let us write this requirement in the following form:

$$\begin{array}{ll} Y_{11}, Y_{12}, \ldots, Y_{1n_1} & \text{are the } n_1 \text{ observations at the } x \text{ value } x_1, \\ Y_{21}, Y_{22}, \ldots, Y_{2n_2} & \text{are the } n_2 \text{ observations at the } x \text{ value } x_2, \\ \cdot & \\ \cdot & \\ \cdot & \\ Y_{k1}, Y_{k2}, \ldots, Y_{kn_k} & \text{are the } n_k \text{ observations at the } x \text{ value } x_k; \end{array} \tag{8.14}$$

$$n_i \geq 2, \qquad i = 1, 2, \ldots, k, \tag{8.15}$$

$$\sum_{i=1}^{k} n_i = n, \tag{8.16}$$

and

$$k \geq 2. \tag{8.17}$$

Here, k denotes the number of distinct *values* of the x_i in our sample.

Equation (8.17) states the previously imposed requirement that there be at least two distinct values of x_i in our sample, whereas equation (8.16) states as before that n is the total number of observations. Our new requirement that there be several observations for each distinct value of x_i is stated in equation (8.15). Equation (8.14) is merely a relabeling of the x and Y values.

In testing the null hypothesis of a linear relationship between x and $E(Y \mid x)$, we shall employ two different estimators of σ^2. The first estimator does not depend upon the linearity assumption, whereas the second does. Moreover, the further the departure of the true relationship from linearity, the larger the expected value of the second estimator. Thus, the ratio of the second estimator to the first, a random variable whose distribution can be found, provides us with a convenient method for testing the null hypothesis—namely, reject H_0 if the observed value of this ratio is too large.

To begin, define

$$\bar{Y}_i = \frac{1}{n_i} \sum_{j=1}^{n_i} Y_{ij}, \qquad \text{for } i = 1, 2, \ldots, k.$$

First note that no matter what the relationship between x and $E(Y \mid x)$, we have, for each i, $E(Y_{ij}) = E(\bar{Y}_i)$ for $j = 1, 2, \ldots, n_i$. Next, observe that one of the assumptions we have been using is that $\mathrm{Var}(Y \mid x) = \sigma^2$ for all x, so, in particular, $\mathrm{Var}(Y_{ij}) = \sigma^2$ for each pair (i, j). Consequently, it follows from Theorem 1 of Chapter 5 that for each i,

$$S_{P_i}^2 \equiv \frac{1}{n_i - 1} \sum_{j=1}^{n_i} (Y_{ij} - \bar{Y}_i)^2$$

is an unbiased estimator of σ^2, for $(n_i - 1)S_{P_i}^2/\sigma^2$ is $\chi_{n_i-1}^2$. Since the k groups of variables $(Y_{11}, \ldots, Y_{1n_1})$; $(Y_{21}, \ldots, Y_{2n_2})$; $\ldots$; $(Y_{k1}, \ldots, Y_{kn_k})$ are independent, it also follows that

$$S_P^2 \equiv \frac{1}{n - k} \sum_{i=1}^{k} \sum_{j=1}^{n_i} (Y_{ij} - \bar{Y}_i)^2 \tag{8.18}$$

is an unbiased estimator of σ^2, because $(n - k)S_P^2/\sigma^2$ is χ_{n-k}^2, as stated at the beginning of the proof of Theorem 1 of Chapter 5.[6]

The estimator S_P^2 is often called the *pooled variance* or the *within-group variance*, for it is a weighted sum of the estimators of σ^2 for the ith group—

[6] Note that this result requires that the variance for each of the k groups be equal.

namely,

$$S_P^2 = \sum_{i=1}^{k} \frac{n_i - 1}{n - k} S_{P_i}^2.$$

Sometimes S_P^2 is called the *explained variance,* for this portion of $\hat{\sigma}^2$, the mean squared deviations of the observations Y_{ij} about the estimated regression line $\hat{\alpha} + \hat{\beta}x$, is due entirely to variations within each of the k groups. Thus, the relationship between x and $E(Y \mid x)$ cannot affect the value of S_P^2 so it does not depend upon the linearity assumption. In particular, S_P^2 is "explained" by the linear model in the sense that no other relationship can reduce it (further).

We now obtain a second estimator for σ^2 that does depend on the assumption of linearity. Under the null hypothesis,

$$E(\bar{Y}_i) = \alpha + \beta x_i = E(\hat{\alpha} + \hat{\beta}x_i),$$

so

$$S_M^2 \equiv \frac{1}{k - 2} \sum_{i=1}^{k} n_i[\bar{Y}_i - (\hat{\alpha} + \hat{\beta}x_i)]^2 \qquad (8.19)$$

is unbiased, as it can be shown[7] that $(k - 2)S_M^2/\sigma^2$ is χ_{k-2}^2. The estimator S_M^2 is often called the *between-group variance,* for the estimated regression coefficients $\hat{\alpha}$ and $\hat{\beta}$ were computed using all n observations, so the deviations of the group means from their computed expected values reflects the extent to which there are differences between the k group variances. Continuing, we have assumed that the k groups do, in fact, have the same variance. Consequently, this (S_M^2) portion of $\hat{\sigma}^2$ is attributable to the linearity assumption, in that it could be reduced further by some other relationship. For this reason, S_M^2 is sometimes called the *unexplained variance* or the *variance due to regression.*

It is interesting to note that the more the true relationship deviates from the assumption of linearity, the larger the (expected) value of S_M^2. Hence, we should reject the null hypothesis when S_M^2 is large. The drawback of this proposal is that the distribution of S_M^2 depends upon the unknown parameter σ^2.

This suggests that we use S_P^2 as our estimate of σ^2. Then, since S_P^2 does not depend upon the null hypothesis, we should reject the null hypothesis whenever S_M^2/S_P^2 is large. Moreover, the distribution of this ratio does not contain any unknown parameters (why?), so this proposal is practicable—provided we can find the distribution of S_M^2/S_P^2.

[7] See pages 333–334 in Hogg and Craig (Reference 17), and note that $\text{Var}(\bar{Y}_i) = \sigma^2/n_i$.

To help us find the distribution of S_M^2/S_P^2 and to further motivate this test of H_0, we now show that

$$n\hat{\sigma}^2 = (n-k)S_P^2 + (k-2)S_M^2. \tag{8.20}$$

To see this, observe that

$$Y_{ij} - (\hat{\alpha} + \hat{\beta}x_i) = (Y_{ij} - \bar{Y}_i) + [\bar{Y}_i - (\hat{\alpha} + \hat{\beta}x_i)]$$

so

$$[Y_{ij} - (\hat{\alpha} + \hat{\beta}x_i)]^2 = (Y_{ij} - \bar{Y}_i)^2 + [\bar{Y}_i - (\hat{\alpha} + \hat{\beta}x_i)]^2 + 2(Y_{ij} - \bar{Y}_i)[\bar{Y}_i - (\hat{\alpha} + \hat{\beta}x_i)].$$

Hence,

$$\begin{aligned}
n\hat{\sigma}^2 &= \sum_{i=1}^{k}\sum_{j=1}^{n_i} [Y_{ij} - (\hat{\alpha} + \hat{\beta}x_i)]^2 \\
&= \sum_{i=1}^{k}\sum_{j=1}^{n_i} (Y_{ij} - \bar{Y}_i)^2 + \sum_{i=1}^{k} n_i[\bar{Y}_i - (\hat{\alpha} + \hat{\beta}x_i)]^2 \\
&\qquad + 2\sum_{i=1}^{k} \left\{[\bar{Y}_i - (\hat{\alpha} + \hat{\beta}x_i)] \sum_{j=1}^{n} (Y_{ij} - \bar{Y}_i)\right\} \\
&= (n-k)S_P^2 + (k-2)S_M^2 + 2\sum_{i=1}^{k} \{[\bar{Y}_i - (\hat{\alpha} + \hat{\beta}x_i)]\cdot 0\} \\
&= (n-k)S_P^2 + (k-2)S_M^2.
\end{aligned}$$

Thus $n\hat{\sigma}^2$, the total variation about the estimated regression line, can be decomposed into the sum of two terms. The first, $(n-k)S_P^2$, is the amount of variation that is explained by the assumption of linearity, while the second, $(k-2)S_M^2$, is the amount that is not explained. Thus, it seems reasonable to reject H_0 whenever the ratio of the unexplained variation to the explained variation is large. This is equivalent to rejecting the null hypothesis whenever the random variable S_R defined by

$$S_R \equiv \frac{\sqrt{S_M^2}}{\sqrt{S_P^2}}$$

is large.

Now we must find the distribution of S_R. First, we can use equation (8.20), the fact that $n\hat{\sigma}^2/\sigma^2$, $(n-k)S_P^2/\sigma^2$, and $(k-2)S_M^2/\sigma^2$ are chi-square random variables, and Theorem 1 on page 309 of Hogg and Craig (Reference 17) to establish the independence of S_M^2 and S_P^2. Using these facts, it follows[8] that S_R is distributed as an *F random variable with* $k-2$

[8] See Theorem 10.4 on page 232 in Mood and Graybill (Reference 18).

and $n - k$ degrees of freedom, written "S_R is $F_{k-2,n-k}$." The F random variable arises frequently, and its cumulative distribution function is given in Table G.

In computing S_P^2 and S_M^2, it is easier to use the formulas

$$S_P^2 = \frac{1}{n-k}\left[\sum_{i=1}^{k}\sum_{j=1}^{n_i} Y_{ij}^2 - \sum_{i=1}^{k}\left(\sum_{j=1}^{n_i} Y_{ij}\right)^2 \Big/ n_i\right] \tag{8.21}$$

and

$$S_M^2 = \frac{1}{k-2}\left[\sum_{i=1}^{k}\left(\sum_{j=1}^{n_i} Y_{ij}\right)^2 \Big/ n_i - \left(\sum_{i=1}^{k}\sum_{j=1}^{n_i} Y_{ij}\right)^2 \Big/ n - \frac{\left(\sum_{i=1}^{k} n_i(x_i - \bar{x})\bar{Y}_i\right)^2}{\sum_{i=1}^{k} n_i(x_i - \bar{x})^2}\right] \tag{8.22}$$

than to use equations (8.18) and (8.19).

▶▶

In Chapter 6 we used the chi-square goodness-of-fit test for testing $H_0: p_1 = p_2 = \cdots = p_k$ when we had independent binomial random variables X_i, where X_i is $B(1, p_i)$, $i = 1, 2, \ldots, k$. Since $E(X_i) = p_i$, we can rewrite the null hypothesis in the form $H_0: \mu_1 = \mu_2 = \cdots = \mu_k$, where $\mu_i \equiv E(X_i)$. In view of the importance of normal random variables, we would also like to be able to test

$$H_0: \mu_1 = \mu_2 = \cdots = \mu_k$$

when the independent random variables Y_{ij} are normal—in particular, when Y_{ij} is $N(\mu_i, \sigma^2)$, $j = 1, 2, \ldots, n_i$, $i = 1, 2, \ldots, k$.

The technique we have used in testing the linearity assumption—which is called *analysis of variance*—can be adapted to test the equality of means of these normal random variables. Here's how: Define S_P^2 as before, but define S_M^2 by

$$S_M^2 = \frac{1}{k-1}\sum_{i=1}^{k} n_i(\bar{Y}_i - \bar{Y})^2, \qquad \text{where } \bar{Y} = \frac{1}{n}\sum_{i=1}^{k} n_i\bar{Y}_i.$$

Then reject H_0 whenever $S_R = \sqrt{S_M^2}/\sqrt{S_P^2}$ is too large; S_R is $F_{k-1,n-k}$.

▶

EXERCISES for Subsection 2.1:

4. Find the maximum-likelihood estimates for α, β, and σ^2 using the data in (a) Exercise 1, (b) Exercise 2, (c) Exercise 3.

5. Show that $\hat{\alpha}$ is a weighted sum of the Y_i. That is, find constants $a_1, a_2, \ldots, a_n$ such that $\hat{\alpha} = \Sigma a_i Y_i$.

6. Verify equations (8.9) and (8.10).

7. Show that

$$\operatorname{Var}(\hat{\alpha}) = \frac{\sigma^2}{n}\left(1 + \frac{n\bar{x}^2}{\Sigma(x_i - \bar{x})^2}\right).$$

Thus, $\operatorname{Var}(\hat{\alpha})$ will be minimized when $\bar{x} = 0$.

8. State a simple condition under which $\hat{\alpha}$ and $\hat{\beta}$ would be independent.

9. Show that $\hat{\alpha}$, $\hat{\beta}$, and $\hat{\sigma}^2$ are sufficient estimators for α, β, and σ^2.

EXERCISES for Subsection 2.2:

10. Find 90 percent confidence intervals for α, for β, and for σ^2 using the data in (a) Exercise 1, (b) Exercise 2, (c) Exercise 3. [*Hint:* See Exercise 4.]

11. Test H_0: $\alpha = 0$ against H_A: $\alpha \neq 0$ at level of significance .10 using the data in (a) Exercise 1, (b) Exercise 2, (c) Exercise 3.

12. Test H_0: $\beta = 0$ against H_A: $\beta > 0$ at level of significance .05 using the data in (a) Exercise 1, (b) Exercise 2, (c) Exercise 3.

EXERCISES for Subsection 2.3:

13. Find a 90 percent prediction interval for Y_0 when
(a) $x_0 = 7$ and $x_0 = 4$ in Exercise 1.
(b) $x_0 = 110$ in Exercise 2.
(c) $x_0 = 9$ and $x_0 = 5$ in Exercise 3.

14. Suppose in the prediction problem considered in Subsection 2.3 that instead of taking a single observation at $x = x_0$ we took a sample of m observations at $x = x_0$. Construct a $100(1 - \delta)$ percent prediction interval for the mean $\bar{Y}_0$ of these m observations. (Of course, our prediction itself is $\hat{\alpha} + \hat{\beta}x_0$.)

Next, suppose that in designing our experiment we are constrained to set $x = x_0$ and $n + m = 2k + 1$, but the exact values of n and m are not specified. Propose and justify heuristically a guideline for choosing n so as to minimize the expected length of the random prediction interval.

15. (Continuation) Find a 90 percent prediction interval for $\bar{Y}_0$ when
    (a) 9 observations are taken at $x_0 = 7$ in Exercise 1.
    (b) 25 observations are taken at $x_0 = 110$ in Exercise 2.
    (c) 5 observations are taken at $x_0 = 5$ in Exercise 3.

16. What effect would increasing the difference between x_0 and $\bar{x}$ have on the expected length of the prediction interval? What does this suggest if our only desire is to have a prediction interval with minimal expected length?

17. In Subsection 2.3 we stated that in forming prediction intervals, we must revise our estimates of α, β, σ^2, and σ_U^2 as we make additional observations. Why is this so? [*Hint:* Suppose that $n = 10$, $\hat{\alpha}_n = 1$, $\hat{\beta}_n = 1$, $\hat{\sigma}_n^2 = 1$, $\Sigma_{i=1}^{10}(x_i - \bar{x})^2 = 20$, and $\delta = .10$. Then suppose 990 additional observations were taken at $x_0 = \bar{x}$ and that $\bar{y}_{1000} = \bar{y}_{10}$. Then $\hat{\beta}_{1000} = \hat{\beta}_{10}$, $\hat{\alpha}_{1000} = \hat{\alpha}_{10}$. However, suppose we also have $\hat{\sigma}_{1000}^2 = \frac{1}{9}\hat{\sigma}_{10}^2$. This would result if the variance of the last 990 observations around $\hat{\alpha}_{10} + \hat{\beta}_{10}x_0$ were slightly less than $\frac{1}{9}\hat{\sigma}_{10}^2$.]

18. Why does the expected length of the prediction interval approach $2\sigma z_{1-(\delta/2)}$ as n approaches infinity? Note that $\sqrt{E(\hat{\sigma}^2)} \neq E(\sqrt{\hat{\sigma}^2})$.

EXERCISES for Subsection 2.4:

19. Test whether or not the linearity assumption is justified in Exercise 3. Use level of significance .05.

20. Verify equations (8.21) and (8.22); then use these equations to verify equation (8.20).

3. THE SIMPLE LINEAR MODEL

Throughout Section 2 it was assumed that the error terms ϵ_i in our simple linear model were normal random variables. We now drop this assumption, and we make no assumptions whatsoever about the form of the distributions of these random variables. We will, of course, continue to assume that they are uncorrelated random variables with mean 0 and variance σ^2.

Since the form of the distributions of the ϵ_i is not known, the maximum-likelihood estimators of α and β cannot be found, so a different method for estimating α and β must be employed. We will use the method of *least-squares*.

In introducing and motivating the method of least-squares, we note that obtaining good estimates of α and β is equivalent to obtaining a good estimate of the true regression line. In turn, a good estimate of the true regression line should fit our data better than any other straight line. Thus, if the data are plotted in a scatter diagram as in Figure 8.3, then we should choose as our estimate that straight line which lies closest to the n sample points.

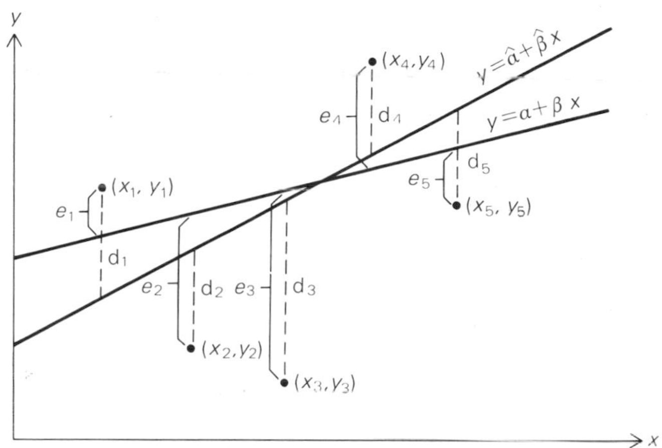

Figure 8.3 Scatter Diagram

We would all agree that the (vertical) distance d_i from the sample point (x_i, y_i) to the estimated regression line $y = \hat{\alpha} + \hat{\beta}x$ is given by the equation

$$d_i = y_i - (\hat{\alpha} + \hat{\beta}x_i).$$

(Notice that d_i is an error term with regard to the estimated regression line, so it is an estimate of the "true" error e_i.) And we would also agree that the line fits the sample point (x_i, y_i) well when d_i is small. But a line that passes through or near certain sample points (has small values for these d_i) may not come close to others (has large values for these d_i), so it is not clear exactly what is meant by the lines fitting all n sample points well. (Of course, $d_1 = d_2 = \cdots = d_n = 0$ is ideal.) In analogy to our use of variance to measure the dispersion of (the values of) a random variable about its mean, we shall use

$$\sum_{i=1}^{n} d_i^2 = \sum_{i=1}^{n} [y_i - (\hat{\alpha} + \hat{\beta}x_i)]^2$$

as our measure of the dispersion of the n sample points about the estimated regression line.

DEFINITION

The *least-squares estimators* of α and β are those random variables $\hat{\alpha}$ and $\hat{\beta}$ that minimize the observed value of

$$\sum_{i=1}^{n} d_i^2 = \sum_{i=1}^{n} [y_i - (\hat{\alpha} + \hat{\beta}x_i)]^2,$$

the sum of squared errors.

It is now easy to establish (see Exercise 22)

THEOREM 2

The least-squares estimators of α and β are

$$\hat{\alpha} = \bar{Y} - \hat{\beta}\bar{x} \quad \text{and} \quad \hat{\beta} = \frac{\sum_{i=1}^{n} (x_i - \bar{x})(Y_i - \bar{Y})}{\sum_{i=1}^{n} (x_i - \bar{x})^2}.$$

Surprisingly, the least-squares estimators are the same as the maximum-likelihood estimators found in Section 2. Thus, when the Y_i are assumed to be normal, the least-squares estimators have all the desirable properties listed in Theorem 1. In particular, they are the minimum-variance unbiased estimators. Unfortunately, if we do not assume that the Y_i are normal, then the least-squares estimators are not the minimum-variance unbiased estimators. They do, however, have minimum variance among the class of unbiased estimators that are linear functions of the Y_i—that is, among the class of unbiased estimators that can be written in the form $\hat{\alpha} = a_0 + \Sigma a_i Y_i$ and $\hat{\beta} = b_0 + \Sigma b_i Y_i$. This result is known as the Gauss-Markov theorem.[9]

Similar least-squares techniques can be used for nonlinear models such as the physics model $h = \frac{1}{2}v^2/g$—here h plays the role of y and v plays the role of x.

EXERCISES:

21. Find the least-squares estimates of α, β, and σ^2 using the data in (a) Exercise 1, (b) Exercise 2, (c) Exercise 3.

22. Show that $\Sigma_{i=1}^{n} d_i = 0$ if $d_i = y_i - (\hat{\alpha} + \hat{\beta}x_i)$ and $\hat{\alpha}$ and $\hat{\beta}$ are the least-squares estimators. Also, show that $\Sigma_{i=1}^{n} d_i = 0$ if $\hat{\beta} = 0$ and $\hat{\alpha} = \bar{Y}$.

[9] Its proof can be found on page 341 of Mood and Graybill (Reference 18).

23. Establish Theorem 2 by setting

$$\frac{\partial}{\partial\hat{\alpha}}\sum_{i=1}^{n} d_i^2 = \frac{\partial}{\partial\hat{\beta}}\sum_{i=1}^{n} d_i^2 = 0.$$

4. A WARNING

In using the methods of this chapter, a large number of tedious arithmetic operations must often be performed—for instance, in computing S_M^2/S_P^2. Consequently, these computations are often performed (with readily available programs) on electronic computers, which are, incidentally, ideally suited for the purpose. Then these computational results are printed out in an exact and scientific-looking format. It is because these otherwise burdensome calculations can be done so readily on the computer and because the printouts look so scientific—not to mention the aura of exactness surrounding the computer itself—that the methods of this chapter are frequently used when a little forethought would reveal that the assumptions underlying the use of these methods are, in fact, totally inappropriate for the problem being considered!

Hopefully, the comments above will be effective in warning potential users of regression models, especially social scientists, to proceed with due caution. Below, we point out two[10] frequently encountered pitfalls in the use of linear regression models.

CAUSE AND EFFECT

If we are able to reject $H_0: \beta = 0$ in favor of, say, $H_A: \beta > 0$, then we say that x and Y are associated or that there is a relationship between x and Y. After having correctly determined that x and Y are associated, people sometimes incorrectly interpret this to mean that changes in the value of the x variable *cause* changes in the value of the Y variable—that is, there is a cause-and-effect relationship between x and Y.

Some of this confusion undoubtedly arises from our labeling x and Y as the independent and dependent variables, respectively, but there is more to it than this. There is a link between association and causation. What then is this link?

First, it is clear upon reflection that if there is causation, then there will always be association. Yet, if there is association, there may not be causation. Why is this true?

[10] Perhaps the spelling should be "too"!

One reason is that association between two variables may be the result of pure chance. For example, there is association between the amount of alcohol consumed in California and the amount of soil erosion in Alaska; yet it is clear that increases in the consumption of alcohol in California will not cause changes in the amount of soil erosion in Alaska or vice versa. A second reason is that association between two variables may be due to the influence of a third, as yet unspecified, variable. For example, the association between the number of aluminum frying pans sold and the number of deaths due to cancer is explained by the concomitant increase in population.

How then can we use statistical techniques to determine whether or not there is causation? The answer is quite simple: we can't!

CORRELATION PROBLEMS

In many, if not most, problems of interest, the independent variable x is a random variable rather than a controllable mathematical variable. When this is the case, we say that we are dealing with a *correlation problem* and not with a regression problem. For example, if x is the gross national product (GNP) and Y is, say, retail sales, then we are dealing with a correlation problem, as the GNP is a random variable and is not within our control. As a rough rule of thumb, problems arising in the physical sciences *tend* to be regression problems, whereas problems arising in the social sciences *tend* to be correlation problems; problems arising in the natural sciences can be of either type.

Because correlation problems are just as important as regression problems, it is imperative that we analyze them too. Fortunately, if we assume that (i) the distribution of the independent variable X is not a function of α, β, or σ^2, and (ii) the distribution of the error term ϵ is independent of x, then all the results of Section 2 remain valid.

To see this, observe from (ii) that

$$\underset{\alpha,\beta,\sigma^2}{L}(\mathbf{x}, \mathbf{y}) = p(\mathbf{x}) \prod_{i=1}^{n} \left(\frac{1}{2\pi\sigma^2}\right)^{n/2} \exp\left\{-\frac{1}{2\sigma^2}(y_i - \alpha - \beta x_i)^2\right\} = p(\mathbf{x}) \cdot \underset{\alpha,\beta,\sigma^2}{L}(\mathbf{y}),$$

where $p(\mathbf{x})$ is the likelihood function of $\mathbf{X}$. By (i), $p(\mathbf{x})$ is not a function of α, β, or σ^2, so the values of α, β, and σ^2 that maximize $\underset{\alpha,\beta,\sigma^2}{L}(\mathbf{y})$ also maximize $\underset{\alpha,\beta,\sigma^2}{L}(\mathbf{x}, \mathbf{y})$. Thus, the maximum-likelihood estimators are as given in Section 2. Moreover, it follows from (ii) that their distributions are also as given in Section 2.

EXERCISES:

24. For which of the following pairs (x, Y) of variables would a regression model be more appropriate than a correlation model?
    (a) SAT test scores and freshman grade point averages.
    (b) Mathematics grades and physics grades (both measured on a scale of 0–100).
    (c) Number of births in year t and number of housing starts in year $t + 3$.
    (d) Density of particles and the speed of diffusion.
    (e) Salesmen's aptitude and salesmen's record.
    (f) Height of cornstalks and yield of the cornstalks.
    (g) Amount of radiation exposure and length of the animal's life.
    (h) Blood pressure and metabolism.
    (i) Farm income and tractor sales.
    (j) Age and weight of trout.
    (k) City population and number of burglaries committed.

9

NONPARAMETRIC TESTS

In the previous chapters of Part II we have learned how to take into account the information contained in random samples in order to draw "good" conclusions about the sampled population. With the exception of Chapter 6, our analysis began with the specification of a particular functional form of the random variables' distribution, and our conclusions were couched in terms of the true values of the distribution's parameters. For example, in Chapter 5 we started by assuming that the random variables of interest were normal with unknown mean and known variance. Our conclusion might have been that $\mu = \mu_0$ or that $\mu > \mu_0$.

In many situations, however, not only do we not know the true values of the parameters, but we do not know the functional form of the distribution. Nevertheless, we need techniques for making statistical inferences and, in particular, for testing statistical hypotheses. The techniques that we shall shortly develop for use in these situations are called *nonparametric* or *distribution-free* methods. The terms indicate that these methods do not depend upon the functional form of the distributions.

1. COMPARING TWO POPULATIONS: THE FISHER-IRWIN TEST

In this and the following two sections we develop three distinct methods for testing whether or not two populations differ with regard to some "performance characteristic." For instance, we might be interested in knowing (1) whether or not no-load mutual funds perform as well (make as much money) as load funds, (2) whether low-protein diets are more successful in reducing cholesterol count than ordinary (low-intake) diets, or

(3) whether the disclosure of a deviation from standard accounting practices will be equally visible to (noticed by) the readers of a financial report if the disclosure is placed in the auditor's report or in a footnote to the financial statements (for example, the balance sheet).

In this section, we treat the special situation in which the performance characteristic is given one of two values. For instance, (1) a mutual fund did or did not recieve a rating of at least zero, (2) a dieter's cholesterol count did or did not fall below 300 milligrams per cubic centimeter, or (3) the reader of the report did or did not notice the auditor's disclosure.

Table 9.1

	Success	*Failure*	*Total*
Population I	x	$n - x$	n
Population II	y	$m - y$	m
Total	$r - x + y$	$N - r$	N

The data resulting from such an experiment can be arranged in a two-by-two table such as Table 9.1. As indicated in Table 9.1, a sample of size n was taken from the first population, and the performance of x of these n individuals was classified as successful whereas the performance of $n - x$ individuals was classified as unsuccessful (that is, a failure). Similarly, y out of the m individuals sampled in the second population performed successfully and $m - y$ performed unsuccessfully. Thus, $r \equiv x + y$ out of the total sample of $N \equiv n + m$ individuals performed successfully and $N - r$ did not.

We wish to test the null hypothesis that there is no difference in the performance characteristic of the two populations against the alternative hypothesis that there is a difference. Under the null hypothesis, an individual's performance is independent of the population from which he came. Consequently, since r out of the N individuals sampled in the two populations performed successfully, it follows that given any s out of the N individuals, the number D of successful performances by these s individuals is a hypergeometric random variable with parameters s, r, and N. In particular, the number X of successful performances by the n individuals in population I is a hypergeometric random variable with parameters

n, r, and N. That is

$$P(X = x) = \frac{\binom{r}{x}\binom{N-r}{n-x}}{\binom{N}{n}}, \quad x = 0, 1, 2, \ldots, \min(r, n). \tag{9.1}$$

Moreover, it is reasonable to reject the null hypothesis in favor of the alternative hypothesis if x, the observed value of X, is either very large or very small. Thus, given the level of significance α, a good rejection region is constructed as follows. Find the largest and smallest integers c_L and c_U that satisfy, respectively,

$$P(X \leq c_L) \leq \frac{\alpha}{2} \tag{9.2}$$

and

$$P(X \geq c_U) \leq \frac{\alpha}{2}. \tag{9.3}$$

Then reject the null hypothesis if $x \geq c_U$ or if $x \leq c_L$.

If the alternative hypothesis had been that performance is better in population I, then we would reject the null hypothesis when $x \geq c$, where c is the smallest integer that satisfies

$$P(X \geq c) \leq \alpha. \tag{9.4}$$

(This test was originally proposed by R. A. Fisher and O. J. Irwin.) Finally, we remark that when appropriate (see Section 5 of Chapter 2) we can approximate X by a binomial random variable—and the binomial by a normal random variable.

EXAMPLE 1. (Mutual Funds)

There has been considerable disagreement among investors concerning the relative performance of two types of mutual funds called load and no-load funds. They differ in that the former charges investors an 8 percent commission whereas the latter charges only a nominal fee. M. C. Jensen[1] made a study evaluating the performance of 115 mutual funds in the twenty-year period 1945–1964. There were 102 load and 13 no-load funds in Jensen's sample. Using a sophisticated technique for evaluating performance (which does not take the commission into account), Jensen found that only 30 of the load funds and 9 of the no-load funds received a

[1] M. C. Jensen, "Problems in Selection of Security Portfolios—The Performance of Mutual Funds in the Period 1945–1964," *Journal of Finance*, 1968, pp. 389–416.

rating of at least zero. Do Jensen's findings reveal a difference between the performance of load and no-load funds?

SOLUTION: Under the null hypothesis that there is no difference between the performance of load and no-load funds, we find (see Table 9.2) that X, the number of no-load funds that received a rating of at least zero, is $H(13, 39, 115)$—that is,

$$P(X = x) = \frac{\binom{39}{x}\binom{76}{13-x}}{\binom{115}{13}}, \qquad x = 0, 1, 2, \ldots, 13.$$[2]

Using $\alpha = .01$, we find (with the aid of a computer) that $c_L = 1$ and $c_U = 9$, so we will reject the null hypothesis if $x \leq 1$ or if $x \geq 9$. [We cannot use the normal approximation to the hypergeometric distribution

Table 9.2

	Rating above zero	*Rating below zero*
No-load funds	9	4
Load funds	30	72

because the sampling fraction $n/N = \frac{13}{115}$ is too large. (See Section 5 of Chapter 2 and Section 2 of Chapter 3.)] Since the observed value of x is 9, we reject the null hypothesis and accept the alternative hypothesis that there is a difference in performance. (Evidently, the 8 percent commission charged by load funds is unjustified—*caveat emptor*!) •

EXAMPLE 2. (Discrimination)

It was reported in the *Los Angeles Times* on August 5, 1969, that a "citizens group charged the city school system with discrimination in the assignment ... of [too many] teachers for gifted children to Anglo areas [the West Side] in preference to predominantly Mexican-American schools on the East Side. . . . West Side schools had 6 teachers for the gifted while the East Side had 2. . . . West Los Angeles has about 1600 identified gifted pupils compared to about 1100 in East Los Angeles." Was the citizens group's charge of discrimination justified? Use $\alpha = .10$.

[2] Also, X is $H(39, 13, 115)$.

SOLUTION: We wish to test the null hypothesis of no discrimination against the alternative hypothesis of discrimination against the Mexican-Americans. Under the null hypothesis of no discrimination, we find (see Table 9.3) that X, the number of teachers assigned to the West Side, is

Table 9.3

	Teachers	*Students*
West Side	6	1600
East Side	2	1100

$H(1606, 8, 2708)$. We will reject the null hypothesis if and only if $x \geq c$, where c is given by equation (9.4). Using $\alpha = .10$, we find (with the aid of a computer) that $c = 7$ [in fact, $P(X \geq 6) = .3007$], so we accept the null hypothesis. •

EXERCISES:

1. The purpose of the study "The Visibility of the Auditor's Disclosure of Deviation from APB Opinion" by Purdy, Smith, and Gray was "to provide one piece of empirical evidence impinging upon the broad question of what constitutes adequate disclosure in published financial statements." More specifically, one of the questions they asked was whether the disclosure of a deviation from standard accounting practices will be equally visible to the readers of financial statements if the disclosure is placed in the auditor's report or in a footnote to the financial statements. Rather than give the entire elaborate procedure they devised for testing the null hypothesis of no difference in visibility against the alternative hypothesis of some difference in visibility, we have distilled the pertinent information. An annual report was given to 133 subjects. Of these, 44 were chosen at random and the annual report presented to this group contained the auditor's disclosure in a footnote, whereas the annual report presented to the other 89 subjects contained the auditor's disclosure in the financial statements. After having read the annual report, the subjects were asked a series of questions of which 6 related to the auditor's disclosure. The number of subjects in the group of 44 and in the other group answering all 6 questions correctly was 6 and 14, respectively.
   (a) Test the null hypothesis using $\alpha = .10$.
   (b) Test the null hypothesis using $\alpha = .10$ and the fact that the number

of subjects in the two groups answering at least 5 questions correctly was 24 and 42, respectively.

2. In a 1954 study of the Salk polio vaccine, 33 of 200,745 inoculated children contracted polio whereas 115 of 201,229 children who were not inoculated contracted polio. Is this evidence that the vaccine is effective? Use $\alpha = .01$. [*Hint:* The necessary computations can be done in just a few minutes.]

3. A manufacturer wanted to know whether his new diet pill was effective in reducing hunger pangs. Therefore, a group of 17 would-be customers were each given what they thought was a diet pill. Unbeknownst to them, 8 were given a placebo (water and flour) instead of the diet pill. Of course the placebo has no physiological effect whatsoever on reducing hunger pangs—only a psychological effect. Six hours later, the dieters were asked if their pill had helped in reducing their hunger pangs. What would the manufacturer have concluded if he used $\alpha = .10$ and 5 of the 9 dieters who took his diet pill and 3 of the 8 dieters who took the placebo responded that the pill had substantially helped to alleviate hunger pangs?

4. Use the Fisher-Irwin test to ascertain the validity of the geneticists' claim in Example 7 of Chapter 5.

5. (Median test) One standard procedure for testing differences in performance is called the *median test*. An observation is labeled successful only if its performance level is among the top half. In particular, if $n + m$ is even, then the $(n + m)/2$ observations with the highest performance level are labeled successful, and the other $(n + m)/2$ are labeled unsuccessful. If $n + m$ is odd, then only $(n + m - 1)/2$ observations are labeled unsuccessful.

   In the example of the 115 mutual funds, the performance of 11 of the 13 no-load funds was successful. Would this result lead you to accept the alternative hypothesis of a difference in the performance of load and no-load mutual funds? Use $\alpha = .01$.

6. Compare the use of the Fisher-Irwin test to the χ^2 test of independence in a two-by-two contingency table. [*Hint:* Recall that if Y is $N(0, 1)$, then Y^2 is χ_1^2.]

2. COMPARING TWO POPULATIONS: THE RANK SUM TEST

In Section 1 we saw how to test the null hypothesis that two populations do not differ with respect to some performance characteristic. This was done for the special situation in which the performance characteristic was given one of only *two* values. We shall now consider the more general situation in which the performance characteristic can take on *any* (numerical) value.

This problem was studied in Chapter 5 under the assumption that the performance characteristic had a normal distribution. Unfortunately, as in the case of Jensen's performance measure for mutual funds, we may not know the distribution of the performance characteristic. We can, however, find the distribution of a random variable W, called the rank sum, which is associated with the performance characteristic. Then W can be used in testing the null hypothesis.

In general, we will have n observations from population I and m observations from population II. That is, we will have independent random samples of size n and m from populations I and II, respectively. The random variable W is found as follows. Make a list of the $n + m$ observations by arranging them in order of magnitude, starting with the smallest value first. Next, record the position of each of the $n + m$ values in the list. The position of each observation is referred to as the *rank* of the observation. Finally, sum the ranks of the observations from population I. This random variable is denoted by W, while w denotes the observed value of W.

For example, if the observations from the two populations are 1.7, −3.2, −1.4, 3.9 and 3.1, 4.1, −2.7, 7.6, 11.5, respectively, then $n = 4$ and $m = 5$. The list and ranks are given in Table 9.4. In this case we find that $w = 1 + 3 + 4 + 6 = 14$.

Table 9.4

−3.2	−2.7	−1.4	1.7	3.1	3.9	4.1	7.6	11.5
1	2	3	4	5	6	7	8	9

It appears that differences between the performance characteristic of the two populations will result in either very large or very small values of W. Consequently, it seems reasonable to reject the null hypothesis in favor of the alternative hypothesis if the observed value w of the random variable W is either very large or very small. Thus, given the level of significance α, a good rejection region is defined by rejecting the null hypothesis when $w \leq c_L$ or $w \geq c_U$, where c_L and c_U are the largest and smallest integers that satisfy, respectively,

$$P(W \leq c_L) \leq \frac{\alpha}{2} \tag{9.5}$$

and

$$P(W \geq c_U) \leq \frac{\alpha}{2}. \tag{9.6}$$

Of course, if the alternative hypothesis had been that performance is better in population I, then we would reject the null hypothesis whenever

$w \geq c$, where c is the smallest integer that satisfies

$$P(W \geq c) \leq \alpha. \tag{9.7}$$

This test was proposed by F. Wilcoxon in 1945, and it is sometimes referred to as the *Wilcoxon two-sample test.*

It now remains to find the distribution of W. Mann and Whitney have shown (see Exercise 13) that when the null hypothesis is true,

$$E(W) = \frac{n(n + m + 1)}{2} \tag{9.8}$$

and

$$\operatorname{Var}(W) = \frac{nm(n + m + 1)}{12}. \tag{9.9}$$

Furthermore, they have shown that W is approximately normally distributed when the null hypothesis is true and both n and m are large. Fortunately, the normal approximation is quite accurate when n and m are both larger than 7, so the distribution of W has, in effect, been found for this case. In the next few paragraphs we shall show how to find the distribution of W when both n and m are not larger than 7.

When the null hypothesis is true, each of the $(n + m)!$ ways of assigning the ranks to the $n + m$ individuals is equally likely. Moreover, given any n of the $n + m$ ranks, there are $n!$ ways of assigning these ranks to the n individuals from population I, and there are $m!$ ways of assigning the m remaining ranks to the m individuals from population II. Thus, given a set of n ranks, the probability that these ranks have been assigned to the individuals from population I is $n!m!/(n + m)!$. Consequently, if we denote by $N(w; n, m)$ the number of sets of n distinct ranks (with each rank no larger than $n + m$) such that the sum of these n ranks is equal to w, then we have

$$P(W = w) = \frac{N(w; n, m)}{\binom{n + m}{n}}$$

and

$$P(W \leq c) = \frac{\sum_{w=0}^{c} N(w; n, m)}{\binom{n + m}{n}}.$$

Thus far, the original problem of finding the distribution of W has been reduced to that of finding $N(w; n, m)$ for each w given values of n and m.

One way of finding $N(w; n, m)$ is simply to list all $\binom{n+m}{n}$ sets of n ranks, sum the ranks in each set, and count the number of sets whose rank sum is equal to w. Needless to say, this is a burdensome task—even when n and m are both less than or equal to 7.

Fortunately, we needn't be bothered with this task, as Table D contains the values of $\sum_{w=0}^{c} N(w; n, m) - n(n+1)/2$ for each relevant value of c and each pair n and m, where $1 \leq n \leq m \leq 7$. Note that in Table D entries are only given for $n \leq m$. This really causes us no trouble in using the table; we simply need refer to the population with the smaller sample size as population I. Also, the distribution of W is symmetrical about $nm/2$.

EXAMPLE 3

Suppose that samples of size 4 and 5 have yielded the data given in Table 9.4, so that the observed value of W is 14. It is easy to show that $N(w; 4, 5) = 1, 1, 2, 3$, and 5 for $w = 10, 11, 12, 13$, and 14, respectively. (Verify this!) Also,

$$\binom{4+5}{4} = 126.$$

Of course, this information can also be found in Table D. In particular, since $n = 4$, $n(n+1)/2 = 10$, so we look in the table in the column headed by $14 - [n(n+1)/2] = 4$ and the row corresponding to $n = 4$ and $m = 5$. There the associated entry is seen to be 12. Thus,

$$P(W \leq 14) = \tfrac{12}{126},$$

so that we would reject the null hypothesis of no difference in performance in favor of the alternative hypothesis that individuals from population II perform better if $\alpha \geq \frac{12}{126}$. •

If in Example 3 we had defined a successful performance to be a value greater than zero, then, using the Fisher-Irwin test given in Section 1, we would have found

$$P(X \leq 3) = \frac{\binom{7}{3}\binom{2}{1}}{\binom{9}{4}} = \frac{5}{9}.$$

Consequently, using the Fisher-Irwin test, we would accept the null hypothesis for values of α less than $\frac{5}{9}$. On the other hand, use of the rank sum test would have led us to reject the null hypothesis for values of α between $\frac{12}{126}$ and $\frac{5}{9}$. This raises the question of which test to use when both are applicable. In general, the rank sum test is preferred, because we can expect it to be more sensitive to differences in performance. The reason is simple. By assigning each observation one of only two values, the Fisher-Irwin test throws away useful information.

EXAMPLE 4. (Mutual Funds)

A closer inspection of Jensen's data reveals the following ranks for the 13 no-load funds:

$$23,\ 47,\ 68,\ 69,\ 77,\ 81,\ 82,\ 83,\ 93,\ 99,\ 102,\ 109,\ 112.$$

Do these data support the null hypothesis of no difference in the performance of load and no-load funds? Use $\alpha = .01$.

SOLUTION: We have $n = 13$, $m = 102$, and $w = 1045$. As both n and m exceed 7, we can use the normal approximation to find $P(W \geq 1045)$. We can conclude from equations (9.7) and (9.8) that $E(W) = 754$ and $\text{Var}(W) = 12{,}818 = (113.2)^2$, so

$$P(W \geq 1045) = P\left(\frac{W - 754}{113.2} \geq \frac{1045 - 754}{113.2}\right) \approx P(Z \geq 2.57) = .0051.$$

Thus, we reject the null hypothesis. •

The rank sum test is, in fact, meant to be applied only when the performance characteristic is a continuous random variable. If, however, we measured the performance characteristic only to the nearest inch, pound, degree, or what have you, then ties in the ranks might occur. Ties could also occur if the performance characteristic were integer-valued (for example, the number of typing errors) or if the performance characteristic were subjective and the experimenter were unable to distinguish between two individuals (for example, two girls might be judged to be equally beautiful).

How then can we incorporate ties in the testing procedure? The simplest method of handling ties is to break them by randomization. That is, if k individuals are tied for ranks $j, j + 1, \ldots, j + k - 1$, then perform an experiment so that each individual has probability $1/k$ of receiving any given one of these ranks (and no two individuals are given the same rank). For example, if two individuals are tied for ranks 5 and 6, then we can randomize by assigning the first individual rank 5 if a fair coin lands heads and rank 6 if it lands tails; the second individual is assigned the remaining rank.

While this method of handling ties is simple and "fair," it is unappealing in that we are adding an extraneous element of randomness to our observations. A better method for handling ties is to assign each individual the average of the ranks for which he is tied. For example, if four individuals are tied for ranks 5, 6, 7, and 8, then we should assign each of these individuals the rank $6\frac{1}{2}$. As before, we let W be the sum of the ranks of the individuals from population I. It is clear upon reflection (see Exercise 14) that we still have $E(W) = n(n + m + 1)/2$ under the null hypothesis. Less obvious, however, is the fact that each set of k tied ranks reduces the variance of W by $nmk(k^2 - 1)/12(n + m)(n + m + 1)$. Luckily, the normal approximation (with the variance adjusted for ties) can still be used if n and m are each larger than 7 and if the variance of W exceeds $(8)8(8 + 8 + 1)/12$.

EXERCISES:

7. Test the null hypothesis that undergraduates and graduates perform equally well in statistics against the alternative hypothesis that there is *some* difference in their performance. Use $\alpha = .10$ and the test scores given in Example 3 of Chapter 6. (There, the first 18 scores listed are for undergraduates.)

8. To find out whether studying for an aptitude test will improve one's score, 5 of 11 students were selected at random and were tutored. The other 6 did not study for the exam. The first 5 students' scores were 930, 1073, 1095, 1120, and 1382 and the other 6 students' scores were 877, 943, 995. 1075, 1105, and 1118. Does studying seem to help? Use $\alpha = .10$.

9. (Rank Correlation) In Section 2.4 of Chapter 8 we considered the problem of measuring how closely two variables are linearly related. There it was necessary to make several assumptions, among them that one of the two variables was normally distributed, in order to test whether the empirical evidence was compatible with the postulated linear relationship. We now present a method for testing the null hypothesis of independence (or, more generally, zero correlation) when it is assumed that the two variables are continuous. We need make no other assumptions about the distributions of the two variables (or about the functional form of their relationship).

   Let $X_1, Y_1; X_2, Y_2; \ldots; X_N, Y_N$ be a random sample of size N when each observation has both an X value and a Y value. Relabel these N observations according to the rank of the X values so that $X_1 < X_2 < \cdots < X_N$. Next, replace each (relabeled) X_i by its rank and each (relabeled) Y_i by its rank. For instance, if we had observed $x_1 = 3$, $y_1 = 7$; $x_2 = 11$, $y_2 = 14$, and $x_3 = 2$, $y_3 = 9$, then after relabeling we would have $x_1 = 2$, $y_1 = 9$; $x_2 = 3$, $y_2 = 7$, $x_3 = 11$, $y_2 = 14$. Using these paired (relabeled) ranks, we compute the ordinary correlation (see Exercise 49 of Chapter 2):

$$S = \frac{\sum_{i=1}^{N} (X_i - \bar{X})(Y_i - \bar{Y})}{\sqrt{\sum_{i=1}^{N} (X_i - \bar{X})^2 \sum_{i=1}^{N} (Y_i - \bar{Y})^2}}.$$

(a) Show that

$$S = \frac{12}{N^3 - N}\left[\sum_{i=1}^{N} iY_i - \frac{N(N+1)^2}{4}\right] = 1 - \frac{6\sum_{i=1}^{N}(i - Y_i)^2}{N^3 - N}.$$

[*Hint:* $\Sigma_{i=1}^{N} i^2 = N(N+1)(2N+1)/6$.]

(b) Show that the distribution of S is independent of the form of the distributions of X and Y under the null hypothesis that X and Y are independent.

(c) Show that under the null hypothesis,

$$E(S) = 0 \quad \text{and} \quad \mathrm{Var}(S) = \frac{1}{N-1}.$$

[*Hint:* Use the form of S involving ΣiY_i.]

Note: Although the exact distribution of S can be found, the normal approximation to S together with part (c) is quite adequate for $N \geq 4$ for small values of α (such as $\alpha = .05$). The use of S in hypothesis testing was introduced in 1904 by C. Spearman and is somewhat more sensitive than the χ^2 test for independence introduced in Chapter 6.

If ties occur in the Y_i's but not in the X_i's, then we should use the "average rank" method for breaking ties. It can be shown (see page 137 of Reference 14) that the expectation of S remains unchanged and the variance is reduced to

$$\frac{1}{N-1}\left[1 - \frac{1}{N(N^2-1)}\sum_{k=1}^{g} \eta_k(\eta_k^2 - 1)\right],$$

where there are g distinct values of the Y_i's and η_k of the Y_i's are tied for the kth distinct value $(1 \leq k \leq g)$.

10. (Continuation) Use the data given in the accompanying table to test the null hypothesis of independence between the number of faculty members in business schools and the number of (a) masters students, (b) Ph.D. students, and (c) masters plus Ph.D. students in business schools. In performing these three tests, use $\alpha = .05$. You will find the random variable of Exercise 3 ideally suited for performing this test. Naturally, we reject H_0 if $s > c$ or if $s < -c$.

Size of Graduate Programs in Business Schools, August 1969

	U.C. Berkeley	*UCLA*	*Carnegie*	*U. of Chicago*	*Columbia*	*Cornell*	*Harvard*	*U. of Indiana*	*MIT*	*Michigan State*	*Northwestern*	*Stanford*	*U. of Washington*
Faculty	60	90	47	75	72	39	160	100	70	96	68	67	71
Masters Students	458	570	77	1623	995	259	1432	500	230	635	980	529	400
Ph.D. Students	128	186	95	106	99	43	238	125	70	272	98	117	97

11. (Continuation) Noting that the ratio of Ph.D. students to faculty is about 1.6 and that the ratio of masters students to faculty is about 8.5, one might infer that the faculty workload commensurate with 5 masters students is about equal to that of 1 Ph.D. student. Let X = number of faculty and let Y = number of masters students plus 5 times the number of Ph.D. students. Using $\alpha = .05$, test whether or not X and Y are independent.

12. (Continuation) In a study of foreign-language teaching techniques, the experimenter, L. Weiss, believed that students of teachers who displayed "flexibility" (for example, did not teach verbatim from the book) would perform better than students of teachers who displayed little flexibility. At the beginning of the semester 10 teachers were ranked according to degree of flexibility. The most flexible teacher received the rank 10, the next most flexible teacher 9, and so on. The average improvement per pupil in the classes of these 10 teachers starting with the lowest teacher's rank was 3.1, 3.4, 5.7, 5.9, 5.6, 6.3, 6.2, 6.4, 6.8, and 8.5. Test the null hypothesis that there is no correlation between teacher flexibility and student performance against the alternative hypothesis that there is a positive correlation. Use $\alpha = .01$.

13. Show that under the null hypothesis,

$$E(W) = \frac{n(n + m + 1)}{2}$$

and

$$\text{Var}(W) = \frac{nm(n + m + 1)}{12}.$$

In so doing you might want to employ Theorem 7 of Chapter 2 and the fact that

$$\Sigma_{j=1}^{k} j = k(k + 1)/2 \quad \text{and} \quad \Sigma_{j=1}^{k} j^2 = k(k + 1)(2k + 1)/6.$$

14. Show that $E(W) = n(n + m + 1)/2$ under the null hypothesis when either randomization or averaging of ranks is used to handle ties.

15. In studying the rank sum test, Mann and Whitney did not employ the random variable W but rather the more complicated and seemingly different random variable U described as follows. Let $X_1, X_2, \ldots, X_n$ and $Y_1, Y_2, \ldots, Y_m$ be the values of the performance characteristic for the first and second populations; and define U to be the total number of the nm pairs (X_i, Y_j) for which X_i is greater than Y_j. Show that

$$U = W - \tfrac{1}{2}n(n + 1).$$

That is, the two random variables are the same—except for a constant. [*Hint:* Note that the number of Y_j less than x_i is $r_i - i$ if $x_1 < x_2 < \cdots < x_n$ and r_i is the rank of x_i.]

3. COMPARING TWO POPULATIONS: THE RUN TEST

In this section we consider a third nonparametric method for testing whether two populations have the same distribution—that is, do not differ with respect to some performance characteristic. Like the rank sum test, the method presented in this section can be used when the performance characteristic can take on any numerical value.

Let n and m be the sample size for populations I and II, respectively. As in the rank sum test, we first make a list of the $n + m$ observations by arranging them in order of magnitude. Next, replace each of the $n + m$ values in the list by either a I or II according to the population with which it is associated. That is, if the observation was taken from population I, then replace it by a I; otherwise, replace it by a II. Each string of I's and each string of II's is called a *run*. That is, a run is a sequence of one or more identical letters that is preceded and followed by a different letter (or preceded by no letter if it is at the beginning or followed by no letter if it is at the end of the sequence of $n + m$ letters). Finally, count the total number of runs. We will denote the total number of runs by R, and we will use r to denote the observed value of R. Note that we will always have $2 \leq r \leq 2 \min\{n, m\} + 1$.

Using the data that generated Table 9.4 in Section 2, we give such a list in Table 9.5. Observe that this list starts with a string of one I. Continuing, we find a string of one II followed by a string of two I's, a string of one II, a string of one I, and finally a string of three II's. Thus, there are 6 runs in all, so $r = 6$.

Table 9.5

−3.2	−2.7	−1.4	1.7	3.1	3.9	4.1	7.6	11.5
I	II	I	I	II	I	II	II	II

If the two populations are so different that the observations from one of the populations are all smaller than the observations from the other population, then we will have $r = 2$. On the other hand, if there is no difference between the two populations, then the I's and II's will tend to be well scattered throughout the list, so that a large value of R will be observed. More generally, differences between the two populations will have a tendency to reduce the observed value of R. Hence, it seems reasonable to reject the null hypothesis of no difference in the performance characteristic in favor of the alternative hypothesis of some difference if the observed value of R is small. Thus, given a level of significance α, a good rejection region is defined by rejecting the null hypothesis when

$r \leq c$, where c is the largest integer such that

$$P(R \leq c) \leq \alpha. \tag{9.10}$$

Before we derive the distribution of R, a few remarks are in order. First, we can show that when the null hypothesis is true,

$$E(R) = \frac{2nm}{n+m} + 1 \tag{9.11}$$

and

$$\operatorname{Var}(R) = \frac{2nm(2nm - n - m)}{(n+m)^2(n+m-1)}. \tag{9.12}$$

Moreover, the distribution of R is approximately normal for large samples,[3] and the normal approximation is sufficiently accurate for practical purposes when both n and m exceed 10. Second, compared to the rank sum test, the run test is, in general, less sensitive to differences in the medians of the two populations. On the other hand, it is in general, more sensitive to differences between the variances and the shapes of the two distributions. Thus, when we are concerned only with testing for differences in the location or median of two populations, we are better off using the rank sum or the median test. (See Exercise 5 for an explanation of the median test.)

Under the null hypothesis, each of the $(n+m)!$ arrangements of the $n + m$ values is equally likely, so each of the $\binom{n+m}{n}$ arrangements of the n I's and m II's is also equally likely. If r is even, say $r = 2k$, then there must be k runs of I's and k runs of II's. Recall from Theorem 10 of Chapter 1 that there are $\binom{k+n-1}{n}$ ways of placing n indistinguishable balls (I's) in k cells (strings). But we seek the number of ways of placing n balls in k cells with the restriction that no cell is empty—that is, no string contains zero I's. Hence, each of the k cells must contain at least one ball. Therefore, if initially one ball is placed in each of the k cells, then the remaining $n - k$ balls can be placed in the k cells without restriction. Now, by Theorem 10 of Chapter 1, this can be done in

$$\binom{k + (n-k) - 1}{n-k} = \binom{n-1}{k-1}$$

ways. Similarly, there are $\binom{m-1}{k-1}$ ways of forming k strings from the

[3] A method for establishing that R is approximately normal for large samples is given in Mood and Graybill (Reference 18), p. 412.

m II's. Finally, since each way of forming k strings of I's is compatible with each way of forming k strings of II's, and since the first run can be either a string of I's or a string of II's, we can conclude that

$$P(R = 2k) = 2\,\frac{\binom{n-1}{k-1}\binom{m-1}{k-1}}{\binom{n+m}{n}}. \tag{9.13}$$

Using similar reasoning, we find

$$P(R = 2k + 1) = \frac{\binom{n-1}{k-1}\binom{m-1}{k} + \binom{n-1}{k}\binom{m-1}{k-1}}{\binom{n+m}{n}}. \tag{9.14}$$

The computation involved in equations (9.13) and (9.14) is easily done by hand if either n or m is small.

EXAMPLE 5

Suppose that samples of size 4 and 5 have yielded the data in Table 9.5, so that the observed value of R is 6. Using (9.13) and (9.14), we easily find $P(R = 9) = \frac{1}{126}$, $P(R = 8) = \frac{8}{126}$, and $P(R = 7) = \frac{18}{126}$, so that

$$P(R \leq 6) = 1 - \tfrac{27}{126} = \tfrac{99}{126}.$$

Thus, we would reject the null hypothesis of no difference in performance in favor of the alternative hypothesis of some difference if $\alpha \geq \frac{99}{126}$. (Compare this with Example 3.) •

EXAMPLE 6. (Mutual Funds)

Using Jensen's data (see Example 4 in Section 2) for comparing load and no-load mutual funds, we find that there are 19 runs. Since n $(= 13)$ and m $(= 102)$ are both larger than 10, we can use the normal distribution to approximate R. From equations (9.11) and (9.12) we find that $E(R) = 24.06$ and $\text{Var}(R) = 4.462 = (2.11)^2$. Hence,

$$P(R \leq 19) = P\left(\frac{R - 24.06}{2.11} \leq \frac{19 - 24.06}{2.11}\right) \approx P(Z \leq -2.39) = .0084.$$

As in Examples 1 and 4, we reject the null hypothesis with $\alpha = .01$. •

EXAMPLE 7. (Auditory Discrimination)

In studying the relationship between auditory discrimination and foreign-language aptitude, Weiss[4] wrote a 29-item test for measuring auditory discrimination. The test lasted about ten minutes and was administered to 56 subjects as follows. A distinct pair or French (pseudo-) words was played on a tape recorder. Next, one of the two words presented was repeated, and the subject was asked to decide whether it was the first or the second word that had been repeated. This was the process used for each of the 29 items. Such tests require careful attention in preparation, for if they are too long, the subjects have a tendency to tire, thus rendering the test unreliable. Do the data below support the null hypothesis that the test was not so long as to tire the subjects? Use $\alpha = .25$.

Item number	1	2	3	4	5	6	7	8	9	10	11	12	13	14	15
Correct responses	24	30	38	28	46	34	46	40	44	34	25	29	50	18	23

Item number	16	17	18	19	20	21	22	23	24	25	26	27	28	29
Correct responses	23	39	32	28	43	26	39	31	35	51	34	47	26	36

SOLUTION: If the null hypothesis that the subjects do not experience fatigue is false, then we should observe a poorer performance by the subjects on the latter part of the test. Therefore, we can test the null hypothesis of no fatigue against the alternative hypothesis that the subjects do tire by dividing the 29 items into two parts corresponding to the first and last items presented. Let us now refer to the first 15 items as population I and the last 14 items as population II. The data, arranged in ascending order of magnitude, are presented in Table 9.6.

Table 9.6

18	23	23	24	25	26	26	28	28	29	30	31	32	34	34
I	II	I	I	I	II	II	I	II	I	I	II	II	I	I
34	35	36	38	39	39	40	43	44	46	46	47	50	51	
II	II	II	I	II	II	I	II	I	I	I	II	I	II	

[4] Louis Weiss, unpublished doctoral dissertation, Stanford University, Stanford, Calif. (1970).

Note that $r = 18$, and it follows from equations (9.11) and (9.12) that $E(R) = 15.48$ and $\text{Var}(R) = 6.97 = (2.64)^2$. Since both n $(= 15)$ and m $(= 14)$ exceed 10, we can use the normal approximation, yielding

$$P(R \leq 18) = P\left(\frac{R - 15.48}{2.64} \leq \frac{18 - 15.48}{2.64}\right) \approx P(Z \leq .95) = .8289,$$

so we accept the null hypothesis. •

We must, as always, be careful in drawing conclusions. The analysis above shows that there is no difference in performance between the first 15 and last 14 items. It could be, however, that there is a fatigue effect that is just offset by a decrease in the intrinsic difficulty of the last 14 questions.

Strictly speaking, the run test is applicable only when the random variable measuring performance is continuous. As in this example, it is often convenient to use it even when the random variable is not continuous or when performance is measured only to the nearest inch, ounce, or whatever. In this case, ties can occur. One method for handling ties is illustrated in Table 9.6.

EXERCISES:

16. Use the run test in Exercise 7.
17. Use the run test in Exercise 8.
18. Consider the regression model and data given in Exercise 3 of Chapter 8. Suppose that the assumption of normal deviations were untenable. Use $\alpha = .05$ and the run test to test the null hypothesis that $\beta = 0$. [*Hint:* Consider the first half of the observations as forming population I.]
19. (Continuation) How would you adjust the procedure given in the exercise above in order to test the null hypothesis that $\beta = -285$?

4. COMPARING TWO POPULATIONS WITH MATCHED PAIRS: THE SIGN TEST

In Sections 1, 2, and 3 we considered the problem of testing whether or not two populations differ with regard to some performance characteristic. In this and in the next section we will consider the closely related problem of determining which of two methods of treatment is preferable with regard to some performance characteristic. Usually, one of the treatments is "no treatment at all." For ease in exposition and to reflect the most common situations and terminology, we will henceforth refer to the group receiving the first treatment as the *experimental group* and to the group receiving the second (or no) treatment as the *control group*.

What distinguishes this problem from the former one is that the individuals under study may not constitute a homogeneous group. Thus, a substantial proportion of the observed differences in the performance characteristic in the two treatment groups could be due to the lack of homogeneity rather than to the treatments themselves. For instance, we might be interested in knowing whether or not fluoridation treatments help prevent tooth decay in children. There are, however, enormous differences in children's eating habits, dental hygiene, and hereditary resistance to tooth decay—not to mention other factors such as age and condition of teeth at the beginning of the treatment period—which together will have considerably more influence upon the amount of decay than the fluoridation treatments.

In order to circumvent the difficulties caused by lack of homogeneity, we do *not* choose individuals at random to be given one or the other of the treatments. Instead, individuals are chosen in *matched pairs* so that the two individuals in each matched pair are as similar to each other as possible. That is, they are matched with respect to many of the factors that in and of themselves account for differences in the performance characteristic. Then one of the individuals in each pair is assigned at random to the control group and the other individual is put in the experimental group.

Suppose then that we have selected N matched pairs from the relevant population (for example, the children), and suppose that we have randomly selected one individual in each pair to receive the treatment. (This random selection can be accomplished by flipping a coin.) We wish to test the null hypothesis that there is no treatment effect against the alternative hypothesis that the treatment has a beneficial effect.

Let X_i and Y_i denote the value of the performance characteristic of the individuals from the ith matched pair in the experimental and control groups, respectively. Since the individuals in each pair were assigned to the experimental group at random, it follows that under the null hypothesis the probability that the difference $X_i - Y_i$ is positive is equal to $\frac{1}{2}$, and $\frac{1}{2}$ also equals the probability that $X_i - Y_i$ is negative for each i, $1 \leq i \leq N$. That is, $P(X_i - Y_i > 0) = \frac{1}{2} = P(X_i - Y_i < 0)$. Whenever the difference is zero, the pair of tied observations will be disregarded (see Exercise 24). (If either X_i or Y_i is a continuous random variable, the probability that $X_i = Y_i$ equals zero.)

Moreover, the differences $X_i - Y_i$ are independent—although not identical—random variables. Consequently, under the null hypothesis, the number B of pairs for which the difference is positive is a binomial random variable with parameters N and $\frac{1}{2}$. That is,

$$P(B = j) = \binom{N}{j}\left(\frac{1}{2}\right)^N, \qquad j = 0, 1, 2, \ldots, N.$$

Now if the treatment does, in fact, have a beneficial effect, then B will have a tendency to be large. Therefore, given the level of significance α, a good rejection region is defined by rejecting the null hypothesis whenever $b \geq c$, where b is the observed value of B and c is the smallest integer for which

$$P(B \geq c) \leq \alpha.$$

This test is known as the *sign test for matched pairs*, and we hasten to point out that it is also applicable when the differences are qualitative rather than quantitative.

EXAMPLE 8. (Fluoridation)

The national dental association decided to investigate what effect, if any, fluoridation treatments, which were then being administered to children in dental offices, had upon tooth decay. As mentioned earlier, environmental and hereditary factors are not only tremendously important in determining the amount of tooth decay, but also extremely variable. Consequently, it was necessary to use matched pairs rather than random selections in designing this experiment. Furthermore, the matches had to be good. This was accomplished by choosing sets of identical twins—which, incidentally, also form ideal pairs for many other experiments. The association picked 144 sets of twins between the ages of 6 and 7 years, and the experimental group was then chosen by a random-selection method. The fluoridation treatments were administered to this group, whereas no special treatments were administered to the members of the control group. Three years later, the children's teeth were examined by a group of dentists who were not told which children had received the treatment. This was done to prevent any *assessment bias* on the part of the examining dentists. It turned out that 89 out of the 144 children in the experimental group exhibited less tooth decay than their twins in the control group. On the basis of this information, what would you report if you were the dental association's statistician? Use $\alpha = .05$.

SOLUTION: Clearly, we can use the sign test for matched pairs to test the null hypothesis of no treatment effect. We will reject the null hypothesis if the observed value of B is very small—as the beneficial effect would be less tooth decay, so negative differences are supportive of the alternative hypothesis. Using the normal approximation to the binomial we have

$$.05 = P(B \leq c) = P\left(\frac{B - 72}{\sqrt{36}} \leq \frac{c - 72}{\sqrt{36}}\right) \approx P\left(Z \leq \frac{c - 72}{6}\right),$$

so $(c - 72)/6 = -1.645$ or $c = 62.13$. Therefore, we reject the null hypothesis, as the observed value of B is $55 \leq 62.13$. •

EXAMPLE 9. (Sound Levels)

After numerous complaints, the F.C.C. decided to investigate a certain TV station to ascertain whether or not this station was guilty of increasing the sound level during paid commercials. The F.C.C. investigator used a highly sensitive instrument to record sound levels for the 27 programs and accompanying commercials that were sampled at various times in the week. The sound level was found to be higher during the accompanying paid commercials for 19 of these 27 programs. What should the investigator have concluded using $\alpha = .01$?

Solution: Let B represent the number of programs for which the sound level was higher during the paid commercial. Then under the null hypothesis B is $B(27, \frac{1}{2})$ so, as can be shown by simple calculations, we should reject the null hypothesis when $b \geq 20$. Since the observed value of B is $19 < 20$, the investigator cannot reject the null hypothesis. [Further complaints led to another investigation in which 1777 programs were sampled. This time the observed value of B was 973 and the null hypothesis was rejected. Action was subsequently taken and the TV station rectified the situation. The number $N = 1777$ ensures that the probability of rejecting the null hypothesis is .95 when $\alpha = .01$ and when the true proportion of programs for which the sound level is higher is .55 or higher. (See Dixon and Massey [Reference 13], p. 338, for further details.)]

•

EXERCISES:

20. Recently, a study was made to uncover the relationship between participation in college sports and income level achieved in later life. A sample of 37 three-year lettermen in football at a particular college was taken. Then each of the football players was matched with another graduate of the college. These pairs were matched for year of graduation, status of father's occupation, and IQ. The survey revealed the following differences in yearly

income 13 years later (letterman's income minus other's):

−972	−4034	−4312	19671	4824	−8770	
−6543	−2184	6704	9767	−7506	2899	
8337	7921	7012	−8499	3888	1950	
6037	−2011	−205	−5177	800	−5526	
1677	4989	8079	−804	−5866	1572	
3260	−5475	1298	1067	−2685	6557	3781

Using $\alpha = .10$, would you conclude that participation in sports has a beneficial effect on income level?

21. There is ample evidence showing that a tendency towards blood clotting can be reduced in people who lower their cholesterol count below 300 milligrams per cubic centimeter. Since blood clots can cause coronary embolisms and strokes, finding an effective means of lowering a high cholesterol count is quite important. Both low-protein diets and ordinary low-intake diets have been proposed as effective for this purpose. In order to determine which is more effective, a group of 472 people with a high cholesterol count was chosen and put in matched pairs. One member of each pair was chosen at random to be given the low-protein diet. The other 236 people were given the low-intake diets. Five months after going on the diets, the entire group had their count taken again. In 127 cases, the person on the low-protein diet had a lower cholesterol count. What conclusion should the experimenters have reached if they used $\alpha = .05$?

22. In a recent study to determine the effectiveness of psychoanalysis, 23 pairs of children classified as psychotic were matched for many relevant factors. One member of each pair was randomly chosen to undergo psychoanalysis. The other member went untreated. Six years later the children were interviewed and a measure—albeit crude—of their mental health was taken. In 12 cases out of the 23, the mental health of the child who received analysis was deemed to be better. Do you accept the alternative hypothesis that psychoanalysis improves mental health? Use $\alpha = .10$.

23. After answering a long battery of questions, 15 matched pairs of students were chosen from a large introductory psychology class and a random assignment to the control and experimental group was made. Half of the members of the class, including all 15 members of the experimental group, were allowed to take their first midterm exam at home, while the other half of the class, including all 15 members of the control group, took their midterm exam in the classroom under the instructor's supervision. It turned out that 11 out of the 15 experimental subjects performed better than their "mates" in the control group. Using $\alpha = .05$, test the null hypothesis that test performance is unaffected by location against the alternative hypothesis

that students perform better if they take their test at home. Also, what reasons might the statistician have had for matching only 15 pairs in this *large* class? (Is this conclusive evidence of cheating or of the benefits of taking a test in familiar and pleasant surroundings?)

24. Suppose that, because measurements are taken only to the nearest $\frac{1}{100}$ inch, in the sign test for matched pairs a large proportion of the differences were zero. Comment on the appropriateness of disregarding these pairs with zero differences.

25. Why are the random variables $X_1 - Y_1, X_2 - Y_2, \ldots, X_N - Y_N$ independent under the null hypothesis? Why are they not identically distributed?

5. COMPARING TWO POPULATIONS WITH MATCHED PAIRS: THE RANK SUM TEST

In the preceding section we saw how to use the sign test for matched pairs to test the null hypothesis of no treatment effect. But the sign test, like the Fisher-Irwin test, assigns to each matched pair one of only two possible values. Thus, it ignores important information that is often available—namely, the magnitudes of the differences. We now present a test that takes this information into account. You will note its similarity to the rank sum test presented in Section 2.

As in Section 4, suppose that $X_i - Y_i$ represents the difference in the values of the performance characteristic for the ith matched pair, and suppose that there are N matched pairs. First, we make a list by arranging these differences in ascending order of absolute magnitude—that is, arrange them in ascending order of magnitude without regard to their signs. Next, assign to each difference its rank in the list. Finally, sum the ranks of the differences that are positive. This random variable is denoted by S^+, and we use s^+ to denote the observed value of S^+. For example, if the observed differences are as given in Table 9.4 of Section 2, then we obtain the list in Table 9.7. In this case we find that $s^+ = 2 + 4 + 6 + 7 + 8 + 9 = 36$.

Table 9.7

−1.4	1.7	−2.7	3.1	−3.2	3.9	4.1	7.6	11.5
1	2	3	4	5	6	7	8	9

Large observed values of S^+ are supportive of the alternative hypothesis of a beneficial treatment effect. This is so because large values of S^+ reflect the fact that not only are many of the differences positive, but also these positive differences tend to be larger in absolute value than do the negative differences. Consequently, it seems reasonable to reject the null hypothesis in favor of the alternative hypothesis if the observed value of the random variable S^+ is very large. Thus, given a level of significance α, a good rejection region is defined by rejecting the null hypothesis when $s^+ \geq c$, where c is the smallest integer such that

$$P(S^+ \geq c) \leq \alpha.$$

This test was also proposed by Wilcoxon in 1945, and it is sometimes referred to as the *Wilcoxon paired-comparison test.*

It now remains only to find the distribution of S^+. To do so we proceed as in Section 2. First, it is easily shown (see Exercises 30 and 31) that when the null hypothesis is true

$$E(S^+) = \frac{N(N+1)}{4} \tag{9.15}$$

and

$$\operatorname{Var}(S^+) = \frac{N(N+1)(2N+1)}{24} \tag{9.16}$$

Furthermore, we can use the normal approximation to S^+ when N is greater than 20. We now show how to find the distribution of S^+ when N is less than or equal to 20.

Under the null hypothesis, each difference is as likely to be positive as negative, without regard to the magnitude of the difference. Moreover, the N differences are independent random variables. Thus, each of the 2^N possible choices of signs for the N ranks has probability $1/2^N$ under the null hypothesis. Consequently, if we denote by $m(s^+; N)$ the number of (the 2^N) subsets of $\{1, 2, \ldots, N\}$ such that the sum of the ranks in each subset equals s^+ (see Table 9.8), then

$$P(S^+ = s^+) = \frac{m(s^+; N)}{2^N}.$$

Also,

$$P(S^+ \leq s^+) = \frac{M(s^+; N)}{2^N},$$

where

$$M(s^+; N) = \sum_{j=0}^{s^+} m(j; N).$$

We have now reduced the original problem of finding the distribution of S^+ to that of finding $M(s^+; N)$. Finding $M(s^+; N)$ is a burdensome task indeed, and the relevant values of $M(s^+; N)$ can be found in Tables E and F. The proper use of these two tables is explained in the next few paragraphs.

Table 9.8

Values of s^+	*Sets of ranks totaling* s^+	*Value of* $m(s^+; 4)$
0	$\varnothing$	1
1	$\{1\}$	1
2	$\{2\}$	1
3	$\{3\}, \{1, 2\}$	2
4	$\{4\}, \{1, 3\}$	2
5	$\{1, 4\}, \{2, 3\}$	2
6	$\{1, 2, 3\}, \{2, 4\}$	2
7	$\{1, 2, 4\}, \{3, 4\}$	2
8	$\{1, 3, 4\}$	1
9	$\{2, 3, 4\}$	1
10	$\{1, 2, 3, 4\}$	1

First note that if we define S^- to be the sum of the ranks of the negative differences, then $S^+ + S^- = 1 + 2 + \cdots + N = N(N + 1)/2$. Therefore, knowing s^- is equivalent to knowing s^+. It also follows that $P(S^+ = i) = P(S^- = i)$ under the null hypothesis, so that $P(S^+ = i) = P(S^+ = N(N + 1)/2 - i)$. In view of this symmetry, our tables need only contain $M(i; N)$ for $i \leq N(N + 1)/4$.

Second, note that for each fixed i with $i \leq N$, $M(i; N)$ does not depend on N (see Exercise 29), so our tables need not contain $M(i; N)$ for $i < N$.

Table E contains the values of $M(N; N)$ and 2^N for $N = 1, 2, \ldots, 20$. Of course, $M(0; N) = 1$ for each N. Table F contains the values of $M(i; N)$ for $N < i \leq N(N + 1)/4$ and $N = 2, 3, \ldots, 20$.

EXAMPLE 10

Using the data contained in Table 9.7, test the null hypothesis of no treatment effect using $\alpha = .10$.

SOLUTION: From Table 9.7 we see that $s^+ = 36$, so $s^- = 8$. Consequently, we can employ Table E to conclude that

$$P(S^+ \geq 36) = P(S^+ \leq 9) = \tfrac{33}{512} < .10,$$

so we reject the null hypothesis of no treatment effect in favor of the alternative hypothesis that there is a treatment effect. •

Had we applied the sign test instead of the rank sum test to the data in Table 9.7, we would have accepted the null hypothesis, because

$$P(B \geq 6) = P(B \leq 3) = \frac{\binom{9}{0} + \binom{9}{1} + \binom{9}{2} + \binom{9}{3}}{2^9} = \frac{130}{512} > .10.$$

This reflects the fact that, in general, the rank sum test is more sensitive and hence preferable to the sign test when both are applicable.

EXAMPLE 11. (Inventory Valuation)

Early in 1968 the accounting firm of Silverberg, Sherman, and Klein, henceforth referred to as the firm, was asked by one of its clients to make an audit of a company with whom the client had recently merged. The merger had become effective on December 31, 1967. Among other things, the firm was requested to check the company's earnings report for the quarter ended March 31, 1968. An important piece of information needed in figuring the first-quarter earnings was the value of inventory on hand on December 31, 1967, and on March 31, 1968. Consequently, the firm took a physical count of the inventory on hand on March 31, 1968. Using the sales records and the physical count taken on March 31, 1968, the firm was able to compute the value of the inventory on hand on December 31, 1967. In particular, for 23 housewares products they found the values listed in Table 9.9. The company had, of course, taken a physical count on December 31, 1967, which yielded the dollar values listed in Table 9.10.

Table 9.9 Firm's December 31 Dollar Valuation of 23 Housewares Products

3,174	2,648	12,687	1,794	4,060	2,434	9,601	14,461
15,604	24,342	6,397	15,425	173	12,571	12,188	14,468
5,716	8,596	6,247	4,600	2,394	12,345	484	

Table 9.10 Company's December 31 Dollar Valuation of 23 Housewares Products

1,380	2,647	11,900	1,804	4,034	2,238	7,120	13,432
15,790	17,242	4,249	14,395	0	15,083	11,850	17,772
5,884	8,912	6,295	4,568	2,396	12,510	498	

The differences between the two valuations are given in Table 9.11. Each entry in Table 9.11 represents the company's valuation minus the firm's valuation. When the two sets of figures were compared, the accountants were startled to observe the large differences in the two valuations. Needless to say, the firm was nervous about attesting to the accuracy of the earnings report, as the total difference in the two valuations accounted for roughly 50 percent of the first-quarter earnings. A rechecking of their figures revealed no errors. (It is important to add that there was never any question of wrongdoing or juggling of figures on the part of the company.) Also, it is well known that even careful counts of inventory can be off by as much as 5 percent. Nevertheless, the firm had to decide whether the observed differences in the evaluation of the inventory were due to chance (that is, to the "minor" counting errors mentioned above) or to one of the counts' being grossly in error. In resolving this question, they wisely called upon a statistician for counsel. What counsel should their statistician have given them?

Table 9.11 Differences in December 31 Dollar Valuations of 23 Housewares Products

−1,794	−1	−787	10	−26	−196	−2,481	−929
186	−7,100	−2,148	−1,030	−173	2,512	−338	3,304
168	316	48	−32	2	165	14	

Solution: The null hypothesis to be tested is that the company's physical count neither overstated nor understated the true value of the inventory. The alternative hypothesis is that the company understated the true value of the inventory. The firm was not concerned with an overstatement, as this would only serve to decrease the quarter earnings. This properly reflects the "conservative" practices of the accounting field.

Although a person's propensity to miscount would have a marked affect on each of the 23 products he helped count, there were enough different people involved in the counting process so that the 23 individual valuations done by the company can be considered independent of one another. Similarly, the 23 valuations done by the firm can be considered independent. Finally, the firm used their own help rather than the company's, so that the two counts for each product are also independent. Thus, under the null hypothesis, the 23 differences are independent—though not necessarily identical—random variables. In view of this, we are justified in using the rank sum test for matched pairs. Rearranging the data in Table 9.11, we observe that

$$s^+ = 2 + 3 + 4 + 7 + 8 + 9 + 11 + 13 + 21 + 22 = 100.$$

Using equations (9.15) and (9.16) and the normal approximation to S^+, we find

$$P(S^+ \leq 100) = P\left(\frac{S^+ - 138}{32.9} \leq \frac{100 - 138}{32.9}\right) \approx P(Z \leq -1.16) = .123.$$

Thus, the probability that the company's count would understate the inventory relative to the firm's count by as much as it did is .123 under the null hypothesis. The accounting firm's statistician advised use of the company's figures after an upward adjustment in the inventory valuation. For further details, see Exercise 28. •

Again, we must consider the possibility of ties. If $X_i - Y_i = 0$, then, as in the sign test, we disregard this matched pair.

Having disregarded such pairs, we are still left with the problem of what to do if there are ties in the magnitudes of the absolute differences. Here, as in the rank sum test, we handle ties by assigning each matched pair the average of the ranks for which it is tied. Moreover, the expected value of S^+ remains unchanged, its variance is reduced by $k(k^2 - 1)/48$ for each set of k tied pairs, and the normal approximation is still valid.

EXERCISES

26. Use the rank sum test for matched pairs to test the null hypothesis in Exercise 20.

27. Use the rank sum test for matched pairs to test the null hypothesis in Exercise 21. Use the fact that the observed values of S^+ was 16,437.

28. (An extension of the rank sum test) In Example 11 we concluded that the company had, in fact, undervalued the inventory. This still left the question of by how much they undervalued it. We can use the rank sum test for matched pairs to test the null hypothesis that the company undervalued the inventory by ϵ percent. This is accomplished by replacing the differences $X_i - Y_i$ by $[1 + (\epsilon/100)]X_i - Y_i$. Perform this test with $\epsilon = 3$ using $\alpha = .05$.

29. Show that $M(i; N) = M(i; i)$ for $N \geq i$.

30. Show that

$$E(S^+) = \frac{N(N + 1)}{4}.$$

To do so, use the facts that $\Sigma_{j=1}^{N} j = N(N + 1)/2$ and that $S^+ = \Sigma_{j=1}^{N} jI_j$, where the indicator random variable I_j equals 1 if the jth rank is positive and 0 if the jth rank is negative.

31. Show that

$$\text{Var}(S^+) = \frac{N(N+1)(2N+1)}{24}.$$

Use the fact that $\sum_{j=1}^{N} j^2 = N(N+1)(2N+1)/6$.

32. Using the data below, find s^+, $E(S^+)$, and $\text{Var}(S^+)$:

Pair	1	2	3	4	5	6	7	8
Difference	−1	0	−3	7	3	0	4	4

6. RUN TESTS FOR RANDOMNESS

Throughout Part II the legitimacy of our inferences has rested upon the fact that the samples we have taken are, in fact, *random samples*—with or without replacement, whichever is appropriate. So if we are to use the classical tools of statistical inference presented in Chapters 4, 5, 6, 7, and 8, we must be able to verify the fact that the samples obtained are indeed random.

In this section we shall introduce two of the many nonparametric procedures for testing whether or not a sample is random. Both of these procedures are based on the concept of runs.

6.1 Runs For Two-Valued Variables

The most natural type of situation in which our first test can be used is one in which there is a sequence of trials—each trial resulting in either success or failure. Under various conditions, binomial, negative binomial, and hypergeometric random variables are used to count the total number of successes in such a sequence of trials. The run test given in Section 3 can be used to test for randomness in situations such as these. We illustrate its use for a hypergeometric situation in Examples 12 and 13 and for a binomial situation in Example 14.

EXAMPLE 12. (Draft Lottery)

On December 1, 1969, the Selective Service people placed 366 allegedly identical blue capsules into an urn. Each capsule contained a slip of paper with one of the days of the year written on it. Then the capsules were withdrawn one by one from the urn. The order of the days drawn determined the vulnerability to the draft of men who were between 19 and 26 years old as of January 1, 1970. In point of fact, the capsules containing the first days of the year were put in the urn before the capsules containing the later dates in the year. Consequently, if the capsules were not mixed very well in the urn, then the draft lottery might not have been random. Using the data in Table 9.12 and $\alpha = .20$, determine whether there is reason to doubt the randomness of this draft lottery.

SOLUTION: Note that we have chosen an extremely large value of α. This is because the error of falsely accepting the null hypothesis of "a fair draft lottery" is so serious. Now, to use the run test for randomness, we can simply identify each of the first 183 days in the year (don't forget February 29) as 0's and identify the last 183 days as 1's. This identification produced the data in Table 9.12. If the alternative hypothesis is true, then there will be a paucity of runs. Hence, we will reject the null hypothesis if the observed value of R is small. Since $n = 183$ and $m = 183$ are both large, we can use the normal approximation to R. From equations (9.11) and (9.12) we have $E(R) = 184$ and $\text{Var}(R) = (9.6)^2$, so, using Table A, we see that we will reject H_0 only if the observed value of R is less than $184 - (9.6)(.84) = 175.94$. But $r = 176$, so we accept H_0. •

Table 9.12

1010111111111011011010110110000001010110111111111
1011010000101001100111000101110101001100000111111
0100111000101111101100101111101100100100111011001 0
1011011101111010110011111110000101101101011011010 0
1111001000000000000010000101011011100100111111100 00
0001101000101101000100000010101111011110100100000 0
0011001011100110000101010011001010000010010100010 0
0100000010000000

EXAMPLE 13. (Lunch Counters)

When taking a seat, say at a lunch counter, people seem to sit apart unless they know each other. Some lunch counters, however, have mostly regular customers who, after a while, get to know each other and therefore sit together. Using the fact that seat numbers 5, 11, 12, and 13 are the only occupied seats at Joe's 20-seat lunch counter, test the null hypothesis that Joe's customers are not mostly regulars against the alternative hypothesis that Joe's has many regular customers. Use $\alpha = .10$.

SOLUTION: Denoting empty and occupied seats by E and O, we see that the people are seated at Joe's lunch counter as shown below:

$$E\ E\ E\ E\ O\ E\ E\ E\ E\ E\ O\ O\ O\ E\ E\ E\ E\ E\ E\ E$$

so there are 5 runs. If the alternative hypothesis is true, then Joe's customers are mostly regulars. Hence, they will sit together, which means there will be few runs. Therefore, we will reject the null hypothesis if the observed value of R is small. Using equations (9.13) and (9.14) with $n = 4$ and $m = 16$ we find that

$$P(R \leq 5) = \frac{2 + 18 + 90 + 360}{4845} = .097 < .10.$$

Thus, we will reject the null hypothesis. •

EXAMPLE 14. (Three Sisters)

Recall the familiar example of the three sisters breaking dishes. In Chapter 4 we tested the null hypothesis that the youngest sister is not clumsy (that is, $p = \frac{1}{3}$). We let X_i be 1 if she broke the ith dish and 0 otherwise. In our probabilistic mechanism, we assumed that $X_1, X_2, \ldots$ is a sequence of independent and identically distributed random variables. Thus, our hypothesis-testing procedure precluded (or ignored) the possibility that, for instance, after breaking the first two dishes she may be nervous, hence more prone to break the third dish, or after breaking the first two dishes she may be very careful, and thus less likely to break the third dish. Now we shall find out whether or not the probabilistic mechanism was, in fact, appropriate.

As stated before, the youngest sister broke the first, second, and fourth dishes. An additional 19 observations revealed that she also broke dishes number 9, 10, 12, 15, 18, 19, 20, and 23. Test the null hypothesis that the X_i's are independent random variables with the same distribution.

SOLUTION: Displaying the data,

$$1\,1\,0\,1\,0\,0\,0\,0\,1\,1\,0\,1\,0\,0\,1\,0\,0\,1\,1\,1\,0\,0\,1$$

we see that there are 13 runs. Since n (= 11) and m (= 12) are larger than 10, we can use the normal distribution to approximate R, the number of runs. From equations (9.11) and (9.12) we find that $E(R) = 12.48$ and $\text{Var}(R) = 5.47 = (2.34)^2$. Hence, it is clear that we will accept the null hypothesis, as the observed value of R is neither too large nor too small. •

6.2 Runs Above and Below the Median

Let $X_1, X_2, \ldots, X_N$ be a sequence of random variables corresponding to the outcomes of N trials. We wish to test the null hypothesis that $X_1, X_2, \ldots, X_N$ constitutes a random sample of size N—that is, the X_i's are independent and have the same distribution. Above, we gave a method for performing this test when the variables could take on only two distinct values. We now present a test for randomness that is appropriate when the X_i's have continuous distributions.

We begin by separating the N observations $x_1, x_2, \ldots, x_N$ into two groups, the a and the b group. If x_i is greater than the median of the sample,[5] then replace x_i by the letter a. If x_i is less than the median of the sample, then replace x_i by the letter b. If N is odd, then there will be one value that is equal to the median of the sample; in this case, replace this value by the letter a. Hence, if N is even, there will be $n \equiv \frac{1}{2}N$ a's and $m \equiv \frac{1}{2}N$ b's. If N is odd, there will be $n \equiv \frac{1}{2}(N + 1)$ a's and $m \equiv \frac{1}{2}(N - 1)$ b's.

Next, let R be the number of runs formed by the n a's and m b's. Then when the null hypothesis is true, the distribution of R is given by equations (9.13) and (9.14) in Section 3.

We will reject the null hypothesis if the observed value of R is either very small or very large. As in Section 3, a very small value of R may indicate a shift in the population median, whereas a very large value of R may indicate that the outcomes are determined by a systematic rather than by a probabilistic mechanism. (If the alternative hypothesis were that the population mean is shifting, then we would reject the null hypothesis only when R is very small.)

EXAMPLE 15

Consider the (infinite) sequence of numbers:

$$\tfrac{1}{2};\ \tfrac{1}{4}, \tfrac{3}{4};\ \tfrac{1}{8}, \tfrac{7}{8}, \tfrac{3}{8}, \tfrac{5}{8};\ \tfrac{1}{16}, \tfrac{15}{16}, \tfrac{3}{16}, \tfrac{13}{16}, \tfrac{5}{16}, \tfrac{11}{16}, \tfrac{7}{16}, \tfrac{9}{16};$$
$$\tfrac{1}{32}, \tfrac{31}{32}, \tfrac{3}{32}, \tfrac{29}{32}, \ldots, \tfrac{15}{32}, \tfrac{17}{32};\ \tfrac{1}{64}, \tfrac{63}{64}, \ldots$$

Next consider the 38 numbers

.91 .48 .40 .32 .24 .92 .94 .96 .98 .00 .34 .97 .60 .23 .83 .43 .37 .69 .01
.33 .55 .87 .19 .24 .11 .98 .85 .72 .59 .66 .99 .88 .77 .66 .55 .44 .33 .22

In each case, decide whether the data could reasonably be held to be a random sample. Use $\alpha = .13$.

SOLUTION: In the first case, note that for each N there are N runs above and below the median in the first N observations. Thus, we reject the null hypothesis. Quite obviously these numbers have been generated by a systematic mechanism.

[5] The median of the sample is any number such that the number of sample values that are strictly greater than the median equals the number of sample values that are strictly less than the median.

In the second case, .57 is the median of the sample, which yields

$$abbbbaaaabbaababbabbbabbbaaaaaaaabbbb,$$

so there are 14 runs. We have $E(R) = 20$ and $\mathrm{Var}(R) = 9.24 = (3.04)^2$, so, using the normal approximation to R, we find that

$$P(20 - 1.51(3.04) \leq R \leq 20 + 1.51(3.04)) = .87.$$

Consequently, we reject the null hypothesis. •

In point of fact, this second set of numbers was also generated by a systematic method. It is interesting to note that had we used the χ^2 goodness-of-fit test to test the null hypothesis that these numbers were the outcomes of independent random variables each having the uniform distribution, then in both cases we would have accepted the null hypothesis and committed a type I error. As made obvious by this example, the χ^2 goodness-of-fit test is insensitive to some systematic effects.

EXERCISES:

33. Use runs above and below the median to test for randomness in the simulation experiment involving the rat in the maze. The needed data are given in Table 2 of Chapter 3. Use $\alpha = .05$, and note that the results of the first 10 observations are to be found in the first row. Compare your result with that found in Example 2 of Chapter 6. Also see Exercise 12 of Chapter 6.

34. In Exercise 2 of Chapter 6 we used a χ^2 goodness-of-fit test to conclude that the Table of Random Digits found on page 100 contains an appropriate percentage of 0's through 9's. Now, check it for randomness by considering the two populations to be the 0-through-4's and 5-through-9's. Use $\alpha = .05$.

35. Use runs above and below the median in order to test the null hypothesis that the degree of difficulty of the questions in the auditory discrimination test is random. The data are given in Example 7. Use $\alpha = .15$.

36. By direct enumeration, determine the exact distribution of the number R of runs above and below the median when $N = 2, 4, 6,$ and 8.

37. Given a positive number x, define x_f to be the fractional part of x. For example, if $x = 2.713$, then $x_f = .713$. Consider the numbers δ_f, $(2\delta)_f$, $(3\delta)_f, \ldots, (n\delta)_f$, and define $N_n(x)$ to be the number of these n fractional numbers that are less than or equal to x. A famous theorem in number theory due to Herman Weyl states that if δ is irrational, then

$$\lim_{n\to\infty} \frac{N_n(x)}{n} = x \qquad \text{for each } x,\ 0 \leq x \leq 1.$$

Consequently, using a χ^2 goodness-of-fit test, we would conclude that the sample $\delta_f, (2\delta)_f, \ldots, (s\delta)_f$ of size s has been generated by a uniform distribution. Why? Would the run test above and below the median also confirm the null hypothesis that these s numbers constitute a random sample of size s from the uniform distribution? Why?

38. Suppose that our data have been rounded to the nearest ounce, that $N = 39$, and that 3 sample values are equal to the median. How would you resolve this in order to use the runs above and below the median test? What if $N = 40$?

39. Another test for randomness based on runs is conducted as follows. Between each pair of observations we place a plus sign if the latter observation is greater than the former and a minus sign otherwise. We then count the number D of runs formed by the plus and minus signs. Do you think $E(D) > E(R)$, where R is the number of runs above and below the median? Why?

40. Suppose that there are just two people at Joe's 20-seat lunch counter. Then the run test for randomness is not sensitive enough to be of use. Rather, we could use as our test statistic the number U of unoccupied seats between the two customers. The alternative hypothesis that they know each other will be accepted if U is small enough. Will you accept the null hypothesis that the seats are chosen at random when $\alpha = .15$ and only seats 4 and 7 are occupied?

41. Let $X_1, X_2, \ldots$ be a sequence of independent random variables with the same probability density function. Let N be the first integer such that $X_1 > X_2 > \cdots > X_{N-1} < X_N$.

    (a) Show that

    $$P(N = n) = \frac{n-1}{n!}, \quad n = 2, 3, \ldots, \qquad E(N) = e,$$

    $$\text{and} \quad \mathrm{Var}(N) = 3e - e^2.$$

    (b) Suggest how N might be used in testing the null hypothesis that the continuous random variables $X_1, X_2, \ldots, X_M$ are independent and identically distributed against the alternative hypothesis that there is a tendency for the means of the X_i's to increase [that is, $E(X_i) < E(X_{i+1})$]. Here, M is the sample size.

7. POWER

As stated in the opening paragraph of this chapter, nonparametric methods should be used when we do not know the functional form of the distribution of the random variables of interest. Thus, the advantage of nonparametric methods over, say those of Chapter 5, is that no questionable (or even unjustified) assumptions about the distributions of the random variables of interest need be made.

On the other hand, we should ask whether we might have attained more power (that is, attained a lower probability of committing a type II error) if we had used a parametric testing procedure. In particular, how does the power of our nonparametric procedure compare with the power of the critical region associated with a t test if we assume that all the random variables of interest are, in fact, independent, normal random variables? We now briefly sketch the answer to this question.

As discussed in Chapter 4, the power of a test procedure against any simple alternative increases as the sample size increases. This leads us to ask: for a given level of significance α, for a given member of the alternative hypothesis, and for a given sample size n used in performing the t test, how large must n_p, the sample size used in our nonparametric procedure, be in order that the two procedures have the same power? The ratio n_p/n is called the *power efficiency* of our nonparametric test. While the power efficiency depends upon α, the particular value of the alternative hypothesis, and n, it usually changes little with α and the particular value of the alternative hypothesis if n is only moderately large. The limit of this ratio as n tends to infinity is called the *asymptotic relative efficiency* (ARE) of our procedure.

If our nonparametric procedure is the rank sum test (or the rank sum test for matched pairs), then the power efficiency is around 95 percent for moderate-sized values of n, and its ARE is[6] $3/\pi \approx 95.5$ percent. Here, the loss in efficiency is a small price to pay if we are uncertain about the assumption that random variables are normally distributed. If our nonparametric procedure is the sign test for matched pairs, then the power efficiency is about 95 percent for $n = 6$, but it declines toward its ARE of $2/\pi \approx 63$ percent as n increases. Here, the loss in efficiency is considerable. Comparing the ARE of the rank sum test for matched pairs with the ARE of the sign test for matched pairs, we see that the rank sum test for matched pairs should be used when both are applicable. A similar comparison would also show that the rank sum test is preferable to the Fisher-Irwin test and to the run test when all three are applicable.

[6] See Mood and Graybill (Reference 18), p. 419–421, and J. Hájek, *Course in Nonparametric Statistics* (San Francisco: Holden-Day, Inc., 1969).

EPILOGUE TO PART II

Like most useful tools, the methods of statistical inference are subject to horrendous misuse. Therefore, we conclude Part II with several examples that illustrate three common misuses of our methods.

DRAWING FALSE CONCLUSIONS

After answering a long battery of questions, 15 matched pairs of students were chosen from a large introductory psychology class. A random assignment to the control and experimental group was made. Half of the members of the class, including all 15 of the members of the experimental group, were allowed to take their first midterm exam at home. The other half of the class, including all 15 members of the control group, took their midterm exam in the classroom under the instructor's supervision. The null hypothesis that exam performance is unaffected by location was then tested against the alternative hypothesis that students perform better if they take their exam at home. It turned out that 11 out of the 15 experimental subjects performed better than their "mates" in the control group. Consequently, the null hypothesis was rejected using $\alpha = .05$ (see Exercise 23 of Chapter 9).

Thus, we know that students perform better if they take their exam at home. However, we do not know whether the better performance is due to cheating or to the effect of taking a test in familiar and pleasant surroundings. Either conclusion would be inappropriate.

In a study of foreign-language teaching techniques, the experimenters, Politzer and Weiss,[1] believed that students of teachers who displayed

[1] Robert L. Politzer and Louis Weiss, "Characteristics and Behaviors of the Successful Foreign Language Teacher," Technical Report No. 5, Stanford Center for Research and Development in Teaching, Stanford, California (1969).

"flexibility" in teaching (for example, did not teach verbatim from the book) would perform better than students of teachers who displayed little flexibility. Throughout the first semester of school, 20 teachers were observed at regular intervals. Based upon these observations, the teachers were ranked according to their flexibility—as judged by the experimenters. At the end of the semester, the experimenters measured the pupil achievement of the 5 most flexible and 5 least flexible teachers. As shown in Exercise 12 of Chapter 9, they found that performance for pupils of the most flexible teachers was far superior to that for pupils of the least flexible teachers. That is, they rejected the null hypothesis of no performance difference in favor of the alternative hypothesis of better performance for pupils of flexible teachers. Therefore, they felt justified in concluding that foreign-language teachers with high flexibility ratings are, in general, more effective than teachers with low flexibility ratings.

A short time after concluding this experiment, Politzer and Weiss made another inspection of their data, which revealed the following information: 78 percent of the students in the classes of the teachers rated most flexible were girls, whereas only 56 percent of the students in the classes of the teachers rated least flexible were girls. They had really discovered that more girls in the classroom means more teacher flexibility. Consequently, they would have come to a false conclusion if they had interpreted their rejection of the null hypothesis to mean that foreign-language teachers who receive low flexibility ratings are ineffective teachers.

Pursuing this matter further in a private communication, Weiss went on to say:

> The fact that there were more girls in the better performing classes led us to suspect that it was *not* the teacher's flexibility that made the difference, but rather the alleged superiority of girls over boys in foreign language study. If this allegation is true, then a class with a very inflexible teacher and a high proportion of girls would rank high in achievement. This is what really bothered us. However, we went one step further in our detective work and investigated the relationship between proportion of girls and teacher flexibility rather than proportion of girls and achievement of the classes. This investigation showed that there is a significant [both practical and statistical!!] positive correlation between proportion of girls and flexibility of teachers. Boys horse around a lot in class, and even the most flexible teacher has to structure her lessons more rigidly if there is to be anything at all going on in class. On the other hand, girls are, for the most part, well behaved and cooperative so it is easy for a teacher to be flexible and sensitive to the needs of the students. Therefore, perhaps we ought to say that the most effective teachers are those with high flexibility ratings due to the happy fact that they have more girls than boys in class. Otherwise, one could infer that the most effective teachers are simply those

teachers who have been accidentally assigned classes with a very high proportion of girls. You can see what an interesting problem this is.

After having read Chapter 9, the reader should quickly observe that, for the purposes of the experiment, the foreign-language students should have been assigned to the classes at random, but with the added condition that each class contain the same percentage of girls. This procedure is called *stratified random sampling.* But, alas, class scheduling and other institutional factors did not permit this.

DOUBLE USE OF INFORMATION

Many problems of hypothesis testing are formulated only after seemingly strange phenomena are observed. In particular, observed values of some random variables may lead us to formulate a null and an alternative hypothesis. If this is the case, then it is *not* permissible to use these observations as the "empirical evidence" in testing the null hypothesis. Instead, we must gather new evidence by taking a new random sample. In short, *we cannot use the same information to both formulate and test hypotheses.* This is illustrated in the following sample.

Consider the familiar example of the three sisters breaking dishes, where we used the level of significance $\alpha = .12$. Suppose that after the first 4 dishes were broken, we noted which sister broke the most dishes, and we tested the null hypothesis H_0 that this sister (who is not necessarily the youngest) is not clumsy against the alternative that she is with $\alpha = .12$. As shown before, we will reject H_0 if and only if she breaks 3 or 4 of the next 4 dishes broken. It is easily seen, however, that the probability that *this* sister broke 3 or 4 of the *first* 4 dishes broken is $\frac{27}{81} = \frac{1}{3}$. Consequently, if we use the number of the first 4 dishes that she broke as the empirical evidence, then the probability that we will reject H_0 when it is true is $\frac{1}{3} > \alpha$. Certainly this example evidences the fact that we cannot use the same information to both formulate and test hypotheses.

This mistake is frequently made by social scientists in their use of regression. These remarks apply also to the problem of estimation.

SAMPLING FROM THE WRONG POPULATION

The following statement from the venerable personal-problems counselor "Dear Abby" appeared in the *Los Angeles Times,* October 29, 1968:

CONFIDENTIAL TO THE TWO SINCERE SERVICEMEN IN THE USMC IN VIETMAN: Sorry, but I'll continue to knock it without trying it.

The official word is as follows: Studies of 2,213 drug addicts treated at the public health service hospital in Lexington, Ky., show that 70.4% of the patients began with marijuana. Does that suggest something to you?

Define the events

$D \equiv$ a U.S. male is a drug addict,
$M \equiv$ a U.S. male's first drug experience was with marijuana.

Abby is trying to say that $P(D \mid M) > P(D)$. To back up her statement she has quoted an estimate of .704 for $P(M \mid D)$. Undoubtedly, she intended to quote an estimate of $P(D \mid M)$. Abby's trouble is that she took a sample (of size 2213) from the population of male drug addicts when she should have taken a sample from the population of males whose first drug experience was marijuana.

Abby is not alone in her mistake of sampling from the wrong population. Perhaps the most famous instance of this mistake was a prediction of the 1932 presidential election. There a survey of readers of *The Literary Digest* magazine was used to estimate the proportion of votes received by Hoover, the Republican candidate. But this magazine's subscribers were mostly Republican, so it should not come as a surprise to learn that their prediction was way off the mark. (In fact, Roosevelt won by a landslide.)

APPENDIX

PRELIMINARIES

1. SUMMATION NOTATION

Suppose we are interested in finding the total time it takes to play five sets of tennis. Then the formula for computing the total time is

total time = time for first set + time for second set + time for third set + time for fourth set + time for fifth set.

Writing the formula in this way is obviously inconvenient. Instead, we may condense it by writing

$$T = x + y + z + u + v,$$

where T represents the total time, x represents the time for the first set, y represents the time for the second set, and so on. But even this notation is inadequate and cumbersome. What would we do if there were 100 sets instead of only five? (Remember, there are but 26 letters in the alphabet.) One solution is to use *subscripts*. For example, let x_1 be the time for the first set, let x_2 be the time for the second set, let x_3 be the time for the third set, and so on. Then the formula becomes

$$T = x_1 + x_2 + x_3 + \cdots + x_{100}.$$

When confronted by such problems, we shall resort to the use of this convenient subscript notation. To save even more time and space, we shall also use the following condensed notation:

$$\sum_{i=1}^{n} x_i \equiv x_1 + x_2 + x_3 + \cdots + x_n,$$

which is read: "the summation of x_i, i going from 1 to n." The symbol x_i is read "x sub i." Thus, instead of the first long formula, we can simply write

$$T = \sum_{i=1}^{100} x_i.$$

At first you may find this notation confusing, but with a little patience your confusion will subside and you will be thankful for the time it saves.

When using summations, we will have occasion to use two easily established rules.

RULE 1. $\sum_{i=1}^{n} (x_i + y_i) = \sum_{i=1}^{n} x_i + \sum_{i=1}^{n} y_i.$

RULE 2. $\sum_{i=1}^{n} kx_i = k \sum_{i=1}^{n} x_i.$

We also will have occasion to use the condensed notation for writing products:

$$\prod_{i=1}^{n} x_i \equiv x_1 \cdot x_2 \cdot \ldots \cdot x_n,$$

which is read: "the product of x_i, i going from 1 to n."

2. SETS

The concept of set is basic to all of mathematics, one branch of which is probability and statistics. We will therefore present a brief resumé of set-theoretic notations which we will use frequently. Since the aim of this book is to present the fundamentals rather than the foundations of probability and statistics, we will not delve into the more complex nature of sets.[1] Instead, we shall regard the word *set* as understood and synonymous with the words "class," "collection," "aggregate," and "ensemble" of objects or things. Some specific examples of sets include:

(i) the pieces of fruit in a bowl,
(ii) the animals in a zoo,
(iii) the members of a football team,
(iv) the students attending a university,
(v) the coins in your coin collection,

[1] The interested reader is referred to Paul Halmos' *Naïve Set Theory* (New York: Van Nostrand-Reinhold Company, 1960).

(vi) the numbers 2, 5, 8,
(vii) the numbers between 0 and 1 inclusive,
(viii) the numbers 1, 2, 3, 4, . . . ,[2]
(ix) Schubert's unfinished symphony.

With these specific examples in mind, the notion of set is now quite simple and readily comprehensible. This intuitive everyday notion of set, however, does not preclude the need for being precise in defining specific sets. For example, consider the set of talented musicians. Deciding which musicians are talented (and therefore in the set) is a matter of opinion, so this set is ambiguously specified—that is, it is not well-defined. This leads to give the following formal

DEFINITION

A *set* is any well-defined collection of distinct objects.

We usually denote sets by capital letters, such as S, A, B, Ω (capital omega). The objects in the set are called *elements* of the set. Usually we denote elements of sets by small letters. If S is a set and s is a member[3] of S, we find it convenient to write

$$s \in S,$$

read "s is a member of S." On the other hand, if s is an element that does *not* belong to S, then we write

$$s \notin S,$$

read "s is not a member of S." In accordance with our naïve conception of a set, we shall require that exactly one of the two possibilities

$$s \in S, \qquad s \notin S$$

holds for an element s and a set S.

There are two ways of defining or specifying a set: the "rule method" and the "roster method." If R denotes a rule or a property that is meaningful (well-defined) for a collection of elements, then we may specify a set S by writing $S = \{x: R(x)\}$, which is read: "the set of x such that $R(x)$"; that is, we may specify a set by stating a characteristic property of its elements. For example,

$$S = \{x: x^2 - 3x + 2 = 0\}.$$

When specifying a set by the roster method, one simply lists all the

[2] This set is called the set of *natural numbers* and is denoted by $\mathfrak{N}$. Note that it has infinitely many objects.

[3] We use the words element and member interchangeably.

members of the set. In this particular case we have $S = \{1, 2\}$. Examples (i)–(v) and (vii) illustrate specification by the rule method whereas (vi), (viii), and (ix) illustrate specification by the roster method. Note that (ix) is a set with one element.

Next consider two sets A and B. If every element of A is an element of B, then we say that A is a *subset* of B or that A is *contained* in B, and we write

$$A \subset B \quad \text{or equivalently} \quad B \supset A,$$

read "A is contained in B." Of course, $B \subset B$. Note that "$\in$" is a relation between an element and a set, whereas "$\subset$" is a relation between two sets. Finally, we say that A *equals* B if they contain the same elements. An immediate consequence of this definition of equality is that $A = B$ if and only if $A \subset B$ and $B \subset A$.

EXAMPLE 1

Consider the following three sets: the set G of all girls, the set S of all students, and the set A of all girl students. These three sets are depicted in Figure A.1 in what is called a Venn diagram. Set G is represented by the contents of the circle, set S by the contents of the triangle, and set A by the common part of the circle and the triangle, which is shaded. Note that $A \subset G$, $A \subset S$, and $G \neq S$. •

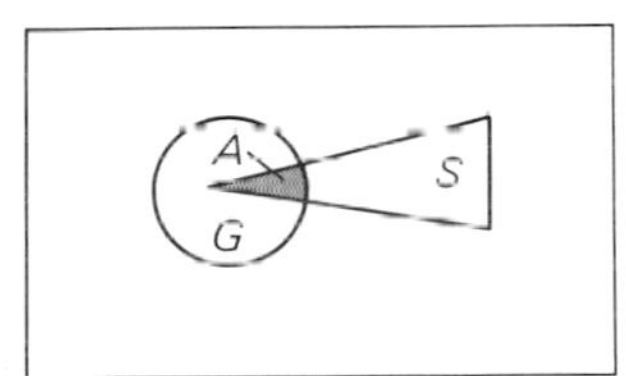

Figure A.1

EXAMPLE 2

If $\mathfrak{N} = \{1, 2, 3, \ldots\}$ denotes the set of natural numbers (positive integers), then the set $C = \{s: s \in \mathfrak{N}, 0 \leq s \leq 3\}$ defined by the rule method may also be written $C = \{1, 2, 3\}$, using the roster method. Also the set $B = \{1, 2\} \subset C$, but the set $D = \{1, 2, 8\}$ is not; that is, $D \not\subset B$. •

EXAMPLE 3

If $A = \{a\}$ and $B = \{a, b\}$, then $a \in B$ and $A \subset B$, but we cannot say $a \subset B$ or $A \in B$. •

We shall always distinguish between the element b and the set $\{b\}$. After all, a box containing one Ritz cracker is not the same as one Ritz cracker.

EXAMPLE 4

Consider the sets $A = \{1, 2, 3\}$ and $B = \{3, 1, 2\}$. Then $A \subset B$ and $B \subset A$, so that $A = B$. Evidently, order is not relevant in specifying sets. •

2.1 Set Operations

We now introduce some methods of constructing new sets from old ones.

DEFINITION

The *intersection* of two sets A and B is the set of all elements that belong to both A and B [see Figure A.2(a)]. We denote this set by $A \cap B$, which is read: "A intersect B." Thus,

$$A \cap B = \{x\colon x \in A \text{ and } x \in B\}.$$

DEFINITION

The *union* of two sets A and B is the set of all elements that belong to A or to B or to both A and B [see Figure A.2(b)]. We denote this set by $A \cup B$, which is read: "A union B." Thus,

$$A \cup B = \{x\colon x \in A \text{ or } x \in B\}.$$

In order to save space, we sometimes mimic the notation used for sums and products and employ a more condensed notation:

$$\bigcup_{i=1}^{n} S_i \equiv S_1 \cup S_2 \cup S_3 \cup \cdots \cup S_n,$$

and

$$\bigcap_{i=1}^{n} S_i \equiv S_1 \cap S_2 \cap S_3 \cap \cdots \cap S_n.$$

When dealing with more than a finite number of sets $S_1, S_2, S_3, \ldots,$ we write

$$\bigcup_{i=1}^{\infty} S_i \quad \text{and} \quad \bigcap_{i=1}^{\infty} S_i$$

for the union of all the S_i and for the intersection of all the S_i, respectively.

A special set we shall encounter is the set with no members, called the *null* or *empty set*, which we denote by $\varnothing$. In the context of set theory, the null set is analogous to the number zero; however, $\varnothing \neq 0$, since it is a set, not a number. If $A \cap B = \varnothing$, then we say that A and B are *disjoint* or *mutually exclusive*. Note that for any sets A and B, it is true that $(A \cap B) \subset A$, that $(A \cap B) \subset B$, and that $A \cap B$ is the largest set contained in both A and B. Also $A \subset (A \cup B)$, $B \subset (A \cup B)$, and $A \cup B$ is the smallest set that contains both A and B.

DEFINITION

The *complement of B in A* is the set of all elements of A that are not in B [see Figure A.2(c)]. We denote this set by $A \sim B$, read "A complement B." Thus,

$$A \sim B = \{x : x \in A \quad \text{and} \quad x \notin B\}.$$

We will always have at hand some *universal set* Ω—that is, all sets we are interested in are subsets of the universal set Ω. Thus, when we speak of the complement of B, written $\tilde{B}$ and read "B complement," we shall mean the set of all elements of Ω that are not in B. Thus,

$$\tilde{B} \equiv \{x : x \in \Omega \quad \text{and} \quad x \notin B\}.$$

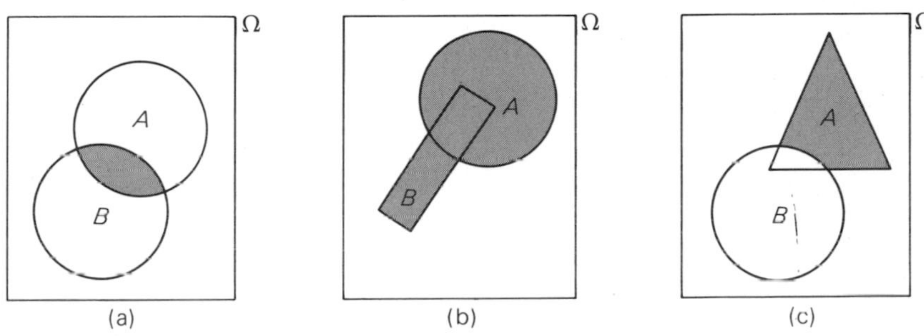

Figure A.2 The Intersection, Union, and Complement of Two Sets

We also use the following

DEFINITION

The sets $S_1, S_2, \ldots, S_n$ are said to *partition* the set Ω if

(a) $S_i \subset \Omega$, $i = 1, 2, \ldots, n$,

(b) $S_i \cap S_j = \emptyset$, $i \neq j$, $i = 1, 2, \ldots, n$ and $j = 1, 2, \ldots, n$, and

(c) $\Omega = \bigcup_{i=1}^{n} S_i$.

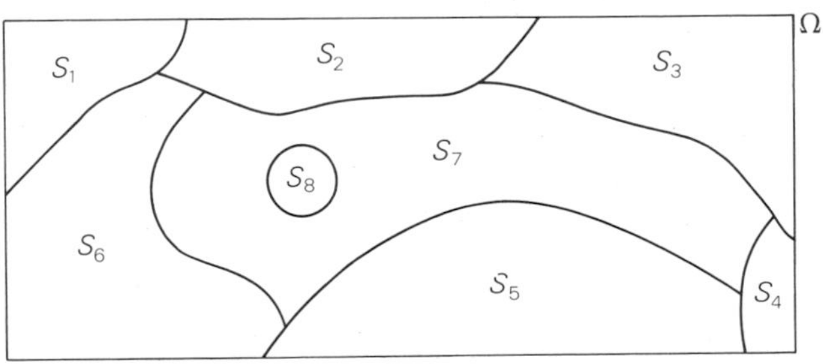

Figure A.3 A Partition of the Set Ω

EXAMPLE 5

Let Ω = set of college students, M = male college students, F = female college students, G = graduate students, U = undergraduates, S_1 = freshmen, S_2 = sophomores, S_3 = juniors, and S_4 = seniors. Then the following are a few of the groups of sets that partition Ω: U and G; M and F; G, S_1, S_2, S_3, and S_4; $M \cap U$, $M \cap G$, $F \cap U$, and $F \cap G$. •

Notice that $M \sim U = M \cap G \neq F \cap U = U \sim M$ in Example 5, so that $A \sim B$ is generally not equal to $B \sim A$. However, we might conjecture, and rightly so, that in general $\widetilde{(A \cap B)} = \tilde{A} \cup \tilde{B}$, since this is true in Example 5. This naturally leads us to ask what general statements (laws) can be made about the equality of unions, intersections, and complements of sets. Below we list a few of these laws. Their proofs are left as exercises.

ALGEBRAIC LAWS OF SET THEORY

Let A, B, and C be subsets of the universal set Ω; then

(a) $A \cup \varnothing = A, \quad A \cap \varnothing = \varnothing;$
(b) $A \cup \Omega = \Omega, \quad A \cap \Omega = A;$
(c) $A \cup \tilde{A} = \Omega, \quad A \cap \tilde{A} = \varnothing;$
(d) $\widetilde{(\tilde{A})} = A;$
(e) $A \cup A = A, \quad A \cap A = A;$
(f) $A \cup B = B \cup A, \quad A \cap B = B \cap A;$
(g) $(A \cup B) \cup C = A \cup (B \cup C), \quad (A \cap B) \cap C = A \cap (B \cap C);$
(h) $A \cup (B \cap C) = (A \cup B) \cap (A \cup C),$
$A \cap (B \cup C) = (A \cap B) \cup (A \cap C).$

DE MORGAN'S LAWS

If A and B are subsets of Ω, then

$$\widetilde{(A \cup B)} = \tilde{A} \cap \tilde{B} \quad \text{and} \quad \widetilde{(A \cap B)} = \tilde{A} \cup \tilde{B}.$$

PROOF. Take $x \in \widetilde{(A \cup B)}$; then $x \notin (A \cup B)$, so $x \notin A$ and $x \notin B$. Equivalently, $x \in \tilde{A}$ and $x \in \tilde{B}$; that is, $x \in \tilde{A} \cap \tilde{B}$. Thus, every member of $\widetilde{(A \cup B)}$ is also a member of $\tilde{A} \cap \tilde{B}$, so

$$\widetilde{(A \cup B)} \subset \tilde{A} \cap \tilde{B}.$$

Next, take $s \in \tilde{A} \cap \tilde{B}$. Then $s \in \tilde{A}$ and $s \in \tilde{B}$. Equivalently, $s \notin A$ and $s \notin B$, so $s \notin A \cup B$—that is, $s \in \widetilde{(A \cup B)}$. Thus, every member of $\tilde{A} \cap \tilde{B}$ is also a member of $\widetilde{(A \cup B)}$, so

$$\tilde{A} \cap \tilde{B} \subset \widetilde{(A \cup B)}.$$

We can now conclude that $(\widetilde{A \cup B}) = \tilde{A} \cap \tilde{B}$, since each is a subset of the other. Similar reasoning establishes the second of De Morgan's laws. ◆

EXERCISES:

1. Give a verbal description of the following sets.
   (a) $A = \{0, 1, 2, 3, 4, 5, 6, 7, 8, 9\}$.
   (b) $B = \{a, e, i, o, u\}$.
   (c) $C = \{$red, green, blue$\}$.
   (d) $D = \{$Eldridge Cleaver, Hubert Humphrey, Richard Nixon, George Wallace$\}$.
   (e) $E = \{$French, Italian, Portuguese, Spanish$\}$.

2. Use the roster method to specify the following sets.
   (a) The Ivy League colleges.
   (b) The consonants in the alphabet.
   (c) The physical sciences.
   (d) The suits in a deck of cards.
   (e) The schools and colleges in the university you attend.

3. Let Ω be a set, and denote by 2^{Ω} the set of all subsets of Ω; 2^{Ω} is called the power set of Ω.
   (a) List all the subsets of $\Omega = \{1, 2, 3\}$ and notice that there $2^3 = 8$ members of 2^{Ω}.
   (b) Use mathematical induction to verify that 2^{Ω} has 2^n elements if Ω has n elements.

4. Let $\Omega = \{x, y, z, u, v\}$, $A = \{x, y, z\}$, $B = \{y, u, v\}$, $C = \{x, z, v\}$. List the elements of the following subsets of Ω:
   (a) $\tilde{A}$.
   (b) $\tilde{B}$.
   (c) $\tilde{C}$.
   (d) $\tilde{B} \cup \tilde{C}$.
   (e) $(\widetilde{B \cap C})$.
   (f) $(B \cap C) \cup (\tilde{B} \cap C)$.
   (g) $\widetilde{\tilde{B}}$.
   (h) $(\widetilde{\tilde{C} \cup \tilde{A}})$.
   (i) $\tilde{A} \cap (\tilde{B} \cup \tilde{C})$.
   (j) $\tilde{B} \cup (\widetilde{B \cap C})$.
   (k) $(A \cup B) \cup (\widetilde{A \cap B})$.
   (l) $(A \cup B) \cap (\tilde{A} \cap \tilde{B})$.
   (m) $(\widetilde{A \cup B}) \cup (\widetilde{\tilde{A} \cap \tilde{B}})$.
   (n) $A \cap C$.

5. Let A be the set of positive even integers, let B be the set of positive integers divisible by 3, let C be the set of positive odd integers, and let $\mathfrak{N}$, the set of positive integers, be the universal set. Describe
   (a) $A \cup C$.
   (b) $A \cap C$.
   (c) $A \cap B$.
   (d) $B \cap C$.
   (e) $A \cap B \cap C$.
   (f) $(A \cup B) \cap C$.
   (g) $A \cup (B \cap C)$.
   (h) $A \sim B$.

(i) $\tilde{B}$.
(j) $\tilde{A}$.
(k) $A \sim C$.
(l) $A \sim \mathfrak{N}$.

6. Let Ω be the set of all people, and let H = the set of all healthy people, S = the set of all sick people, R = the set of all rich people, P = the set of all poor people, I = the set of all intelligent people. Give verbal descriptions of the following statements.
   (a) $H \subset P$.
   (b) $H \cap P \neq \emptyset$.
   (c) $H \cap R = \emptyset$.
   (d) $S \cap P = \emptyset$.
   (e) $\tilde{H} \cap (I \cup P) = \emptyset$.
   (f) $I \cap S \subset P$.
   (g) $P \subset I$.
   (h) $P = H$.
   (i) $R = S \cup (I \cap H)$.
   (j) $H \cup S = \Omega$.
   (k) $H \cap S = \emptyset$.

7. Establish the algebraic laws of set theory.

8. Prove that $A \subset B$ if and only if $A \cap B = A$.

9. Show that the sets $A \cap B$ and $A \sim B$ partition A and the sets E and $\tilde{E}$ partition Ω.

10. Show that $\tilde{B} \subset \tilde{A}$ if and only if $A \subset B$.

11. Let $|A|$ be the number of elements in A for each subset A of $\Omega = \{1, 2, \ldots, n\}$. State each of the following conditions in terms of $|A|$.
   (a) $A = \emptyset$.
   (b) $A \cap B = \emptyset$.
   (c) $A = \Omega$.
   (d) $A \cup B = \Omega$.
   (e) $A \subset B$.
   (f) $A = B$.
   (g) $A \neq \emptyset$.
   (h) $A \cup \tilde{A} = \Omega$.

3. FUNCTIONS[4]

We now turn to a discussion of the fundamental notion of function or mapping. As we shall soon see, a function is only a special kind of set.

To the mathematician of a century ago the word "function" ordinarily meant a definite formula, such as

$$f(x) = x^2 + 3x - 5,$$

which associates to each real number x another real number $f(x)$. As mathematics developed, however, it became increasingly clear that the requirement that a function be a formula was unduly restrictive and that a more general definition was needed. In our study of probability, we too will need a more general definition, so without further ado we plunge in and give our formal

[4] Much of this discussion follows along the lines of Robert Bartles' excellent book *The Elements of Real Analysis* (New York: John Wiley & Sons, Inc., 1964).

DEFINITION

A *function* is a set f of ordered pairs[5] with the property that if (a, b) and (a, b') are elements of f, then $b = b'$; that is, a cannot occur as the first member of two different ordered pairs. The set of all elements that occur as first members of elements of f is called the *domain* of f and will be denoted by $D(f)$, while the set of all elements that occur as second members of elements of f is called the *range* of f and will be denoted by $R(f)$. Thus,

$$D(f) = \{a\colon \text{there is some } b \text{ with } (a, b) \in f\}$$

and

$$R(f) = \{b\colon \text{there is some } a \text{ with } (a, b) \in f\}.$$

If (a, b) is an element of a function f, then it is customary to write

$$b = f(a)$$

instead of $(a, b) \in f$, and we often refer to the element b as the value of f at the point a. Thus, a function is a rule of correspondence that associates with each element x in a set called the domain a unique element $f(x)$ in a set called the range. Notice that there need not be a unique element of the domain associated with each element of the range; that is, we may have $(a, b) \in f$, $(a', b) \in f$, and $a \neq a'$.

EXAMPLE 6

Consider the following sets of ordered pairs:

$$\begin{aligned} F &= \{(1, 1), (2, 4), (3, 9)\}, \\ X &= \{(\text{heads}, 1), (\text{tails}, 0)\}, \\ C &= \{(7, \text{win}), (11, \text{win}), (2, \text{lose}), (3, \text{lose}), (12, \text{lose})\}, \\ E &= \{(1, 0), (0, 1), (-1, 0), (0, -1)\}. \end{aligned}$$

Notice that each set but E is a function. The domain of X and the range of C are not real numbers, and C does not associate a unique element of its domain with each element of its range. [The function F corresponds to the formula $F(x) = x^2$, while X and C correspond to tossing a coin and playing craps.] •

One way of visualizing a function is as a *graph* (see Figure A.4). Another way that is important and widely used is as a *table* (see Figure A.5). Yet a third way is as a *transformation*. In this phraseology, when $(a, b) \in f$, we literally think of f as taking the element a from $D(f)$ and "transforming" or "mapping" it into an element b of $R(f)$ (see Figure A.6).

[5] An *ordered pair* is a pair of objects in which order is important—for example, $(4, 7) \neq (7, 4)$. (Of course $\{4, 7\} = \{7, 4\}$.) Also, in contrast to most other sets, it is common to denote functions by lower-case letters.

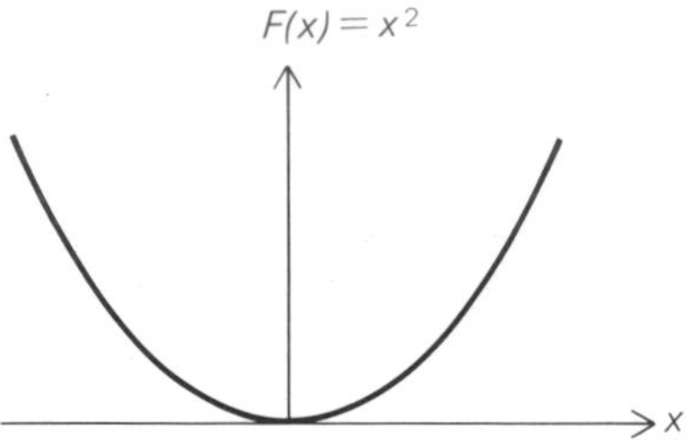

Figure A.4 A Function as a Graph

Player	*Points scored*
Alcindor	32
Allen	19
Lynn	9
Shackelford	13
Warren	10

Figure A.5 A Function as a Table

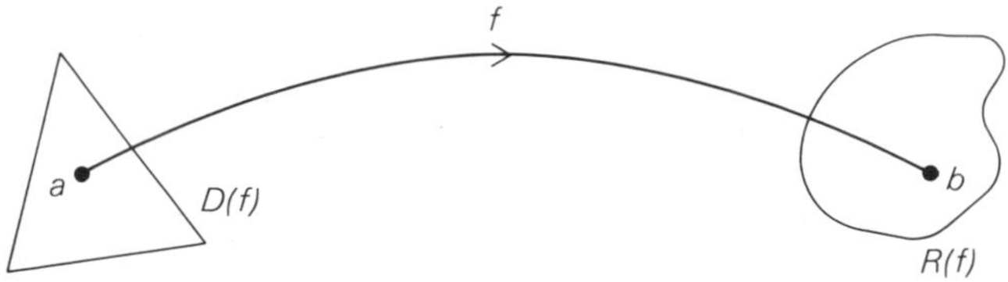

Figure A.6 A Function as a Transformation

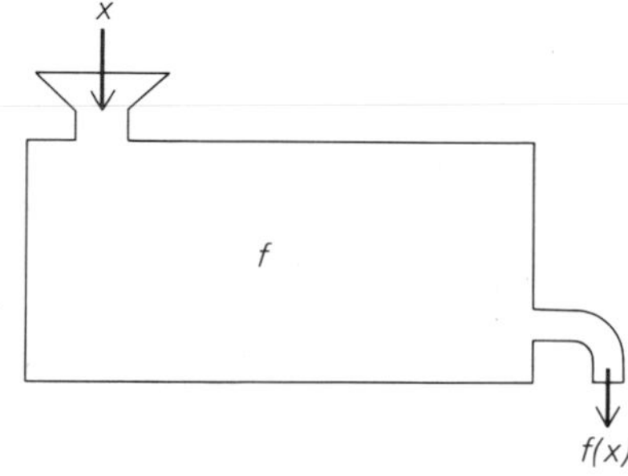

Figure A.7 A Function as a Machine

Perhaps the best way to visualize a function is to picture a *machine* that will accept elements of $D(f)$ as inputs and yield corresponding elements of $R(f)$ as outputs (see Figure A.7). If we take an element a from $D(f)$ and drop it into f, then out comes the corresponding value $f(a)$. If we drop a different element c of $D(f)$ into f, we get $f(c)$—which may or may not differ from $f(a)$. If we try to insert something that does not belong to $D(f)$ into f, we find that this machine f does not accept it.

This last visualization makes clear the distinction between f and $f(a)$: the first is the machine, the second is the output of the machine when we put a into it. Certainly it is useful to distinguish between a machine and its outputs. (Only a fool would confuse a sausage grinder with a ground sausage!)

EXAMPLE 7

All of the following formulas define functions with domain the real numbers and range contained in the real numbers. (When the range of a function is contained in the real numbers, we say that the function is *real-valued*.)

$$f(x) = 1,$$

$$g(x) = 2x^2 + 4,$$

$$h(x) = \begin{cases} x^2, & \text{if } x > 0, \\ 1, & \text{if } x \leq 0, \end{cases}$$

$$k(x) = \begin{cases} 1, & \text{if } x \text{ is rational}, \\ 0, & \text{if } x \text{ is irrational}. \end{cases}$$

Notice that it is not possible to draw a graph corresponding to k. Nevertheless k is a function. •

EXAMPLE 8

Let P be the function whose domain is the set of all subsets of $\{1, 2, 3\}$, and define P by $P(E) = \frac{1}{3}$ times the number of points in E, where E is a subset of $\{1, 2, 3\}$. Thus,

$$P(\varnothing) = 0, \quad P(\{1\}) = P(\{2\}) = P(\{3\}) = \tfrac{1}{3}, \quad P(\{1, 2, 3\}) = 1,$$
$$P(\{1, 2\}) = P(\{1, 3\}) = P(\{2, 3\}) = \tfrac{2}{3}. \bullet$$

Such real-valued functions defined on a class of sets will be of considerable interest to us, especially in Chapter 1.

EXAMPLE 9. (Absolute Values)

One function of special interest that we will use from time to time is the absolute-value function A defined by $A(x) = x$ if $x \geq 0$ and $A(x) = -x$ if $x < 0$. The domain of A is the set of real numbers and the range of A is the set of nonnegative real numbers. Owing to its special importance, it is common to write $|x|$, read "the absolute value of x," instead of $A(x)$. •

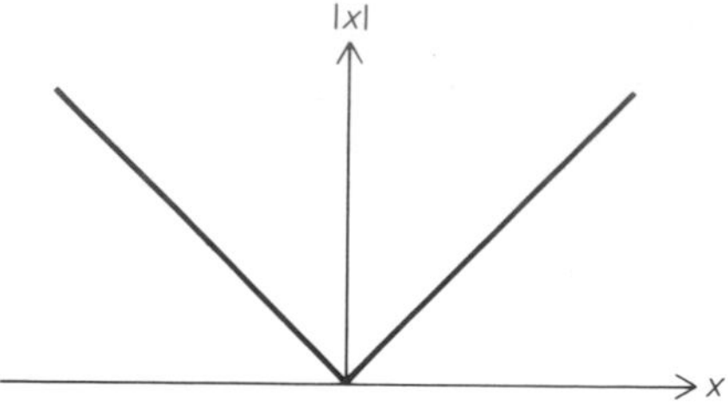

Figure A.8 The Absolute-Value Function

▶▶

Just as we were able to construct new sets from old ones, we can also construct new functions from old ones. Similar to the union and intersection of sets, we have addition and multiplication of functions.

If f and g are real-valued functions both having domain D, then we define the functions $f + g$ and $f \cdot g$ with domain D by

$$(f + g)(x) = f(x) + g(x) \quad \text{and} \quad (f \cdot g)(x) = f(x) \cdot g(x) \quad \text{for each } x \in D.$$

Thus, if $f(x) = x^2$ and $g(x) = 2$ for each $x \geq 0$, then $(f + g)(x) = 2 + x^2$ and $(f \cdot g)(x) = 2x^2$ for each $x \geq 0$.

If $R(g) \subset D(f)$, then we may define the *composition of f with g*, written $f \circ g$ or $f(g)$, by

$$(f \circ g)(x) = f(g(x)) \qquad \text{for each } x \in D(g).$$

Note that $D(f \circ g) = D(g)$ and $R(f \circ g) \subset R(f)$. Thus, if f and g are defined as above, then $(f \circ g)(x) = 4$ and $(g \circ f)(x) = 2$ for each $x \geq 0$.

▶

EXERCISES:

12. Explain why the set E in Example 6 is not a function.

13. Find the domains and ranges of the functions given in Example 7.

14. Let $\mathcal{R}$ be the set of real numbers and consider the set C defined by

$$C = \{(x, y) : x^2 + y^2 = 1, x \in \mathcal{R}, y \in \mathcal{R}\}.$$

Is this set a function? What about the sets

$$D = \{(x, y) : 2x + y = 1, x \in \mathcal{R}, y \in \mathcal{R}\}?$$
$$E = \{(x, -x) : x \in \mathcal{R}\}? \quad \text{and} \quad F = \{(x, x^2) : x \in \mathcal{R}\}?$$

15. Into what number is the number 5 mapped by the functions

$$g(x) = x^2 + 4, \quad x \in \mathcal{R}, \qquad h(x) = \frac{1}{(1 - x^2)}, \quad x \in \mathcal{R} \sim \{-1, 1\}?$$

What is the range of these two functions?

16. (The generation gap) Let $C = \{$grandmother, grandfather, mother, father, daughter, son$\}$ ($C = \{GM, GF, M, F, D, S\}$) and consider a real-valued function g defined on the set of ordered pairs from C. We give eight of the thirty values of this function. Complete the definition of this function in an appropriate manner.

$$g((S, D)) = g((D, S)) = 6, \qquad g((F, M)) = g((M, F)) = 4,$$
$$g((D, GM)) = g((S, GM)) = 4, \qquad g((GM, GF)) = g((GF, GM)) = 2.$$

17. Let f be a function whose domain is the set of real numbers such that $f(x) = x$ if $x \geq 0$ and $f(x) = -x$ if $x < 0$.
    (a) Draw a picture (graph) of the function f.
    (b) What is the range of f?
    (c) Let $g(x) = -f(x)$ for each real number x. Describe $g + f$, $g \cdot f$, and $g \circ f$ find their ranges.

18. (Countable sets) A set A is said to be *countable* if there is a function f that has domain $\mathfrak{N}$ the set of positive integers, that has range A, and that maps different members of $\mathfrak{N}$ into different members of A—that is, if $n \neq m$, then $f(n) \neq f(m)$. Show that the following sets are countable:
    (a) $\mathfrak{N}$ = set of positive integers.
    (b) $2\mathfrak{N}$ = set of even positive integers.
    (c) I = set of all integers.
    (d) Q = set of rational numbers.

REFERENCES

ANALYSIS

1. Bartle, R. C., *The Elements of Real Analysis.* New York: John Wiley & Sons, Inc., 1964.
2. Royden, H. L., *Real Analysis.* New York: The Macmillan Company, 1963.

PROBABILITY THEORY

3. Dwass, M., *First Steps in Probability.* New York: McGraw-Hill, Inc., 1967.
4. Feller, W., *An Introduction to Probability Theory and Its Applications,* Vol. I, 3d ed. New York: John Wiley & Sons, Inc., 1968.
5. Mosteller, F., R. Rourke, and G. Thomas, Jr., *Probability with Statistical Applications.* Reading, Mass.: Addison-Wesley Publishing Company, Inc., 1961.
6. Thompson, W. A., Jr., *Applied Probability.* New York: Holt, Rinehart and Winston, Inc., 1969.
7. Parzen, E., *Modern Probability Theory and Its Applications.* New York: John Wiley & Sons, Inc., 1960.

REGRESSION

8. Draper, N. R., and H. Smith, *Applied Regression Analysis.* New York: John Wiley & Sons, Inc., 1966.

SAMPLING

9. Cochran, W. G., *Sampling Techniques,* 2d ed. New York: John Wiley & Sons, Inc., 1953.
10. Hansen, M. H., W. H. Hurwitz, and W. G. Madow, *Sample Survey Methods and Theory,* Vols. I and II. New York: John Wiley & Sons, Inc., 1953.

SET THEORY

11. Halmos, P. *Naïve Set Theory.* New York: Van Nostrand-Reinhold Company, Inc., 1960.

STATISTICS

12. Bowker, A. H., and G. J. Lieberman, *Engineering Statistics*, 2d ed. Englewood Cliffs, N.J.: Prentice-Hall, Inc., 1971.
13. Dixon, W. J., and F. J. Massey, Jr., *Introduction to Statistical Analysis*, 3d ed. New York: McGraw-Hill, Inc., 1969.
14. Hájek, J., *Course in Nonparametric Statistics*. San Francisco: Holden-Day, Inc., 1969.
15. Hays, W. L., and R. L. Winkler, *Statistics: Probability, Inference, and Decision*, Vol. I. New York: Holt, Rinehart and Winston, Inc., 1970.
16. Hodges, J. L., Jr., and E. L. Lehmann, *Basic Concepts of Probability and Statistics*. San Francisco: Holden-Day, Inc., 1964.
17. Hogg, R. V., and A. T. Craig, *Introduction to Mathematical Statistics*, 2d ed. New York: The Macmillan Company, 1965.
18. Mood, A. E., and F. A. Graybill, *Introduction to the Theory of Statistics*, 2d ed. New York: McGraw-Hill, Inc., 1963.
19. Siegel, W., *Nonparametric Statistics for the Behavioral Sciences*. New York: McGraw-Hill, Inc., 1956.
20. Wallis, A. W., and H. V. Roberts, *Statistics: A New Approach*. New York: The Free Press, 1956.

TABLES

Table A Cumulative Normal Distribution*

$[\Phi(z_\alpha) = P(Z \leq z_\alpha) = \alpha$; e.g., $P(Z \leq 2) = .9772]$

z	.00	.01	.02	.03	.04	.05	.06	.07	.08	.09
.0	.5000	.5040	.5080	.5120	.5160	.5199	.5239	.5279	.5319	.5359
.1	.5398	.5438	.5478	.5517	.5557	.5596	.5636	.5675	.5714	.5753
.2	.5793	.5832	.5871	.5910	.5948	.5987	.6026	.6064	.6103	.6141
.3	.6179	.6217	.6255	.6293	.6331	.6368	.6406	.6443	.6480	.6517
.4	.6554	.6591	.6628	.6664	.6700	.6736	.6772	.6808	.6844	.6879
.5	.6915	.6950	.6985	.7019	.7054	.7088	.7123	.7157	.7190	.7224
.6	.7257	.7291	.7324	.7357	.7389	.7422	.7454	.7486	.7517	.7549
.7	.7580	.7611	.7642	.7673	.7704	.7734	.7764	.7794	.7823	.7852
.8	.7881	.7910	.7939	.7967	.7995	.8023	.8051	.8078	.8106	.8133
.9	.8159	.8186	.8212	.8238	.8264	.8289	.8315	.8340	.8365	.8389
1.0	.8413	.8438	.8461	.8485	.8508	.8531	.8554	.8577	.8599	.8621
1.1	.8643	.8665	.8686	.8708	.8729	.8749	.8770	.8790	.8810	.8830
1.2	.8849	.8869	.8888	.8907	.8925	.8944	.8962	.8980	.8997	.9015
1.3	.9032	.9049	.9066	.9082	.9099	.9115	.9131	.9147	.9162	.9177
1.4	.9192	.9207	.9222	.9236	.9251	.9265	.9279	.9292	.9306	.9319
1.5	.9332	.9345	.9357	.9370	.9382	.9394	.9406	.9418	.9429	.9441
1.6	.9452	.9463	.9474	.9484	.9495	.9505	.9515	.9525	.9535	.9545
1.7	.9554	.9564	.9573	.9582	.9591	.9599	.9608	.9616	.9625	.9633
1.8	.9641	.9649	.9656	.9664	.9671	.9678	.9686	.9693	.9699	.9706
1.9	.9713	.9719	.9726	.9732	.9738	.9744	.9750	.9756	.9761	.9767
2.0	.9772	.9778	.9783	.9788	.9793	.9798	.9803	.9808	.9812	.9817
2.1	.9821	.9826	.9830	.9834	.9838	.9842	.9846	.9850	.9854	.9857
2.2	.9861	.9864	.9868	.9871	.9875	.9878	.9881	.9884	.9887	.9890
2.3	.9893	.9896	.9898	.9901	.9904	.9906	.9909	.9911	.9913	.9916
2.4	.9918	.9920	.9922	.9925	.9927	.9929	.9931	.9932	.9934	.9936
2.5	.9938	.9940	.9941	.9943	.9945	.9946	.9948	.9949	.9951	.9952
2.6	.9953	.9955	.9956	.9957	.9959	.9960	.9961	.9962	.9963	.9964
2.7	.9965	.9966	.9967	.9968	.9969	.9970	.9971	.9972	.9973	.9974
2.8	.9974	.9975	.9976	.9977	.9977	.9978	.9979	.9979	.9980	.9981
2.9	.9981	.9982	.9982	.9983	.9984	.9984	.9985	.9985	.9986	.9986
3.0	.9987	.9987	.9987	.9988	.9988	.9989	.9989	.9989	.9990	.9990
3.1	.9990	.9991	.9991	.9991	.9992	.9992	.9992	.9992	.9993	.9993
3.2	.9993	.9993	.9994	.9994	.9994	.9994	.9994	.9995	.9995	.9995
3.3	.9995	.9995	.9995	.9996	.9996	.9996	.9996	.9996	.9996	.9997
3.4	.9997	.9997	.9997	.9997	.9997	.9997	.9997	.9997	.9997	.9998

z	1.282	1.645	1.960	2.326	2.576	3.090	3.291	3.891	4.417
$\Phi(z)$	.90	.95	.975	.99	.995	.999	.9995	.99995	.999995

Table B Cumulative Chi-Square Distributions*

$[P(\chi_n^2 \leq \chi_n^2(\alpha)) = \alpha;$ e.g., $P(\chi_7^2 \leq 14.1) = .95]$

d.f. †n \ α	.005	.010	.025	.050	.100	.250	.500	.750	.900	.950	.975	.990	.995
1	.00004	.00016	.001	.0039	.0158	.102	.455	1.32	2.71	3.84	5.02	6.63	7.88
2	.0100	.0201	.0506	.103	.211	.575	1.39	2.77	4.64	5.99	7.38	9.21	10.6
3	.0717	.115	.216	.352	.584	1.21	2.37	4.11	6.25	7.81	9.35	11.3	12.8
4	.207	.297	.484	.711	1.06	1.92	3.36	5.39	7.78	9.49	11.1	13.3	14.9
5	.412	.554	.831	1.15	1.61	2.67	4.35	6.63	9.24	11.1	12.8	15.1	16.7
6	.676	.872	1.24	1.64	2.20	3.45	5.35	7.81	10.6	12.6	14.4	16.8	18.5
7	.989	1.24	1.69	2.17	2.83	4.25	6.35	9.04	12.0	14.1	16.0	18.5	20.3
8	1.34	1.65	2.18	2.73	3.49	5.07	7.34	10.2	13.4	15.5	17.5	20.1	22.0
9	1.73	2.09	2.70	3.33	4.17	5.90	8.34	11.4	14.7	16.9	19.0	21.7	23.6
10	2.16	2.56	3.25	3.94	4.87	6.74	9.34	12.5	16.0	18.3	20.5	23.2	25.2
11	2.60	3.05	3.82	4.57	5.58	7.58	10.3	13.7	17.3	19.7	21.9	24.7	26.8
12	3.07	3.57	4.40	5.23	6.30	8.44	11.3	14.8	18.5	21.0	23.3	26.2	28.3
13	3.57	4.11	5.01	5.89	7.04	9.30	12.3	16.0	19.8	22.4	24.7	27.7	29.8
14	4.07	4.66	5.63	6.57	7.79	10.2	13.3	17.1	21.1	23.7	26.1	29.1	31.3
15	4.60	5.23	6.26	7.26	8.55	11.0	14.3	18.2	22.3	25.0	27.5	30.6	32.8
16	5.14	5.81	6.91	7.96	9.31	11.9	15.3	19.4	23.5	26.3	28.8	32.0	34.3
17	5.70	6.41	7.56	8.67	10.1	12.8	16.3	20.5	24.8	27.6	30.2	33.4	35.7
18	6.26	7.01	8.23	9.39	10.9	13.7	17.3	21.6	26.0	28.9	31.5	34.8	37.2
19	6.84	7.03	8.91	10.1	11.7	14.6	18.3	22.7	27.2	30.1	32.9	36.2	38.6
20	7.43	8.26	9.59	10.9	12.4	15.5	19.3	23.8	28.4	31.4	34.2	37.6	40.0
21	8.03	8.90	10.3	11.6	13.2	16.3	20.3	24.9	29.6	32.7	35.5	38.9	41.4
22	8.64	9.54	11.0	12.3	14.0	17.2	21.3	26.0	30.8	33.9	36.8	40.3	42.8
23	9.26	10.2	11.7	13.1	14.8	18.1	22.3	27.1	32.0	35.2	38.1	41.6	44.2
24	9.89	10.9	12.4	13.8	15.7	19.0	23.3	28.2	33.2	36.4	39.4	43.0	45.6
25	10.5	11.5	13.1	14.6	16.5	19.9	24.3	29.3	34.4	37.7	40.6	44.3	46.9
26	11.2	12.2	13.8	15.4	17.3	20.8	25.3	30.4	35.6	38.9	41.9	45.6	48.3
27	11.8	12.9	14.6	16.2	18.1	21.7	26.3	31.5	36.7	40.1	43.2	47.0	49.6
28	12.5	13.6	15.3	16.9	18.9	22.7	27.3	32.6	37.9	41.3	44.5	48.3	51.0
29	13.1	14.3	16.0	17.7	19.8	23.6	28.3	33.7	39.1	42.6	45.7	49.6	52.3
30	13.8	15.0	16.8	18.5	20.6	24.5	29.3	34.8	40.3	43.8	47.0	50.9	53.7

* Thin table is abridged from "Tables of percentage points of the incomplete beta function and of the chi-square distribution," *Biometrika*, Vol. 32 (1941). It is here published with the kind permission of the author, Catherine M. Thompson, and the editor of *Biometrika*.

† d.f. = degrees of freedom.

Table C Cumulative Student's-t Distributions*

$[P(T_n \leq t_n(\alpha)) = \alpha$; e.g., $P(T_4 \leq 2.132) = .95]$

d.f.† n \ α	.75	.90	.95	.975	.99	.995	.9995
1	1.000	3.078	6.314	12.706	31.821	63.657	636.619
2	.816	1.886	2.920	4.303	6.965	9.925	31.598
3	.765	1.638	2.353	3.182	4.541	5.841	12.941
4	.741	1.533	2.132	2.776	3.747	4.604	8.610
5	.727	1.476	2.015	2.571	3.365	4.032	6.859
6	.718	1.440	1.943	2.447	3.143	3.707	5.959
7	.711	1.415	1.895	2.365	2.998	3.499	5.405
8	.706	1.397	1.860	2.306	2.896	3.355	5.041
9	.703	1.383	1.833	2.262	2.821	3.250	4.781
10	.700	1.372	1.812	2.228	2.764	3.169	4.587
11	.697	1.363	1.796	2.201	2.718	3.106	4.437
12	.695	1.356	1.782	2.179	2.681	3.055	4.318
13	.694	1.350	1.771	2.160	2.650	3.012	4.221
14	.692	1.345	1.761	2.145	2.624	2.977	4.140
15	.691	1.341	1.753	2.131	2.602	2.947	4.073
16	.690	1.337	1.746	2.120	2.583	2.921	4.015
17	.689	1.333	1.740	2.110	2.567	2.898	3.965
18	.688	1.330	1.734	2.101	2.552	2.878	3.922
19	.688	1.328	1.729	2.093	2.539	2.861	3.883
20	.687	1.325	1.725	2.086	2.528	2.845	3.850
21	.686	1.323	1.721	2.080	2.518	2.831	3.819
22	.686	1.321	1.717	2.074	2.508	2.819	3.792
23	.685	1.319	1.714	2.069	2.500	2.807	3.767
24	.685	1.318	1.711	2.064	2.492	2.797	3.745
25	.684	1.316	1.708	2.060	2.485	2.787	3.725
26	.684	1.315	1.706	2.056	2.479	2.779	3.707
27	.684	1.314	1.703	2.052	2.473	2.771	3.690
28	.683	1.313	1.701	2.048	2.467	2.763	3.674
29	.683	1.311	1.699	2.045	2.462	2.756	3.659
30	.683	1.310	1.697	2.042	2.457	2.750	3.646
40	.681	1.303	1.684	2.021	2.423	2.704	3.551
60	.679	1.296	1.671	2.000	2.390	2.660	3.460
120	.677	1.289	1.658	1.980	2.358	2.617	3.373
∞	.674	1.282	1.645	1.960	2.326	2.576	3.291

* This table is abridged from the "Statistical Tables" of R. A. Fisher and Frank Yates published by Oliver & Boyd, Ltd., Edinburgh and London, 1938. It is here published with the kind permission of the authors and their publishers.

† d.f. = degrees of freedom.

Table D Wilcoxon Two-Sample Distribution*

Values of $\sum_{w=0}^{c} N(w; n, m) - \frac{n(n+1)}{2}$

n	m	$\binom{n+m}{n}$	0	1	2	3	4	5	6	7	8	9	10	11	12	13	14	15	16	17	18	19	20
3	3	20	1	2	4	7	10	13	16	18	19	20											
3	4	35	1	2	4	7	11	15	20	24	28	31	33	34	35								
4	4	70	1	2	4	7	12	17	24	31	39	46	53	58	63	66	68	69	70				
3	5	56	1	2	4	7	11	16	22	28	34	40	45	49	52	54	55	56					
4	5	126	1	2	4	7	12	18	26	35	46	57	69	80	91	100	108	114	119	122	124	125	126
5	5	252	1	2	4	7	12	19	28	39	53	69	87	106	126	146	165	183	199	213	224	233	240
3	6	84	1	2	4	7	11	16	23	30	38	46	54	61	68	73	77	80	82	83	84		
4	6	210	1	2	4	7	12	18	27	37	50	64	80	96	114	130	146	160	173	183	192	198	203
5	6	462	1	2	4	7	12	19	29	41	57	76	99	124	153	183	215	247	279	309	338	363	386
6	6	924	1	2	4	7	12	19	30	43	61	83	111	143	182	224	272	323	378	433	491	546	601
3	7	120	1	2	4	7	11	16	23	31	40	50	60	70	80	89	97	104	109	113	116	118	119
4	7	330	1	2	4	7	12	18	27	38	52	68	87	107	130	153	177	200	223	243	262	278	292
5	7	792	1	2	4	7	12	19	29	42	59	80	106	136	171	210	253	299	347	396	445	493	539
6	7	1716	1	2	4	7	12	19	30	44	63	87	118	155	201	253	314	382	458	539	627	717	811
7	7	3432	1	2	4	7	12	19	30	45	65	91	125	167	220	283	358	445	545	657	782	918	1064
3	8	165	1	2	4	7	11	16	23	31	41	52	64	76	89	101	113	124	134	142	149	154	158
4	8	495	1	2	4	7	12	18	27	38	53	70	91	114	141	169	200	231	264	295	326	354	381
5	8	1287	1	2	4	7	12	19	29	42	60	82	110	143	183	228	280	337	400	466	536	607	680
6	8	3003	1	2	4	7	12	19	30	44	64	89	122	162	213	272	343	424	518	621	737	860	994
7	8	6135	1	2	4	7	12	19	30	45	66	93	129	174	232	302	388	489	609	746	904	1080	1277
8	8	12870	1	2	4	7	12	19	30	45	67	95	133	181	244	321	418	534	675	839	1033	1254	1509

* This table is reprinted with permission from J. L. Hodges, Jr. and E. L. Lehmann, *Basic Concepts of Probability and Statistics* (San Francisco: Holden-Day, Inc., 1964).

Table E Values of $M(N; N)$ and 2^N for Use in the Wilcoxon Paired-Comparison Test*

N	$M(N;N)$	2^N	N	$M(N;N)$	2^N	N	$M(N;N)$	2^N
1	2	2	8	25	256	15	137	32768
2	3	4	9	33	512	16	169	65536
3	5	8	10	43	1024	17	207	131072
4	7	16	11	55	2048	18	253	262144
5	10	32	12	70	4096	19	307	524288
6	14	64	13	88	8192	20	371	1048576
7	19	128	14	110	16384			

* This table is reprinted with permission from J. L. Hodges, Jr. and E. L. Lehmann, *Basic Concepts of Probability and Statistics* (San Francisco: Holden-Day, Inc., 1964).

Table F Values of $M(i; N)$ with $N < i$ for Use in the Wilcoxon Paired-Comparison Test*

$i - N$ \ N	3	4	5	6	7	8	9	10	11	12	13	14	15	16	17	18	19	20
1	6	9	13	18	24	32	42	54	69	87	109	136	168	206	252	306	370	446
2	7	11	16	22	30	40	52	67	85	107	134	166	204	250	304	368	444	533
3	8	13	19	27	37	49	64	82	104	131	163	201	247	301	365	441	530	634
4		14	22	32	44	59	77	99	126	158	196	242	296	360	436	525	629	751
5		15	25	37	52	70	92	119	151	189	235	289	353	429	518	622	744	886
6		16	27	42	60	82	109	141	179	225	279	343	419	508	612	734	876	1,041
7			29	46	68	95	127	165	211	265	329	405	494	598	720	862	1,027	1,219
8			30	50	76	108	146	192	246	310	386	475	579	701	843	1,008	1,200	1,422
9			31	54	84	121	167	221	285	361	450	551	676	818	983	1,175	1,397	1,653
10			32	57	91	135	188	252	328	417	521	643	785	950	1,142	1,364	1,620	1,916
11				59	98	148	210	285	374	478	600	742	907	1,099	1,321	1,577	1,873	2,213
12				61	101	161	253	320	423	545	687	852	1,044	1,266	1,522	1,818	2,158	2,548
13				62	109	174	256	356	476	617	782	974	1,196	1,452	1,748	2,088	2,478	2,926
14				63	114	186	279	394	532	695	886	1,108	1,364	1,660	2,000	2,390	2,838	3,350
15				64	118	197	302	433	591	779	999	1,254	1,550	1,890	2,280	2,728	3,240	3,825
16					121	207	324	472	653	868	1,120	1,414	1,753	2,143	2,591	3,103	3,688	4,356
17					123	216	345	512	717	962	1,251	1,587	1,975	2,422	2,934	3,519	4,187	4,947
18					125	224	366	552	783	1,062	1,391	1,774	2,218	2,728	3,312	3,980	4,740	5,604
19					126	231	385	591	851	2,166	1,539	1,976	2,481	3,062	3,728	4,487	5,351	6,333
20					127	237	403	630	920	1,274	1,697	2,192	2,766	3,427	4,183	5,045	6,026	7,139
21					128	242	420	668	989	1,387	1,863	2,423	3,074	3,823	4,680	5,658	6,769	8,028
22						246	435	704	1,059	1,502	2,037	2,669	3,404	4,251	5,222	6,328	7,584	9,008
23						249	448	739	1,128	1,620	2,219	2,929	3,757	4,714	5,810	7,059	8,478	10,084
24						251	460	772	1,197	1,741	2,408	3,203	4,135	5,212	6,447	7,856	9,455	11,264
25						253	470	803	1,265	1,863	2,603	3,492	4,536	5,746	7,136	8,721	10,520	12,557
26						254	479	832	1,331	1,986	2,805	3,794	4,961	6,318	7,878	9,658	11,681	13,968
27						255	487	859	1,395	2,110	3,012	4,109	5,411	6,928	8,675	10,673	12,941	15,506
28						256	493	883	1,457	2,233	3,233	4,437	5,881	7,576	9,531	11,766	14,306	17,180
29							498	905	1,516	2,355	3,438	4,776	6,380	8,265	10,445	12,942	15,783	18,997
30							502	925	1,572	2,476	3,656	5,126	6,901	8,993	11,420	14,206	17,377	20,966

* This table is reprinted with permission from J. L. Hodges, Jr. and E. L. Lehmann, *Basic Concepts of Probability and Statistics* (San Francisco: Holden-Day, Inc., 1964).

Table G The 95th Percentile for the F Distribution*
[example: $P(F_{5,8} \leq 3.69) = .95$]

DEGREES OF FREEDOM FOR NUMERATOR (columns); DEGREES OF FREEDOM FOR DENOMINATOR (rows)

	1	2	3	4	5	6	7	8	9	10	12	15	20	24	30	40	60	120	∞
1	161	200	216	225	230	234	237	239	241	242	244	246	248	249	250	251	252	253	254
2	18.5	19.0	19.2	19.2	19.3	19.3	19.4	19.4	19.4	19.4	19.4	19.4	19.4	19.5	19.5	19.5	19.5	19.5	19.5
3	10.1	9.55	9.28	9.12	9.01	8.94	8.89	8.85	8.81	8.79	8.74	8.70	8.66	8.64	8.62	8.59	8.57	8.55	8.53
4	7.71	6.94	6.59	6.39	6.26	6.16	6.09	6.04	6.00	5.96	5.91	5.86	5.80	5.77	5.75	5.72	5.69	5.66	5.63
5	6.61	5.79	5.41	5.19	5.05	4.95	4.88	4.82	4.77	4.74	4.68	4.62	4.56	4.53	4.50	4.46	4.43	4.40	4.37
6	5.99	5.14	4.76	4.53	4.39	4.28	4.21	4.15	4.10	4.06	4.00	3.94	3.87	3.84	3.81	3.77	3.74	3.70	3.67
7	5.59	4.74	4.35	4.12	3.97	3.87	3.79	3.73	3.68	3.64	3.57	3.51	3.44	3.41	3.38	3.34	3.30	3.27	3.23
8	5.32	4.46	4.07	3.84	3.69	3.58	3.50	3.44	3.39	3.35	3.28	3.22	3.15	3.12	3.08	3.04	3.01	2.97	2.93
9	5.12	4.26	3.86	3.63	3.48	3.37	3.29	3.23	3.18	3.14	3.07	3.01	2.94	2.90	2.86	2.83	2.79	2.75	2.71
10	4.96	4.10	3.71	3.48	3.33	3.22	3.14	3.07	3.02	2.98	2.91	2.85	2.77	2.74	2.70	2.66	2.62	2.58	2.54
11	4.84	3.98	3.59	3.36	3.20	3.09	3.01	2.95	2.90	2.85	2.79	2.72	2.65	2.61	2.57	2.53	2.49	2.45	2.40
12	4.75	3.89	3.49	3.26	3.11	3.00	2.91	2.85	2.80	2.75	2.69	2.62	2.54	2.51	2.47	2.43	2.38	2.34	2.30
13	4.67	3.81	3.41	3.18	3.03	2.92	2.83	2.77	2.71	2.67	2.60	2.53	2.46	2.42	2.38	2.34	2.30	2.25	2.21
14	4.60	3.74	3.34	3.11	2.96	2.85	2.76	2.70	2.65	2.60	2.53	2.46	2.39	2.35	2.31	2.27	2.22	2.18	2.13
15	4.54	3.68	3.29	3.06	2.90	2.79	2.71	2.64	2.59	2.54	2.48	2.40	2.33	2.29	2.25	2.20	2.16	2.11	2.07
16	4.49	3.63	3.24	3.01	2.85	2.74	2.66	2.59	2.54	2.49	2.42	2.35	2.28	2.24	2.19	2.15	2.11	2.06	2.01
17	4.45	3.59	3.20	2.96	2.81	2.70	2.61	2.55	2.49	2.45	2.38	2.31	2.23	2.19	2.15	2.10	2.06	2.01	1.96
18	4.41	3.55	3.16	2.93	2.77	2.66	2.58	2.51	2.46	2.41	2.31	2.27	2.19	2.15	2.11	2.06	2.02	1.97	1.92
19	4.38	3.52	3.13	2.90	2.74	2.63	2.54	2.48	2.42	2.38	2.31	2.23	2.16	2.11	2.07	2.03	1.98	1.93	1.88
20	4.35	3.49	3.10	2.87	2.71	2.60	2.51	2.45	2.39	2.35	2.28	2.20	2.12	2.08	2.04	1.99	1.95	1.90	1.84
21	4.32	3.47	3.07	2.84	2.68	2.57	2.49	2.42	2.37	2.32	2.25	2.18	2.10	2.05	2.01	1.96	1.92	1.87	1.81
22	4.30	3.44	3.05	2.82	2.66	2.55	2.46	2.40	2.31	2.30	2.23	2.15	2.07	2.03	1.98	1.94	1.89	1.84	1.78
23	4.28	3.42	3.03	2.80	2.64	2.53	2.44	2.37	2.32	2.27	2.20	2.13	2.05	2.01	1.96	1.91	1.86	1.81	1.76
24	4.26	3.40	3.01	2.78	2.62	2.51	2.42	2.36	2.30	2.25	2.18	2.11	2.03	1.98	1.94	1.89	1.84	1.79	1.73
25	4.24	3.39	2.99	2.76	2.60	2.49	2.40	2.34	2.28	2.24	2.16	2.09	2.01	1.96	1.92	1.87	1.82	1.77	1.71
30	4.17	3.32	2.92	2.69	2.53	2.42	2.33	2.27	2.21	2.16	2.09	2.01	1.93	1.89	1.84	1.79	1.74	1.68	1.62
40	4.08	3.23	2.84	2.61	2.45	2.34	2.25	2.18	2.12	2.08	2.00	1.92	1.84	1.79	1.74	1.69	1.64	1.58	1.51
60	4.00	3.15	2.76	2.53	2.37	2.25	2.17	2.10	2.04	1.99	1.92	1.84	1.75	1.70	1.65	1.59	1.53	1.47	1.93
120	3.92	3.07	2.68	2.45	2.29	2.18	2.09	2.02	1.96	1.91	1.83	1.75	1.66	1.61	1.55	1.50	1.43	1.35	1.25
∞	3.84	3.00	2.60	2.37	2.21	2.10	2.01	1.94	1.88	1.83	1.75	1.67	1.57	1.52	1.46	1.39	1.32	1.22	1.00

Interpolation should be performed using reciprocals of the degrees of freedom.

* By permission of Prof. E. S. Pearson from M. Merrington, C. M. Thompson, "Tables of percentage points of the inverted beta (F) distribution," *Biometrika*, Vol. 33 (1943), p. 73.

Table H Binomial Probabilities with $p = .5$ Entries of $P(X = j)$ When X Is $B(n, \frac{1}{2})$*

n	j	$p = .5$
2	0	.2500
	1	.5000
3	0	.1250
	1	.3750
4	0	.0625
	1	.2500
	2	.3750
5	0	.0312
	1	.1562
	2	.3125
6	0	.0156
	1	.0938
	2	.2344
	3	.3125
7	0	.0078
	1	.0547
	2	.1641
	3	.2734
8	0	.0039
	1	.0312
	2	.1094
	3	.2188
	4	.2734
9	0	.0020
	1	.0176
	2	.0703
	3	.1641
	4	.2461
10	0	.0010
	1	.0098
	2	.0439
	3	.1172
	4	.2051
	5	.2461
11	0	.0005
	1	.0054
	2	.0269
	3	.0806
	4	.1611
	5	.2256
12	0	.0002
	1	.0029
	2	.0161
	3	.0537
	4	.1208
	5	.1934
	6	.2256
13	0	.0001
	1	.0016
	2	.0095
	3	.0349
	4	.0873
	5	.1571
	6	.2095
14	0	.0001
	1	.0009
	2	.0056
	3	.0222
	4	.0611
	5	.1222
	6	.1833
	7	.2095
15	0	.0000
	1	.0005
	2	.0032
	3	.0139
	4	.0417
	5	.0916
	6	.1527
	7	.1964
16	0	.0000
	1	.0002
	2	.0018
	3	.0085
	4	.0278
	5	.0667
	6	.1222
	7	.1746
	8	.1964
17	0	.0000
	1	.0001
	2	.0010
	3	.0052
	4	.0182
	5	.0472
	6	.0944
	7	.1484
	8	.1855
18	0	.0000
	1	.0001
	2	.0006
	3	.0031
	4	.0117
	5	.0327
	6	.0708
	7	.1214
	8	.1669
	9	.1855
19	1	.0000
	2	.0003
	3	.0018
	4	.0074
	5	.0222
	6	.0518
	7	.0961
	8	.1442
	9	.1762
20	1	.0000
	2	.0002
	3	.0011
	4	.0046
	5	.0148
	6	.0370
	7	.0739
	8	.1201
	9	.1602
	10	.1762
21	1	.0000
	2	.0001
	3	.0006
	4	.0029
	5	.0097
	6	.0259
	7	.0554
	8	.0970
	9	.1402
	10	.1682
22	1	.0000
	2	.0001
	3	.0004
	4	.0017
	5	.0063
	6	.0178
	7	.0407
	8	.0762
	9	.1186
	10	.1542
	11	.1682
23	2	.0000
	3	.0002
	4	.0011
	5	.0040
	6	.0120
	7	.0292
	8	.0584
	9	.0974
	10	.1364
	11	.1612
24	2	.0000
	3	.0001
	4	.0006
	5	.0025
	6	.0080
	7	.0206
	8	.0438
	9	.0779
	10	.1169
	11	.1488
	12	.1612
25	2	.0000
	3	.0001
	4	.0004
	5	.0016
	6	.0053
	7	.0143
	8	.0322
	9	.0609
	10	.0974
	11	.1328
	12	.1550
26	3	.0000
	4	.0002
	5	.0010
	6	.0034
	7	.0098
	8	.0233
	9	.0466
	10	.0792
	11	.1151
	12	.1439
	13	.1550
27	3	.0000
	4	.0001
	5	.0006
	6	.0022
	7	.0066
	8	.0165
	9	.0349
	10	.0629
	11	.0971
	12	.1295
	13	.1494
28	3	.0000
	4	.0001
	5	.0004
	6	.0014
	7	.0044
	8	.0116
	9	.0257
	10	.0489
	11	.0800
	12	.1133
	13	.1395
	14	.1494
29	4	.0000
	5	.0002
	6	.0009
	7	.0029
	8	.0080
	9	.0187
	10	.0373
	11	.0644
	12	.0967
	13	.1264
	14	.1445
30	4	.0000
	5	.0001
	6	.0006
	7	.0019
	8	.0055
	9	.0133
	10	.0280
	11	.0509
	12	.0806
	13	.1115
	14	.1354
	15	.1445

* This table is reprinted with permission from J. L. Hodges, Jr. and E. L. Lehmann, *Basic Concepts of Probability and Statistics* (San Francisco: Holden-Day, Inc., 1964).

Table I Square Roots

n	$\sqrt{n}$	$\sqrt{10n}$	n	$\sqrt{n}$	$\sqrt{10n}$	n	$\sqrt{n}$	$\sqrt{10n}$
1	1.0000	3.1623	34	5.8310	18.439	67	8.1854	25.884
2	1.4142	4.4721	35	5.9161	18.708	68	8.2462	26.077
3	1.7321	5.4772	36	6.0000	18.974	69	8.3066	26.268
4	2.0000	6.3246	37	6.0828	19.235	70	8.3666	26.458
5	2.2361	7.0711	38	6.1644	19.494	71	8.4261	26.646
6	2.4495	7.7460	39	6.2450	19.748	72	8.4853	26.833
7	2.6458	8.3666	40	6.3246	20.000	73	8.5440	27.019
8	2.8284	8.9443	41	6.4031	20.248	74	8.6023	27.203
9	3.0000	9.4868	42	6.4807	20.494	75	8.6603	27.386
10	3.1623	10.000	43	6.5574	20.736	76	8.7178	27.568
11	3.3166	10.488	44	6.6332	20.976	77	8.7750	27.749
12	3.4641	10.954	45	6.7082	21.213	78	8.8318	27.928
13	3.6056	11.402	46	6.7823	21.448	79	8.8882	28.107
14	3.7417	11.832	47	6.8557	21.679	80	8.9443	28.284
15	3.8730	12.247	48	6.9282	21.909	81	9.0000	28.460
16	4.0000	12.649	49	7.0000	22.136	82	9.0554	28.636
17	4.1231	13.038	50	7.0711	22.361	83	9.1104	28.810
18	4.2426	13.416	51	7.1414	22.583	84	9.1652	28.983
19	4.3589	13.784	52	7.2111	22.804	85	9.2195	29.155
20	4.4721	14.142	53	7.2801	23.022	86	9.2736	29.326
21	4.5826	14.491	54	7.3485	23.238	87	9.3274	29.496
22	4.6904	14.832	55	7.4162	23.452	88	9.3808	29.665
23	4.7958	15.166	56	7.4833	23.664	89	9.4340	29.833
24	4.8990	15.492	57	7.5498	23.875	90	9.4868	30.000
25	5.0000	15.811	58	7.6158	24.083	91	9.5394	30.166
26	5.0990	16.125	59	7.6811	24.290	92	9.5917	30.332
27	5.1962	16.432	60	7.7460	24.495	93	9.6437	30.496
28	5.2915	16.733	61	7.8102	24.698	94	9.6954	30.659
29	5.3852	17.029	62	7.8740	24.900	95	9.7468	30.822
30	5.4772	17.321	63	7.9373	25.100	96	9.7980	30.984
31	5.5678	17.607	64	8.0000	25.298	97	9.8489	31.145
32	5.6569	17.889	65	8.0623	25.495	98	9.8995	31.305
33	5.7446	18.166	66	8.1240	25.690	99	9.9499	31.464

This table has added flexibility illustrated as follows:

$$\sqrt{6.3} = \sqrt{630/100} = \sqrt{630}/\sqrt{100} = 25.1/10 = 2.51.$$

INDEXES

SUBJECT INDEX

EXAMPLE INDEX

SYMBOL INDEX

GREEK ALPHABET

It is often convenient to make use of the Greek alphabet, although it is not necessary to learn the *order* of this alphabet.

A	α	alpha	N	ν	nu
B	β	beta	Ξ	ξ	xi
Γ	γ	gamma	O	o	omicron
Δ	δ	delta	Π	π	pi
E	ϵ	epsilon	P	ρ	rho
Z	ζ	zeta	Σ	σ	sigma
H	η	eta	T	τ	tau
Θ	θ	theta	Υ	υ	upsilon
I	ι	iota	Φ	φ	phi
K	κ	kappa	X	χ	chi
Λ	λ	lambda	Ψ	ψ	psi
M	μ	mu	Ω	ω	omega